Student's Solutions Manual

to accompany

Elementary Number Theory and Its Applications

Fifth Edition

Bart Goddard

Kenneth H. Rosen
AT&T Labs

Boston   San Francisco   New York
London   Toronto   Sydney   Tokyo   Singapore   Madrid
Mexico City   Munich   Paris   Cape Town   Hong Kong   Montreal

Reproduced by Pearson Addison-Wesley from electronic files supplied by the author.

Copyright © 2005 Pearson Education, Inc.
Publishing as Pearson Addison-Wesley, 75 Arlington Street, Boston, MA 02116.

All rights reserved. No part of this publication may be reproduced, stored in a retrieval system, or transmitted, in any form or by any means, electronic, mechanical, photocopying, recording, or otherwise, without the prior written permission of the publisher. Printed in the United States of America.

ISBN   0-321-26840-7

3 4 5 6 OPM 08 07 06

Contents

Chapter 1. The Integers
    1.1. Numbers, Sequences, and Sums   1
    1.2. Sums and Products   4
    1.3. Mathematical Induction   5
    1.4. The Fibonacci Numbers   7
    1.5. Divisibility   10

Chapter 2. Integer Representations and Operations
    2.1. Representations of Integers   13
    2.2. Computer Operations with Integers   15
    2.3. Complexity and Integer Operations   16

Chapter 3. Primes and Greatest Common Divisors
    3.1. Prime Numbers   19
    3.2. The Distribution of Primes   20
    3.3. Greatest Common Divisors   22
    3.4. The Euclidean Algorithm   24
    3.5. The Fundamental Theorem of Arithmetic   27
    3.6. Factorization Methods and the Fermat Numbers   32
    3.7. Linear Diophantine Equations   34

Chapter 4. Congruences
    4.1. Introduction to Congruences   37
    4.2. Linear Congruences   40
    4.3. The Chinese Remainder Theorem   42
    4.4. Solving Polynomial Congruences   44
    4.5. Systems of Linear Congruences   45
    4.6. Factoring Using the Pollard Rho Method   47

Chapter 5. Applications of Congruences
    5.1. Divisibility Tests   49
    5.2. The Perpetual Calendar   50
    5.3. Round-Robin Tournaments   51
    5.4. Hashing Functions   53
    5.5. Check Digits   54

Chapter 6. Some Special Congruences
    6.1. Wilson's Theorem and Fermat's Little Theorem   57
    6.2. Pseudoprimes   59
    6.3. Euler's Theorem   59

Chapter 7. Multiplicative Functions
    7.1. The Euler Phi-Function   61
    7.2. The Sum and Number of Divisors   64
    7.3. Perfect Numbers and Mersenne Primes   67
    7.4. Mobius Inversion   69

Chapter 8. Cryptology

8.1. Character Ciphers	71
8.2. Block and Stream Ciphers	71
8.3. Exponentiation Ciphers	75
8.4. Public Key Cryptography	75
8.5. Knapsack Ciphers	75
8.6. Cryptographic Protocols and Applications	76

Chapter 9. Primitive Roots

9.1. The Order of an Integer and Primitive Roots	79
9.2. Primitive Roots for Primes	80
9.3. The Existence of Primitive Roots	81
9.4. Index Arithmetic	83
9.5. Primality Tests Using Orders of Integers and Primitive Roots	85
9.6. Universal Exponents	85

Chapter 10. Applications of Primitive Roots and the Order of an Integer

10.1. Pseudorandom Numbers	89
10.2. The ElGamal Cryptosystem	90
10.3. An Application to the Splicing of Telephone Cables	91

Chapter 11. Quadratic Residues

11.1. Quadratic Residues and Nonresidues	93
11.2. The Law of Quadratic Reciprocity	96
11.3. The Jacobi Symbol	100
11.4. Euler Pseudoprimes	102
11.5. Zero-Knowledge Proofs	102

Chapter 12. Decimal Fractions and Continued Fractions

12.1. Decimal Fractions	105
12.2. Finite Continued Fractions	107
12.3. Infinite Continued Fractions	108
12.4. Periodic Continued Fractions	110
12.5. Factoring Using Continued Fractions	112

Chapter 13. Some Nonlinear Diophantine Equations

13.1. Pythagorean Triples	115
13.2. Fermat's Last Theorem	116
13.3. Sums of Squares	118
13.4. Pell's Equation	120

Chapter 14. The Gaussian Integers

14.1. Gaussian Integers and Gaussian Primes	121
14.2. Greatest Common Divisors and Unique Factorization	124
14.3. Gaussian Integers and Sums of Squares	128

Appendices

Appendix A. Axioms for the Set of Integers	131
Appendix B. Binomial Coefficients	133

CHAPTER 1

The Integers

1.1. Numbers, Sequences, and Sums

1.1.1. a. The set of integers greater than 3 is well-ordered. Every subset of this set is also a subset of the set of positive integers, and hence must have a least element.

b. The set of even positive integers is well-ordered. Every subset of this set is also a subset of the set of positive integers, and hence must have a least element.

c. The set of positive rational numbers is not well-ordered. This set does not have a least element. If a/b were the least positive rational number then $a/(b+a)$ would be a smaller positive rational number, which is a contradiction.

d. The set of positive rational numbers of the form $a/2$ is well-ordered. Consider a subset of numbers of this form. The set of numerators of the numbers in this subset is a subset of the set of positive integers, so it must have a least element b. Then $b/2$ is the least element of the subset.

e. The set of nonnegative rational numbers is not well-ordered. The set of positive rational numbers is a subset with no least element, as shown in part c.

1.1.3. Suppose that x and y are rational numbers. Then $x = a/b$ and $y = c/d$, where $a, b, c,$ and d are integers with $b \neq 0$ and $d \neq 0$. Then $xy = (a/b) \cdot (c/d) = ac/bd$ and $x + y = a/b + c/d = (ad + bc)/bd$ where $bd \neq 0$. Since both $x + y$ and xy are ratios of integers, they are both rational.

1.1.5. Suppose that $\sqrt{3}$ were rational. Then there would exist positive integers a and b with $\sqrt{3} = a/b$. Consequently, the set $S = \{k\sqrt{3} \mid k \text{ and } k\sqrt{3} \text{ are positive integers}\}$ is nonempty since $a = b\sqrt{3}$. Therefore, by the well-ordering property, S has a smallest element, say $s = t\sqrt{3}$. We have $s\sqrt{3} - s = s\sqrt{3} - t\sqrt{3} = (s-t)\sqrt{3}$. Since $s\sqrt{3} = 3t$ and s are both integers, $s\sqrt{3} - s = (s-t)\sqrt{3}$ must also be an integer. Furthermore, it is positive, since $s\sqrt{3} - s = s(\sqrt{3} - 1)$ and $\sqrt{3} > 1$. It is less than s since $s = t\sqrt{3}$, $s\sqrt{3} = 3t$, and $\sqrt{3} < 3$. This contradicts the choice of s as the smallest positive integer in S. It follows that $\sqrt{3}$ is irrational.

1.1.7. a. Since $0 \leq 1/4 < 1$, we have $[1/4] = 0$.

b. Since $-1 \leq -3/4 < 0$, we have $[-3/4] = -1$.

c. Since $3 \leq 22/7 < 4$, we have $[22/7] = 3$.

d. Since $-2 \leq -2 < -1$, we have $[-2] = -2$.

e. We compute $[1/2 + [1/2]] = [1/2 + 0] = [1/2] = 0$.

f. We compute $[-3 + [-1/2]] = [-3 - 1] = [-4] = -4$.

1.1.9. a. Since $[8/5] = 1$, we have $\{8/5\} = 8/5 - [8/5] = 8/5 - 1 = 3/5$.

b. Since $[1/7] = 0$, we have $\{1/7\} = 1/7 - [1/7] = 1/7 - 0 = 1/7$.

    c. Since $[-11/4] = -3$, we have $\{-11/4\} = -11/4 - [-11/4] = -11/4 - (-3) = 1/4$.

    d. Since $[7] = 7$, we have $\{7\} = 7 - [7] = 7 - 7 = 0$.

1.1.11. If x is an integer, then $[x] + [-x] = x - x = 0$. Otherwise, $x = z + r$, where z is an integer and r is a real number with $0 < r < 1$. In this case, $[x] + [-x] = [z + r] + [-z - r] = z + (-z - 1) = -1$.

1.1.13. We have $[x] \leq x$ and $[y] \leq y$. Adding these two inequalities gives $[x] + [y] \leq x + y$. Hence $[x + y] \geq [[x] + [y]] = [x] + [y]$.

1.1.15. Let $x = a + r$ and $y = b + s$, where a and b are integers and r and s are real numbers such that $0 \leq r, s < 1$. Then $[xy] = [ab + as + br + sr] = ab + [as + br + sr]$, whereas $[x][y] = ab$. Thus we have $[xy] \geq [x][y]$. If x and y are both negative, then $[xy] \leq [x][y]$. If one of x and y is positive and the other negative, then the inequality could go either direction. For examples take $x = -1.5, y = 5$ and $x = -1, y = 5.5$. In the first case we have $[-1.5 \cdot 5] = [-7.5] = -8 > [-1.5][5] = -2 \cdot 5 = -10$. In the second case we have $[-1 \cdot 5.5] = [-5.5] = -6 < [-1][5.5] = -1 \cdot 5 = -5$.

1.1.17. Let $x = [x] + r$. Since $0 \leq r < 1$, $x + \frac{1}{2} = [x] + r + \frac{1}{2}$. If $r < \frac{1}{2}$, then $[x]$ is the integer nearest to x and $[x + \frac{1}{2}] = [x]$ since $[x] \leq x + \frac{1}{2} = [x] + r + \frac{1}{2} < [x] + 1$. If $r \geq \frac{1}{2}$, then $[x] + 1$ is the integer nearest to x (choosing this integer if x is midway between $[x]$ and $[x+1]$) and $[x + \frac{1}{2}] = [x] + 1$ since $[x] + 1 \leq x + r + \frac{1}{2} < [x] + 2$.

1.1.19. Let $x = k + \epsilon$ where k is an integer and $0 \leq \epsilon < 1$. Further, let $k = a^2 + b$, where a is the largest integer such that $a^2 \leq k$. Then $a^2 \leq k = a^2 + b \leq x = a^2 + b + \epsilon < (a+1)^2$. Then $[\sqrt{x}] = a$ and $[\sqrt{[x]}] = [\sqrt{k}] = a$ also, proving the theorem.

1.1.21. a. Since the difference between any two consecutive terms of this sequence is 8, we may compute the nth term by adding 8 to the first term $n - 1$ times. That is, $a_n = 3 + (n-1)8 = 8n - 5$.

    b. For each n, we have $a_n - a_{n-1} = 2^{n-1}$, so we may compute the nth term of this sequence by adding all the powers of 2, up to the $(n-1)$th, to the first term. Hence $a_n = 5 + 2 + 2^2 + 2^3 + \cdots + 2^{n-1} = 5 + 2^n - 2 = 2^n + 3$.

    c. The nth term of this sequence appears to be zero, unless n is a perfect square, in which case the term is 1. If n is not a perfect square, then $[\sqrt{n}] < \sqrt{n}$, where $[x]$ represents the greatest integer function. If n is a perfect square, then $[\sqrt{n}] = \sqrt{n}$. Therefore, $[[\sqrt{n}]/\sqrt{n}]$ equals 1 if n is a perfect square and 0 otherwise, as desired.

    d. This is a Fibonacci-like sequence, with $a_n = a_{n-1} + a_{n-2}$, for $n \geq 3$, and $a_1 = 1$, and $a_2 = 3$.

1.1.23. Three possible answers are $a_n = 2^{n-1}$, $a_n = (n^2 - n + 2)/2$, and $a_n = a_{n-1} + 2a_{n-2}$.

1.1.25. This set is exactly the sequence $a_n = n - 100$, and hence is countable.

1.1.27. One way to show this is to imitate the proof that the set of rational numbers is countable, replacing a/b with $a + b\sqrt{2}$. Another way is to consider the function $f(a + b\sqrt{2}) = 2^a 3^b$ which is a one-to-one map of this set into the rational numbers, which is known to be countable.

1.1.29. Suppose $\{A_i\}$ is a countable collection of countable sets. Then each A_i can be represented by a sequence, as follows:

$$\begin{aligned} A_1 &= a_{11} \quad a_{12} \quad a_{13} \quad \ldots \\ A_2 &= a_{21} \quad a_{22} \quad a_{23} \quad \ldots \\ A_3 &= a_{31} \quad a_{32} \quad a_{33} \quad \ldots \\ &\vdots \end{aligned}$$

Consider the listing $a_{11}, a_{12}, a_{21}, a_{13}, a_{22}, a_{31}, \ldots$, in which we first list the elements with subscripts adding to 2, then the elements with subscripts adding to 3 and so on. Further, we order the elements with subscripts adding to k in order of the first subscript. Form a new sequence c_i as follows. Let $c_1 = a_1$. Given that c_n is determined, let c_{n+1} be the next element in the listing which is different from each c_i with $i = 1, 2, \ldots, n$. Then this sequence is exactly the elements of $\bigcup_{i=1}^{\infty} A_i$, which is therefore countable.

1.1.31. a. Note that $\sqrt{3} = 1.73 \approx 7/4$, so we might guess that $\sqrt{3} - 7/4 \approx 0$. If we multiply through by 4 we find that $|4\sqrt{3} - 7| = 0.07\ldots < 1/10$. So we may take $a = 4 \leq 10$ and $b = 7$.

b. It is helpful to keep the decimal expansions of the multiples of $1/7$ in mind in these exercises. Here $\sqrt[3]{3} = 1.442\ldots$ and $3/7 = 0.428\ldots$ so that we have $\sqrt[3]{3} \approx 10/7$. Then as in part a., we investigate $|7\sqrt[3]{3} - 10| = 0.095\ldots < 1/10$. So we may take $a = 7 \leq 10$ and $b = 10$.

c. Since $\pi^2 = 9.869\ldots$ and $6/7 = 0.857\ldots$, we have that $\pi^2 \approx 69/7$, so we compute $|7\pi^2 - 69| = 0.087\ldots < 1/10$. So we may take $a = 7 \leq 10$ and $b = 69$.

d. Since $e^3 = 20.0855\ldots$ we may take $a = 1$ and $b = 20$ to get $|1e^3 - 20| = 0.855\ldots < 1/10$.

1.1.33. The number α must lie in some interval of the form $r/k \leq \alpha < (r+1)/k$. If we divide this interval into equal halves, then α must lie in one of the halves, so either $r/k \leq \alpha < (2r+1)/2k$ or $(2r+1)/2k \leq \alpha < (r+1)/k$. In the first case we have $|\alpha - r/k| < 1/2k$, so we take $u = r$. In the second case we have $|\alpha - (r+1)/k| < 1/2k$, so we take $u = r+1$.

1.1.35. First we have $|\sqrt{2} - 1/1| = 0.414\ldots < 1/1^2$. Second, Exercise 30, part a., gives us $|\sqrt{2} - 7/5| < 1/50 < 1/5^2$. Third, observing that $3/7 = 0.428\ldots$ leads us to try $|\sqrt{2} - 10/7| = 0.014\ldots < 1/7^2 = 0.0204\ldots$. Fourth, observing that $5/12 = 0.4166\ldots$ leads us to try $|\sqrt{2} - 17/12| = 0.00245\ldots < 1/12^2 = 0.00694\ldots$.

1.1.37. We may assume that b and q are positive. Note that if $q > b$, we have $|p/q - a/b| = |pb - aq|/qb \geq 1/qb > 1/q^2$. Therefore, solutions to the inequality must have $1 \leq q \leq b$. For a given q, there can be only finitely many p such that the distance between the rational numbers a/b and p/q is less than $1/q^2$ (indeed there is at most one.) Therefore there are only finitely many p/q satisfying the inequality.

1.1.39. a. Since $n3$ is an integer for all n, so is $[n3]$, so the first ten terms of the spectrum sequence are 3, 6, 9, 12, 15, 18, 21, 24, 27, 30.

b. The sequence for $n\sqrt{3}$, rounded, is 1.732, 3.464, 5.196, 6.928, 8.660, 10.39, 12.12, 13.86, 15.59, 17.32. When we apply the floor function to these numbers we get 1, 3, 5, 6, 8, 10, 12, 13, 15, 17 for the spectrum sequence.

c. The sequence for $n(3+\sqrt{3})/2$, rounded, is 2.366, 4.732, 7.098, 9.464, 11.83, 14.20, 16.56, 18.93, 21.29, 23.66. When we apply the floor function to these numbers we get 2, 4, 7, 9, 11, 14, 16, 18, 21, 23 for the spectrum sequence.

d. The sequence for $n\pi$, rounded is 3.142, 6.283, 9.425, 12.57, 15.71, 18.85, 21.99, 25.13, 28.27, 31.42. When we apply the floor function to these numbers we get 3, 6, 9, 12, 15, 18, 21, 25, 28, 31, for the spectrum sequence.

1.1.41. Assume that $1/\alpha + 1/\beta = 1$. Note first that for all integers n and m, $m\alpha \neq n\beta$, for otherwise, we solve the equations $m\alpha = n\beta$ and $1/\alpha + 1/\beta = 1$ and get rational solutions for α and β, a contradiction. Therefore the sequences $m\alpha$ and $n\beta$ are disjoint.

For an integer k, define $N(k)$ to be the number of elements of the sequences $m\alpha$ and $n\beta$ which are less than k. Now $m\alpha < k$ if and only if $m < k/\alpha$, so there are exactly $[k/\alpha]$ members of the sequence $m\alpha$ less than k. Likewise, there are exactly $[k/\beta]$ members of the sequence $n\beta$ less than k. So we have $N(k) = [k/\alpha] + [k/\beta]$. By definition of the greatest integer function, we have $k/\alpha - 1 < [k/\alpha] < k/\alpha$ and $k/\beta - 1 < [k/\beta] < k/\beta$, where the inequalities are strict because the numbers are irrational. If we add these inequalities we get $k/\alpha + k/\beta - 2 < N(k) < k/\alpha + k/\beta$ which simplifies to $k - 2 < N(k) < k$. Since $N(k)$ is an integer, we conclude that $N(k) = k - 1$. This shows that there is exactly one member of the union of the sequences $m\alpha$ and $n\beta$ in each interval of the form $k - 1 \leq x < k$, and therefore, when we apply the floor function to each member, exactly one will take on the value k.

Conversely, suppose that α and β are irrational numbers such that $1/\alpha + 1/\beta \neq 1$. If $1/\alpha + 1/\gamma = 1$ then we know from the first part of the theorem that the spectrum sequences for α and γ partition the positive integers. By Exercise 40., we know that the spectrum sequences for β and γ are different, so the sequences for α and β can not partition the positive integers.

1.1.43. Assume that there are only finitely many Ulam numbers. Let the two largest Ulam numbers be u_{n-1} and u_n. Then the integer $u_n + u_{n-1}$ is an Ulam number larger than u_n. It is the unique sum of two Ulam numbers since $u_i + u_j < u_n + u_{n-1}$ if $j < n$ or $j = n$ and $i < n - 1$.

1.2. Sums and Products

1.2.1. a. We have $\sum_{j=1}^{5} j^2 = 1^2 + 2^2 + 3^2 + 4^2 + 5^2 = 55$.

b. We have $\sum_{j=1}^{5}(-3) = (-3) + (-3) + (-3) + (-3) + (-3) = -15$.

c. We have $\sum_{j=1}^{5} 1/(j+1) = 1/2 + 1/3 + 1/4 + 1/5 + 1/6 = 29/20$.

1.2.3. a. We use the formula from Example 1.15 as follows. We evaluate the sum $\sum_{j=0}^{8} 2^j = 2^9 - 1 = 511$ as in Example 1.17. Then we have $\sum_{j=1}^{8} 2^j = \sum_{j=0}^{8} 2^j - 2^0 = 510$.

b. We could proceed as in part (a), or we may do the following: $\sum_{j=1}^{8} 5(-3)^j = \sum_{j=0}^{7} 5(-3)^{j+1}$
$= \sum_{j=0}^{7} -15(-3)^j$. We may apply the formula in Example 1.15 to this last sum, with $a = -15$, $n = 7$ and $r = -3$, to get the sum equal to $\dfrac{-15(-3)^8 - (-15)}{-3 - 1} = 24600$.

c. We manipulate the sum as in part b., so we can apply the formula from Example 1.15. $\sum_{j=1}^{8} 3(-1/2)^j = \sum_{j=0}^{7} 3(-1/2)^{j+1} = \sum_{j=0}^{7} (-3/2)(-1/2)^j = \dfrac{(-3/2)(-1/2)^8 - (-3/2)}{-1/2 - 1} = -\dfrac{255}{256}$.

1.2.5. The sum $\sum_{k=1}^{n} [\sqrt{k}]$ counts 1 for every value of k with $\sqrt{k} \geq 1$. There are n such values of k in the range $k = 1, 2, 3, \ldots, n$. It counts another 1 for every value of k with $\sqrt{k} \geq 2$. There are $n - 3$ such values in the range. The sum counts another 1 for each value of k with $\sqrt{k} \geq 3$. There are $n - 8$ such values in the range. In general, for $m = 1, 2, 3, \ldots, [\sqrt{n}]$ the sum counts a 1 for each value of k with $\sqrt{k} \geq m$, and there are $n - (m^2 - 1)$ values in the range. Therefore $\sum_{k=1}^{n} [\sqrt{k}] = \sum_{m=1}^{[\sqrt{n}]} n - (m^2 - 1) = [\sqrt{n}](n+1) - \sum_{m=1}^{[\sqrt{n}]} m^2 = [\sqrt{n}](n+1) - ([\sqrt{n}]([\sqrt{n}]+1)(2[\sqrt{n}]+1))/6$.

1.2.7. We see that $t_n = \sum_{j=1}^{n} j = \sum_{j=1}^{n}(n - j + 1)$. Thus, $2t_n = \sum_{j=1}^{n} j + \sum_{j=1}^{n}(n - j + 1) = \sum_{j=1}^{n}(n+1) = n(n+1)$.

1.2.9. From Exercise 8, we have $p_n = \sum_{k=1}^{n}(3k - 2) = 3\sum_{k=1}^{n} k - 2\sum_{k=1}^{n} 1 = 3n(n+1)/2 - 2n = (3n^2 - n)/2$. On the other hand, $t_{n-1} + n^2 = (n-1)n/2 + n^2 = (3n^2 - n)/2$, which is the same as above.

1.2.11. a. Consider a regular heptagon which we border successively by heptagons with $3, 4, 5, \ldots$ on each side. Define the *heptagonal numbers* $s_1, s_2, s_3, \ldots, s_k, \ldots$ to be the number of dots contained in the k nested heptagons.

b. First note that $s_1 = 1$. To get a recursive relationship we consider $s_k - s_{k-1}$, which counts the dots added to the $(k-1)$st heptagon to obtain the kth heptagon. To do this, we must add 5 sides of k dots each, but 4 of the dots belong to two sides. Therefore $s_k - s_{k-1} = 5k - 4$. A closed formula is then given by adding these differences together: $s_k = \sum_{i=1}^{k}(5i - 4) = 5t_k - 4k = 5k(k+1)/2 - 4k = (5k^2 - 3k)/2$.

1.2.13. We continue with the formula from Exercise 12. $T_n = \sum_{k=1}^{n} t_k = \sum_{k=1}^{n} k(k+1)/2$. Exploiting the same technique as in Example 1.19, we consider $(k+1)^3 - k^3 = 3k^2 + 3k + 1 = 3(k^2 + k) + 1$ and solve for $k^2 + k$ to get $k^2 + k = (k+1)^3 - k^3/3 - (1/3)$. Then $T_n = (1/2)\sum_{k=1}^{n} k(k+1) = (1/6)\sum_{k=1}^{n}((k+1)^3 - k^3) - (1/6)\sum_{k=1}^{n} 1$. The first sum is telescoping and the second sum is trivial, so we have $T_n = (1/6)((n+1)^3 - 1^3) - (1/6) = (n^3 + 3n^2 + 2n)/6$.

1.2.15. Each of these four quantities are products of 100 integers. The largest product is 100^{100}, since it is the product of 100 factors of 100. The second largest is $100!$ which is the product of the integers $1, 2, \ldots, 100$, and each of these terms is less or equal to 100. The third largest is $(50!)^2$ which is the product of $1^2, 2^2, \ldots, 50^2$, and each of these factors j^2 is less than $j(50 + j)$, whose product is $100!$. The smallest is 2^{100} which is the product of 100 2's.

1.2.17. $\sum_{k=1}^{n}\left(\frac{1}{k(k+1)}\right) = \sum_{k=1}^{n}\left(\frac{1}{k} - \frac{1}{k+1}\right)$. Let $a_j = 1/(j+1)$. Notice that this is a telescoping sum, and using the notation in the text preceding Example 1.15, we have $\sum_{k=1}^{n}\left(\frac{1}{k(k+1)}\right) = \sum_{j=1}^{n} a_{j-1} - a_j = -(a_n - a_0) = 1 - 1/(n+1) = n/(n+1)$.

1.2.19. We sum both sides of the identity $(k+1)^3 - k^3 = 3k^2 + 3k + 1$ from $k = 1$ to $k = n$. $\sum_{k=1}^{n}((k+1)^3 - k^3) = (n+1)^3 - 1$, since the sum is telescoping. $\sum_{k=1}^{n}(3k^2 + 3k + 1) = 3(\sum_{k=1}^{n} k^2) + 3(\sum_{k=1}^{n} k) + \sum_{k=1}^{n} 1 = 3(\sum_{k=1}^{n} k^2) + 3n(n+1)/2 + n$. As these two expressions are equal, solving for $\sum_{k=1}^{n} k^2$, we find that $\sum_{k=1}^{n} k^2 = (n/6)(2n+1)(n+1)$.

1.2.21. a. $10! = (7!)(8 \cdot 9 \cdot 10) = (7!)(720) = (7!)(6!)$.

b. $10! = (7!)(6!) = (7!)(5!) \cdot 6 = (7!)(5!)(3!)$.

c. $16! = (14!)(15 \cdot 16) = (14!)(240) = (14!)(5!)(2!)$.

d. $9! = (7!)(8 \cdot 9) = (7!)(6 \cdot 6 \cdot 2) = (7!)(3!)(3!)(2!)$

1.2.23. Assume that $x \leq y$. Then $z! = x! + y! \leq y! + y! = 2(y!)$. Since $z > y$ we have $z! \geq (y+1)y!$. This implies that $y + 1 \leq 2$. Hence the only solution with x, y, and z positive integers is $x = y = 1$ and $z = 2$.

1.3. Mathematical Induction

1.3.1. For $n = 1$ we have $1 < 2^1 = 2$. This is the basis case. Now assume $n < 2^n$. We then have $n + 1 < 2^n + 1 < 2^n + 2^n = 2^{n+1}$. This completes the inductive step and the proof by mathematical induction.

1.3.3. For the basis step we have $\sum_{k=1}^{1}\frac{1}{k^2} = 1 \leq 2 - \frac{1}{1} = 1$. For the inductive step, we assume that $\sum_{k=1}^{n}\frac{1}{k^2} \leq 2 - \frac{1}{n}$. Then, $\sum_{k=1}^{n+1}\frac{1}{k^2} = \sum_{k=1}^{n}\frac{1}{k^2} + \frac{1}{(n+1)^2} \leq 2 + \frac{1}{(n+1)^2}$ by the induction hypothesis. This is less than

$2 - \frac{1}{n+1} + \frac{1}{(n+1)^2} = 2 - \frac{1}{n+1}(1 - \frac{1}{n+1}) \leq 2 - \frac{1}{n+1}$, as desired.

1.3.5. We see that $\mathbf{A} = \begin{pmatrix} 1 & 1 \\ 0 & 1 \end{pmatrix}, \mathbf{A}^2 = \begin{pmatrix} 1 & 2 \\ 0 & 1 \end{pmatrix}, \mathbf{A}^3 = \mathbf{A}^2 \mathbf{A} = \begin{pmatrix} 1 & 3 \\ 0 & 1 \end{pmatrix}$ and so on. We conjecture that $\mathbf{A}^n = \begin{pmatrix} 1 & n \\ 0 & 1 \end{pmatrix}$. To prove this by mathematical induction we first note that the basis step follows since $\mathbf{A} = \begin{pmatrix} 1 & 1 \\ 0 & 1 \end{pmatrix}$. Next, we assume that $\mathbf{A}^n = \begin{pmatrix} 1 & n \\ 0 & 1 \end{pmatrix}$. Then $\mathbf{A}^{n+1} = \mathbf{A}^n \mathbf{A} = \begin{pmatrix} 1 & n \\ 0 & 1 \end{pmatrix} \begin{pmatrix} 1 & 1 \\ 0 & 1 \end{pmatrix} = \begin{pmatrix} 1 & n+1 \\ 0 & 1 \end{pmatrix}$.

1.3.7. For the basis step, we have $\sum_{j=1}^{1} j^2 = 1 = 1(1+1)(2 \cdot 1 + 1)/6$. For the inductive step, we assume that $\sum_{j=1}^{n} j^2 = n(n+1)(2n+1)/6$. Then, $\sum_{j=1}^{n+1} j^2 = \sum_{j=1}^{n} j^2 + (n+1)^2 = n(n+1)(2n+1)/6 + (n+1)^2 = (n+1)(n(2n+1)/6 + n+1) = (n+1)(2n^2 + 7n + 6)/6 = (n+1)(n+2)[2(n+1) + 1]/6$, as desired.

1.3.9. For the basis step, we have $\sum_{j=1}^{1} j(j+1) = 2 = 1(2)(3)/3$. Assume it is true for n. Then $\sum_{j=1}^{n+1} j(j+1) = n(n+1)(n+2)/3 + (n+1)(n+2) = (n+1)(n+2)(n/3 + 1) = (n+1)(n+2)(n+3)/3$.

1.3.11. We have $\prod_{j=1}^{n} 2^j = 2^{\sum_{j=1}^{n} j} = 2^{n(n+1)/2}$ since $\sum_{j=1}^{n} j = \frac{n(n+1)}{2}$.

1.3.13. We will prove this using mathematical induction. We see that $12 = 4 \cdot 3$. Now assume that postage of n cents can be formed, with $n = 4a + 5b$, where a and b are nonnegative integers. To form $n+1$ cents postage, if $a > 0$ we can replace a 4-cent stamp with a 5-cent stamp; that is, $n+1 = 4(a-1) + 5(b+1)$. If no 4-cent stamps are present, then all 5-cent stamps were used. It follows that there must be at least three 5-cent stamps and these can be replaced by four 4-cent stamps; that is, $n+1 = 4(a+4) + 5(b-3)$.

1.3.15. We use mathematical induction. The inequality is true for $n = 0$ since $H_{2^0} = H_1 = 1 \geq 1 = 1 + 0/2$. Now assume that the inequality is true for n, that is, $H_{2^n} \geq 1 + n/2$. Then $H_{2^{n+1}} = \sum_{j=1}^{2^n} 1/j + \sum_{j=2^n+1}^{2^{n+1}} 1/j \geq H_{2^n} + \sum_{j=2^n+1}^{2^{n+1}} 1/2^{n+1} \geq 1 + n/2 + 2^n \cdot 1/2^{n+1} = 1 + n/2 + 1/2 = 1 + (n+1)/2$. This completes the inductive proof.

1.3.17. For the basis step, we have $(2 \cdot 1)! = 2 < 2^{2 \cdot 1}(1!)^2 = 4$. For the inductive step, we assume that $(2n)! < 2^{2n}(n!)^2$. Then $[2(n+1)]! = (2n)!(2n+1)(2n+2) < 2^{2n}(n!)^2(2n+1)(2n+2) < 2^{2n}(n!)^2(2n+2)^2 = 2^{2(n+1)}[(n+1)!]^2$, as desired.

1.3.19. Let A be such a set. Define B as $B = \{x - k + 1 \mid x \in A \text{ and } x \geq k\}$. Since $x \geq k$, B is a set of positive integers. Since $k \in A$ and $k \geq k$, $k - k + 1 = 1$ is in B. Since $n+1$ is in A whenever n is, $n+1-k+1$ is in B whenever $n-k+1$ is. Thus B satisfies the hypothesis for mathematical induction, i.e. B is the set of positive integers. Mapping B back to A in the natural manner, we find that A contains the set of integers greater than or equal to k.

1.3.21. For the basis step, we have $4^2 = 16 < 24 = 4!$. For the inductive step, we assume that $n^2 < n!$. Then, $(n+1)^2 = n^2 + 2n + 1 < n! + 2n + 1 < n! + 3n < n! + n! = 2n! < (n+1)n! = (n+1)!$, as desired.

1.3.23. We use the second principle of mathematical induction. For the basis step, if the puzzle has only one piece, then it is assembled with exactly 0 moves. For the induction step, assume that all puzzles with $k \leq n$ pieces require $k - 1$ moves to assemble. Suppose it takes m moves to assemble a puzzle with $n+1$ pieces. Then the m move consists of joining two blocks of size a and b, respectively, with $a + b = n + 1$. But by the induction hypothesis, it requires exactly $a - 1$ and $b - 1$ moves to assemble each of these blocks. Thus, $m = (a-1) + (b-1) + 1 = a + b + 1 = n + 1$. This completes the induction.

1.3.25. Suppose that $f(n)$ is defined recursively by specifying the value of $f(1)$ and a rule for finding $f(n+1)$ from $f(n)$. We will prove by mathematical induction that such a function is well-defined. First, note that $f(1)$ is well-defined since this value is explicitly stated. Now assume that $f(n)$ is well-defined. Then

1.3.27. We have $g(1) = 2, g(2) = 2^{g(1)} = 4, g(3) = 2^{g(2)} = 2^4 = 16$, and $g(4) = 2^{g(3)} = 2^{16} = 65536$.

1.3.29. We use the second principle of mathematical induction. The basis step consists of verifying the formula for $n = 1$ and $n = 2$. For $n = 1$ we have $f(1) = 1 = 2^1 + (-1)^1$ and for $n = 2$ we have $f(2) = 5 = 2^2 + (-1)^2$. Now assume that $f(k) = 2^k + (-1)^k$ for all positive integers k with $k < n$ where $n > 2$. By the induction hypothesis it follows that $f(n) = f(n-1) + 2f(n-2) = (2^{n-1} + (-1)^{n-1}) + 2(2^{n-2} + (-1)^{n-2}) = (2^{n-1} + 2^{n-1}) + (-1)^{n-2}(-1 + 2) = 2^n + (-1)^n$. This finishes the proof.

1.3.31. We use the second principle of mathematical induction. We see that $a_0 = 1 \leq 3^0 = 1$, $a_1 = 3 \leq 3^1 = 3$, and $a_2 = 9 \leq 3^2 = 9$. These are the basis cases. Now assume that $a_k \leq 3^k$ for all integers k with $0 \leq k < n$. It follows that $a_n = a_{n-1} + a_{n-2} + a_{n-3} \leq 3^{n-1} + 3^{n-2} + 3^{n-3} = 3^{n-3}(1 + 3 + 9) = 13 \cdot 3^{n-3} < 27 \cdot 3^{n-3} = 3^n$. The induction argument is complete.

1.3.33. Let P_n be the statement for n. Then P_2 is true, since we have $(a_1 + a_2)/2)^2 - a_1 a_2 = (a_1 - a_2/2)^2 \geq 0$. Assume P_n is true. Then by P_2, for $2n$ positive real numbers $a_1, \ldots, a_{2n}$ we have $a_1 + \cdots + a_{2n} \geq 2(\sqrt{a_1 a_2} + \sqrt{a_3 a_4} + \cdots + \sqrt{a_{2n-1} a_{2n}})$. Apply P_n to this last expression to get $a_1 + \cdots + a_{2n} \geq 2n(a_1 a_2 \cdots a_{2n})^{1/2n}$ which establishes P_n for $n = 2^k$ for all k. Again, assume P_n is true. Let $g = (a_1 a_2 \cdots a_{n-1})^{1/(n-1)}$. Applying P_n, we have $a_1 + a_2 + \cdots + a_{n-1} + g \geq n(a_1 a_2 \cdots a_{n-1} g)^{1/n} = n(g^{n-1} g)^{1/n} = ng$. Therefore, $a_1 + a_2 + \cdots + a_{n-1} \geq (n-1)g$ which establishes P_{n-1}. Thus P_{2^k} is true and P_n implies P_{n-1}. This establishes P_n for all n.

1.3.35. Note that since $0 < p < q$ we have $0 < p/q < 1$. The proposition is trivially true if $p = 1$. We proceed by strong induction on p. Let p and q be given and assume the proposition is true for all rational numbers between 0 and 1 with numerators less than p. To apply the algorithm, we find the unit fraction $1/s$ such that $1/(s-1) > p/q > 1/s$. When we subtract, the remaining fraction is $p/q - 1/s = (ps - q)/qs$. On the other hand, if we multiply the first inequality by $q(s-1)$ we have $q > p(s-1)$ which leads to $p > ps - q$, which shows that the numerator of p/q is strictly greater than the numerator of the remainder $(ps - q)/qs$ after one step of the algorithm. By the induction hypothesis, this remainder is expressible as a sum of unit fractions, $1/u_1 + \cdots + 1/u_k$. Therefore $p/q = 1/s + 1/u_1 + \cdots + 1/u_k$ which completes the induction step.

1.4. The Fibonacci Numbers

1.4.1. a. We have $f_1 = 1$, $f_2 = 1$, and $f_n = f_{n-1} + f_{n-2}$ for $n \geq 3$. Hence $f_3 = f_2 + f_1 = 1 + 1 = 2$, $f_4 = f_3 + f_2 = 2 + 1 = 3$, $f_5 = 3 + 2 = 5$, $f_6 = 5 + 3 = 8$, $f_7 = 8 + 5 = 13$, $f_8 = 13 + 8 = 21$, $f_9 = 21 + 13 = 34$, and $f_{10} = 34 + 21 = 55$.

b. We continue beyond part (a) finding that $f_{11} = f_{10} + f_9 = 55 + 34 = 89$, $f_{12} = 89 + 55 = 144$, and $f_{13} = 144 + 89 = 233$.

c. We continue beyond part (b) finding that $f_{14} = f_{13} + f_{12} = 233 + 144 = 377$, and $f_{15} = 377 + 233 = 610$.

d. We continue beyond part (c) finding that $f_{16} = 610 + 377 = 987$, $f_{17} = 987 + 610 = 1597$, and $f_{18} = 1597 + 987 = 2584$.

e. We continue beyond part (d) finding that $f_{19} = 2584 + 1597 = 4181$, $f_{20} = 4181 + 2584 = 6765$.

f. We continue beyond part (e) finding that $f_{21} = 6765 + 4181 = 10946$, $f_{22} = 10946 + 6765 = 17711$, $f_{23} = 17711 + 10946 = 28657$, $f_{24} = 28657 + 17711 = 46368$, and $f_{25} = 46368 + 28657 = 75025$.

1.4.3. Note that from the Fibonacci identity, whenever n is a positive integer, $f_{n+2} - f_n = f_{n+1}$. Then we have $2f_{n+2} - f_n = f_{n+2} + (f_{n+2} - f_n) = f_{n+2} + f_{n+1} = f_{n+3}$. If we add f_n to both sides of this equation,

1.4.5. For $n = 1$ we have $f_{2 \cdot 1} = 1 = 1^2 + 2 \cdot 1 \cdot 0 = f_1^2 + 2f_0 f_1$, and for $n = 2$, we have $f_{2 \cdot 2} = 3 = 1^2 + 2 \cdot 1 \cdot 1 = f_2^2 + 2f_1 f_2$. So the basis step holds for strong induction. Assume, then that $f_{2n-4} = f_{n-2}^2 + 2f_{n-3}f_{n-2}$ and $f_{2n-2} = f_{n-1}^2 + 2f_{n-2}f_{n-1}$. Now compute $f_{2n} = f_{2n-1} + f_{2n-2} = 2f_{2n-2} + f_{2n-3} = 3f_{2n-2} - f_{2n-4}$. Now we may substitute in our induction hypotheses to set this last expression equal to $3f_{n-1}^2 + 6f_{n-2}f_{n-1} - f_{n-2}^2 - 2f_{n-3}f_{n-2} = 3f_{n-1}^2 + 6(f_n - f_{n-1})f_{n-1} - (f_n - f_{n-1})^2 - 2(f_{n-1} - f_{n-2})(f_n - f_{n-1}) = -2f_{n-1}^2 + 6f_n f_{n-1} - f_n^2 + 2f_n(f_n - f_{n-1}) - 2f_{n-1}(f_n - f_{n-1}) = f_n^2 + 2f_{n-1}f_n$ which completes the induction step.

1.4.7. Note that $f_1 = 1 = f_2$, $f_1 + f_3 = 3 = f_4$, and $f_1 + f_3 + f_5 = 8 = f_6$ so we conjecture that $f_1 + f_3 + f_5 + \cdots + f_{2n-1} = f_{2n}$. We prove this by induction. The basis step is checked above. Assume that our formula is true for n, and consider $f_1 + f_3 + f_5 + \cdots + f_{2n-1} + f_{2n+1} = f_{2n} + f_{2n+1} = f_{2n+2}$, which is the induction step. Therefore the formula is correct.

1.4.9. First suppose $n = 2k$ is even. Then $f_n - f_{n-1} + \cdots + (-1)^{n+1}f_1 = (f_{2k} + f_{2k-1} + \cdots + f_1) - 2(f_{2k-1} + f_{2k-3} + \cdots + f_1) = (f_{2k+2} - 1) - 2(f_{2k})$ by the formulas in Example 1.23 and Exercise 3. This last equals $(f_{2k+2} - f_{2k}) - f_{2k} - 1 = f_{2k+1} - f_{2k} - 1 = f_{2k-1} - 1 = f_{n-1} - 1$. Now suppose $n = 2k+1$ is odd. Then $f_n - f_{n-1} + \cdots + (-1)^{n+1} = f_{2k+1} - (f_{2k} - f_{2k-1} + \cdots - (-1)^{n+1}f_1) = f_{2k+1} - (f_{2k-1} - 1)$ by the formula just proved for the even case. This last equals $(f_{2k+1} - f_{2k-1}) + 1 = f_{2k} + 1 = f_{n-1} + 1$. We can unite the formulas for the odd and even cases by writing the formula as $f_{n-1} - (-1)^n$.

1.4.11. We can construct an induction proof similar to the ones in Exercises 5 and 10, or we may proceed as follows. From Exercise 5, we have $f_{2n} = f_n^2 + 2f_{n-1}f_n = f_n(f_n + f_{n-1} + f_{n-1}) = (f_{n+1} - f_{n-1})(f_{n+1} + f_{n-1}) = f_{n+1}^2 - f_{n-1}^2$, which is the desired identity.

1.4.13. We proceed by mathematical induction. For the basis step, $\sum_{j=1}^{1} f_j^2 = f_1^2 = f_1 f_2$. To make the inductive step we assume that $\sum_{j=1}^{n} f_j^2 = f_n f_{n+1}$. Then $\sum_{j=1}^{n+1} f_j^2 = \sum_{j=1}^{n} f_j^2 + f_{n+1}^2 = f_n f_{n+1} + f_{n+1}^2 = f_{n+1}f_{n+2}$.

1.4.15. From Exercise 13, we have $f_{n+1}f_n - f_{n-1}f_{n-2} = (f_1^2 + \cdots + f_n^2) - (f_1^2 + \cdots f_{n-2}^2) = f_n^2 + f_{n-1}^2$. The identity in Exercise 10 shows that this is equal to f_{2n-1} when n is a positive integer, and in particular when n is greater than 2.

1.4.17. For fixed m, we proceed by induction on n. The basis step is $f_{m+1} = f_m f_2 + f_{m-1}f_1 = f_m \cdot 1 + f_{m-1} \cdot 1$ which is true. Assume the identity holds for $1, 2, \ldots, k$. Then $f_{m+k} = f_m f_{k+1} + f_{m-1}f_k$ and $f_{m+k-1} = f_m f_k + f_{m-1}f_{k-1}$. Adding these equations gives us $f_{m+k} + f_{m+k-1} = f_m(f_{k+1} + f_k) + f_{m-1}(f_k + f_{k-1})$. Applying the recursive definition yields $f_{m+k+1} = f_m f_{k+2} + f_{m-1}f_{k+1}$, which is precisely the identity.

1.4.19. A few trial cases lead us to conjecture that $\sum_{i=1}^{n} L_i = L_{n+2} - 3$. We prove that this formula is correct by induction. The basis step is $L_1 = 1$ and $L_3 - 3 = 4 - 3 = 1$, which checks. Assume that the formula holds for n and compute $\sum_{i=1}^{n+1} L_i = \sum_{i=1}^{n} L_i + L_{n+1} = L_{n+2} - 3 + L_{n+1}$ by the induction hypothesis. This last equals $(L_{n+2} + L_{n+1}) - 3 = L_{n+3} - 3$, which completes the induction step.

1.4.21. A few trial cases lead us to conjecture that $\sum_{i=1}^{n} L_{2i} = L_{2n+1} - 1$. We prove that this formula is correct by induction. The basis step is $L_2 = 3 = L_3 - 1$. Assume that the formula holds for n and compute $\sum_{i=1}^{n+1} L_{2i} = \sum_{i=1}^{n} L_{2i} + L_{2n+2} = L_{2n+1} - 1 + L_{2n+2} = L_{2n+3} - 1$, which completes the induction step.

1.4.23. We proceed by induction. The basis step is $L_1^2 = 1 = L_1 L_2 - 2 = 1 \cdot 3 - 2$. Assume the formula holds for n and consider $\sum_{i=1}^{n+1} L_i^2 = \sum_{i=1}^{n} L_i^2 + L_{n+1}^2 = L_n L_{n+1} - 2 + L_{n+1}^2 = L_{n+1}(L_n + L_{n+1}) - 2 = L_{n+1}L_{n+2} - 2$, which completes the induction step.

1.4.25. For the basis step, we check that $L_1 f_1 = 1 \cdot 1 = 1 = f_2$ and $L_2 f_2 = 3 \cdot 1 = 3 = f_4$. Assume the identity is true for all positive integers up to n. Then we have $f_{n+1}L_{n+1} = (f_{n+2} - f_n)(f_{n+2} - f_n)$ from Exercise 16. This equals $f_{n+2}^2 - f_n^2 = (f_{n+1} + f_n)^2 - (f_{n-1} - f_{n-2})^2 = f_{n+1}^2 + 2f_{n+1}f_n + f_n^2 - f_{n-1}^2 - 2f_{n-1}f_{n-2} - f_{n-2}^2 = (f_{n+1}^2 - f_{n-1}^2) + (f_n^2 - f_{n-2}^2) + 2(f_{n+1}f_n - f_{n-1}f_{n-2}) = (f_{n+1} - f_{n-1})(f_{n+1} + f_{n-1}) + (f_n -$

$f_{n-2})(f_n + f_{n-2}) + 2(f_{2n-1})$, where the last parenthetical expression is obtained from Exercise 8. This equals $f_n L_n + f_{n-1}L_{n-1} + 2f_{2n-1}$. Applying the induction hypothesis yields $f_{2n} + f_{2n-2} + 2f_{2n-1} = (f_{2n} + f_{2n-1}) + (f_{2n-1} + f_{2n-2}) = f_{2n+1} + f_{2n} = f_{2n+2}$, which completes the induction.

1.4.27. We prove this by induction on n. Fix m a positive integer. If $n = 2$, then for the basis step we need to show that $L_{m+2} = f_{m+1}L_2 + f_m L_1 = 3f_{m+1} + f_m$, for which we will use induction on m. For $m = 1$ we have $L_3 = 4 = 3 \cdot f_2 + f_1$ and for $m = 2$ we have $L_4 = 7 = 3 \cdot f_3 + f_2$, so the basis step for m holds. Now assume that the basis step for n holds for all values of m less than and equal to m. Then $L_{m+3} = L_{m+2} + L_{m+1} = 3f_{m+1} + f_m + 3f_m + f_{m-1} = 3f_{m+2} + f_{m+1}$, which completes the induction step on m and proves the basis step for n. To prove the induction step on n, we compute $L_{m+n+1} = L_{m+n} + L_{m+n-1} = (f_{m+1}L_n + f_m L_{n-1}) + (f_{m+1}L_{n-1} + f_m L_{n-2}) = f_{m+1}(L_n + L_{n-1}) + f_m(L_{n-1} + L_{n-2}) = f_{m+1}L_{n+1} + f_m L_n$, which completes the induction on n and proves the identity.

1.4.29. We find that $50 = 34 + 13 + 3 = f_9 + f_7 + f_4$, $85 = 55 + 21 + 8 + 1 = f_{10} + f_8 + f_6 + f_2$, $110 = 89 + 21 = f_{11} + f_8$ and $200 = 144 + 55 + 1 = f_{12} + f_{10} + f_2$. In each case, we used the "greedy" algorithm, always subtracting the largest possible Fibonacci number from the remainder.

1.4.31. We proceed by mathematical induction. The basis steps ($n = 2$ and 3) are easily seen to hold. For the inductive step, we assume that $f_n \leq \alpha^{n-1}$ and $f_{n-1} \leq \alpha^{n-2}$. Now, $f_{n+1} = f_n + f_{n-1} \leq \alpha^{n-1} + \alpha^{n-2} = \alpha^n$, since α satisfies $\alpha^n = \alpha^{n-1} + \alpha^{n-2}$.

1.4.33. Using Theorem 1.3 and the notation therein, we have $\alpha^2 = \alpha + 1$ and $\beta^2 = \beta + 1$, since they are roots of $x^2 - x - 1 = 0$. Then we have $f_{2n} = (\alpha^{2n} - \beta^{2n})/\sqrt{5} = (1/\sqrt{5})((\alpha + 1)^n - (\beta + 1)^n) = (1/\sqrt{5})\left(\sum_{j=0}^n \binom{n}{j}\alpha^j - \sum_{j=0}^n \binom{n}{j}\beta^j\right) = (1/\sqrt{5})\sum_{j=0}^n \binom{n}{j}(\alpha^j - \beta^j) = \sum_{j=1}^n \binom{n}{j}f_j$ since the first term is zero in the penultimate sum.

1.4.35. On one hand, $\det(\mathbf{F}^n) = \det(\mathbf{F})^n = (-1)^n$. On the other hand,
$$\det\begin{pmatrix} f_{n+1} & f_n \\ f_n & f_{n-1} \end{pmatrix} = f_{n+1}f_{n-1} - f_n^2.$$

1.4.37. We use the relationship $f_n = f_{n+2} - f_{n+1}$ to extend the definition to include negative indices. Thus, $f_0 = 0, f_{-1} = 1, f_{-2} = -1, f_{-3} = 2, f_{-4} = -3, f_{-5} = 5, f_{-6} = -8, f_{-7} = 13, f_{-8} = -21, f_{-9} = 34, f_{-10} = -55$.

1.4.39. The square has area 64 square units, while the rectangle has area 65 square units. This corresponds to the identity in Exercise 7, which tells us that $f_7 f_5 - f_6^2 = 1$. Notice that the slope of the hypotenuse of the triangular piece is $3/8$, while the slope of the top of the trapezoidal piece is $2/5$. We have $2/5 - 3/8 = 1/40$. Thus, the "diagonal" of the rectangle is really a very skinny parallelogram of area 1, hidden visually by the fact that the two slopes are nearly equal.

1.4.41. We solve the equation $r^2 - r - 1 = 0$ to discover the roots $r_1 = (1 + \sqrt{5})/2$ and $r_2 = (1 - \sqrt{5})/2$. Then according to the theory in the paragraph above, $f_n = C_1 r_1^n + C_2 r_2^n$. For $n = 0$ we have $0 = C_1 r_1^0 + C_2 r_2^0 = C_1 + C_2$. For $n = 1$ we have $1 = C_1 r_1 + C_2 r_2 = C_1(1 + \sqrt{5})/2 + C_2(1 - \sqrt{5})/2$. Solving these two equations simultaneously yields $C_1 = 1/\sqrt{5}$ and $C_2 = -1/\sqrt{5}$. So the explicit formula is $f_n = (1/\sqrt{5})r_1^n - (1/\sqrt{5})r_2^n = (r_1^n - r_2^n)/\sqrt{5}$.

1.4.43. We seek to solve the recurrence relation $L_n = L_{n-1} + L_{n-1}$ subject to the initial conditions $L_1 = 1$ and $L_2 = 3$. We solve the equation $r^2 - r - 1 = 0$ to discover the roots $\alpha = (1 + \sqrt{5})/2$ and $\beta = (1 - \sqrt{5})/2$. Then according to the theory in the paragraph above Exercise 31, $L_n = C_1 \alpha^n + C_2 \beta^n$. For $n = 1$ we have $L_1 = 1 = C_1 \alpha + C_2 \beta$. For $n = 2$ we have $3 = C_1 \alpha^2 + C_2 \beta^2$. Solving these two equations simultaneously yields $C_1 = 1$ and $C_2 = 1$. So the explicit formula is $L_n = \alpha^n + \beta^n$.

1.4.45. First check that $\alpha^2 = \alpha + 1$ and $\beta^2 = \beta + 1$. We proceed by induction. The basis steps are $(1/\sqrt{5})(\alpha - \beta) = (1/\sqrt{5})(\sqrt{5}) = 1 = f_1$ and $(1/\sqrt{5})(\alpha^2 - \beta^2) = (1/\sqrt{5})((1 + \alpha) - (1 + \beta)) = (1/\sqrt{5})(\alpha - \beta) = 1 =$

f_2. Assume the identity is true for all positive integers up to n. Then $f_{n+1} = f_n + f_{n-1} = (1/\sqrt{5})(\alpha^n - \beta^n) + (1/\sqrt{5})(\alpha^{n-1} - \beta^{n-1}) = (1/\sqrt{5})(\alpha^{n-1}(\alpha+1) - \beta^{n-1}(\beta+1)) = (1/\sqrt{5})(\alpha^{n-1}(\alpha^2) - \beta^{n-1}(\beta^2)) = (1/\sqrt{5})(\alpha^{n+1} - \beta^{n+1})$, which completes the induction.

1.5. Divisibility

1.5.1. We find that $3 \mid 99$ since $99 = 3 \cdot 33$, $5 \mid 145$ since $145 = 5 \cdot 29$, $7 \mid 343$ since $343 = 7 \cdot 49$, and $888 \mid 0$ since $0 = 888 \cdot 0$.

1.5.3. a. Yes, $0 = 7 \cdot 0$.

    b. Yes, $707 = 7 \cdot 101$.

    c. By the division algorithm, we have $1717 = 245 \cdot 7 + 2$. Since the remainder is nonzero, we know that $7 \nmid 1717$.

    d. By the division algorithm, we have $123321 = 17617 \cdot 7 + 2$. Since the remainder is nonzero, we know that $7 \nmid 123321$.

    e. By the division algorithm, we have $-285714 = -40817 \cdot 7 + 5$. Since the remainder is nonzero, we know that $7 \nmid -285714$.

    f. By the division algorithm, we have $-430597 = -61514 \cdot 7 + 1$. Since the remainder is nonzero, we know that $7 \nmid -430597$.

1.5.5. a. We have $100 = 5 \cdot 17 + 15$, so the quotient is 5 and the remainder is 15.

    b. We have $289 = 17 \cdot 17$, so the quotient is 17 and the remainder is 0.

    c. We have $-44 = -3 \cdot 17 + 7$, so the quotient is -3 and the remainder is 7.

    d. We have $-100 = -6 \cdot 17 + 2$, so the quotient is -6 and the remainder is 2.

1.5.7. By hypothesis we know $b = ra$ and $d = sc$, for some r and s. Thus $bd = rs(ac)$ and $ac \mid bd$.

1.5.9. If $a \mid b$, then $b = na$ and $bc = n(ca)$, i.e. $ac \mid bc$. Now, suppose $ac \mid bc$. Thus $bc = nac$ and, as $c \neq 0$, $b = na$, i.e., $a \mid b$.

1.5.11. By definition, $a \mid b$ if and only if $b = na$ for some integer n. Then raising both sides of this equation to the kth power yields $b^k = n^k a^k$ whence $a^k \mid b^k$.

1.5.13. Let a and b be odd, and c even. Then $ab = (2x+1)(2y+1) = 4xy + 2x + 2y + 1 = 2(2xy + x + y) + 1$, so ab is odd. On the other hand, for any integer n, we have $cn = (2z)n = 2(zn)$ which is even.

1.5.15. By the division algorithm, $a = bq + r$, with $0 \leq r < b$. Thus $-a = -bq - r = -(q+1)b + b - r$. If $0 \leq b - r < b$ then we are done. Otherwise $b - r = b$, or $r = 0$ and $-a = -qb + 0$.

1.5.17. a. The division algorithm covers the case when b is positive. If b is negative, then we may apply the division algorithm to a and $|b|$ to get a quotient q and remainder r such that $a = q|b| + r$ and $0 \leq r < |b|$. But since b is negative, we have $a = q(-b) + r = (-q)b + r$, as desired.

    b. We have $17 = -7(-2) + 3$. Here $r = 3$.

1.5.19. By the division algorithm, let $m = qn + r$, with $0 \leq r < n - 1$ and $q = [m/n]$. Then $[(m+1)/n] = [(qn + r + 1)/n] = [q + (r+1)/n] = q + [(r+1)/n]$ as in Example 1.31. If $r = 0, 1, 2, \ldots, n-2$, then $m \neq kn - 1$ for any integer k and $1/n \leq (r+1)/n < 1$ and so $[(r+1)/n] = 0$. In this case, we have

1.5. DIVISIBILITY

$[(m+1)/n] = q + 0 = [m/n]$. On the other hand, if $r = n - 1$, then $m = qn + n - 1 = n(q+1) - 1 = nk - 1$, and $[(r+1)/n] = 1$. In this case, we have $[(m+1)/n] = q + 1 = [m/n] + 1$.

1.5.21. The positive integers divisible by the positive integer d are those integers of the form kd where k is a positive integer. The number of these that are less than x is the number of positive integers k with $kd \leq x$, or equivalently with $k \leq x/d$. There are $[x/d]$ such integers.

1.5.23. There are $[1000/7] - [100/7] = 142 - 14 = 128$ integers between 100 and 1000 that are divisible by 7. There are $[1000/49] - [100/49] = 20 - 2 = 18$ integers between 100 and 1000 that are divisible by 49.

1.5.25. Using the Principle of Inclusion-Exclusion, the answer is $1000 - ([1000/3] + [1000/5] + [1000/7]) + ([1000/15] + [1000/21] + [1000/35]) - ([1000/105]) = 1000 - (333 + 200 + 142) + (66 + 47 + 28) - 9 = 462$.

1.5.27. Let w be the weight of a letter in ounces. Note that the function $-[-x]$ rounds x up to the least integer less than or equal to x. (That is, it's the equivalent of the ceiling function.) The cost of mailing a letter weighing w ounces is, then, 33 cents plus 22 cents for each ounce or part thereof more than 1, so we need to round $w - 1$ up to the next integer. So the cost is $c(w) = 33 - [1 - w]23$ cents. Suppose that $33 - [1 - w]22 = 145$. then $-[1-w]22 = 145 - 33 = 112$ which is not a multiple of 22, so no letter can cost $1.45. Suppose that $33 - [1 - w]22 = 231$. then $-[1 - w]22 = 231 - 33 = 198 = 9 \cdot 22$. Then $[1 - w] = -9$, so $-9 \leq 1 - w < -8$, or $9 < w \leq 10$. So a letter weight at least 9 ounces but less than 10 ounces would cost $2.31.

1.5.29. Multiplying two integers of this form gives us $(4n + 1)(4m + 1) = 16mn + 4m + 4n + 1 = 4(4mn + m + n) + 1$. Similarly, $(4n + 3)(4m + 3) = 16mn + 12m + 12n + 9 = 4(4mn + 3m + 3n + 2) + 1$.

1.5.31. Every odd integer may be written in the form $4k + 1$ or $4k + 3$. Observe that $(4k + 1)^4 = 16^2 k^4 + 4(4k)^3 + 6(4k)^2 + 4(4k) + 1 = 16(16k^4 + 16k^3 + 6k^2 + k) + 1$. Proceeding further, $(4k + 3)^4 = (4k)^4 + 12(4k)^3 + 54(4k)^2 + 108(4k) + 3^4 = 16(16k^4 + 48k^3 + 54k^2 + 27k + 5) + 1$.

1.5.33. Of any consecutive three integers, one is a multiple of three. Also, at least one is even. Therefore, the product is a multiple of $2 \cdot 3 = 6$.

1.5.35. For the basis step note that $0^3 + 1^3 + 2^3 = 9$ is a multiple of 9. Suppose that $n^3 + (n+1)^3 + (n+2)^3 = 9k$ for some integer k. Then $(n+1)^3 + (n+2)^3 + (n+3)^3 = n^3 + (n+1)^3 + (n+2)^3 + (n+3)^3 - n^3 = 9k + n^3 + 9n^2 + 27n + 27 - n^3 = 9k + 9n^2 + 27n + 27 = 9(k + n^2 + 3n + 3)$ which is a multiple of 9.

1.5.37. We proceed by mathematical induction. The basis step is clear. Assume that only f_{4n}'s are divisible by 3 for $f_i, i \leq 4k$. Then, as $f_{4k+1} = f_{4k} + f_{4k-1}$, $3 \mid f_{4k}$ and $3 \mid f_{4k+1}$ gives us the contradiction $3 \mid f_{4k-1}$. Thus $3 \nmid f_{4k+1}$. Continuing on, if $3 \mid f_{4k}$ and $3 \mid f_{4k+2}$, then $3 \mid f_{4k+1}$, which contradicts the statement just proved. If $3 \mid f_{4k}$ and $3 \mid f_{4k+3}$, then since $f_{4k+3} = 2f_{4k+1} + f_{4k}$, we again have a contradiction. But, as $f_{4k+4} = 3f_{4k+1} + 2f_{4k}$, and $3 \mid f_{4k}$ and $3 \mid 3 \cdot f_{4k+1}$, we see that $3 \mid f_{4k+4}$.

1.5.39. First note that for $n > 5$, $5f_{n-4} + 3f_{n-5} = 2f_{n-4} + 3(f_{n-4} + f_{n-5}) = 2f_{n-4} + 3f_{n-3} = 2(f_{n-4} + f_{n-3}) + f_{n-3} = 2f_{n-2} + f_{n-3} = f_{n-2} + f_{n-2} + f_{n-3} = f_{n-2} + f_{n-1} = f_n$, which proves the first identity. Now note that $f_5 = 5$ is divisible by 5. Suppose that f_{5n} is divisible by 5. From the identity above $f_{5n+5} = 5f_{5n+5-4} + 3f_{5n+5-5} = 5f_{5n+1} + 3f_{5n}$, which is divisible by 5 since $5f_{5n+1}$ is a multiple of 5 and, by the induction hypothesis, so is f_{5n}. This completes the induction.

1.5.41. Iterating the transformation T starting with 39 we find that $T(39) = 59$; $T(59) = 89$; $T(89) = 134$; $T(134) = 67$; $T(67) = 101$; $T(101) = 152$; $T(152) = 76$; $T(76) = 38$; $T(38) = 19$; $T(19) = 29$; $T(29) = 44$; $T(44) = 22$; $T(22) = 11$; $T(11) = 17$; $T(17) = 26$; $T(26) = 13$; $T(13) = 20$; $T(20) = 10$; $T(10) = 5$; $T(5) = 8$; $T(8) = 4$; $T(4) = 2$; $T(2) = 1$.

1.5.43. We prove this using the second principle of mathematical induction. Since $T(2) = 1$, the Collatz conjecture is true for $n = 2$. Now assume that the conjecture holds for all integers less that n. By assumption there is an integer k such that k iterations of the transformation T, starting at n, produces an integer m less than n. By the inductive hypothesis there is an integer l such that iterating T l times starting at m

produces the integer 1. Hence iterating T $k+l$ times starting with n leads to 1. This finishes the proof.

1.5.45. We first show that $(2+\sqrt{3})^n + (2-\sqrt{3})^n$ is an even integer. By the binomial theorem it follows that $(2+\sqrt{3})^n + (2-\sqrt{3})^n = \sum_{j=0}^{n} \binom{n}{j} 2^j \sqrt{3}^{n-j} + \sum_{j=0}^{n} \binom{n}{j} 2^j (-1)^{n-j} \sqrt{3}^{n-j} = 2(2^n + \binom{n}{2} 3 \cdot 2^{n-2} + \binom{n}{4} 3^2 \cdot 2^{n-4} + \cdots) = 2l$ where l is an integer. Next, note that $(2-\sqrt{3})^n < 1$. We see that $[(2+\sqrt{3})^2] = (2+\sqrt{3})^n + (2-\sqrt{3})^n - 1$. It follows that $[(2+\sqrt{3})^2]$ is odd.

CHAPTER 2

Integer Representations and Operations

2.1. Representations of Integers

2.1.1. We have $1999 = 7 \cdot 285 + 4, 285 = 7 \cdot 40 + 5$, and $40 = 7 \cdot 5 + 5$, and $5 = 7 \cdot 0 + 5$. The sequence of remainders gives the base 7 digits. Hence $(1999)_{10} = (5554)_7$. We have $(6105)_7 = 6 \cdot 7^3 + 1 \cdot 7^2 + 0 \cdot 7 + 5 = (2112)_{10}$.

2.1.3. We have $(10101111)_2 = (175)_{10}$, and $(999)_{10} = (1111100111)_2$.

2.1.5. We group together blocks of four binary digits starting from the right. We have $(0101)_2 = (5)_{16}$, $(1111)_2 = (F)_{16}, (1000)_2 = (8)_{16}$. Hence $(100011110101)_2 = (8F5)_{16}$. Likewise, $(1110)_2 = (E)_{16}$, $(0100)_2 = (4)_{16}$, and $(0111)_2 = (7)_{16}$. Therefore, $(11101001110)_2 = (74E)_{16}$.

2.1.7. This is because we are using the blocks of three digits as one "digit," which has 1000 possible values.

2.1.9. We find that $(101001)_{-2} = 1(-2)^5 + 0(-2)^4 + 1 \cdot (-2)^3 + 0(-2)^2 + 0(-2)^1 + 1(-2)^0 = -39$ and $(12012)_{-3} = 1(-3)^4 + 2(-3)^3 + 0(-3)^2 + 1(-3)^1 + 2(-3)^0 = 26$.

2.1.11. If m is any integer weight less than 2^k, then by Theorem 1.10, m has a base two expansion $m = a_{k-1}2^{k-1} + a_{k-2}2^{k-2} + \cdots + a_1 2^1 + a_0 2^0$, where each a_i is 0 or 1. The 2^i weight is used if and only if $a_i = 1$.

2.1.13. Let w be the weight to be measured. By Exercise 10, w has a unique balanced ternary expansion. Place the object in pan 1. If $e_i = 1$ then place a weight of 3^i into pan 2. If $e_i = -1$ then place a weight of 3^i in pan 1. If $e_i = 0$ then do not use the weight of 3^i. Now the pans will be balanced.

2.1.15. To convert a number from base r to base r^n, take the number in blocks of size n. To go the other way, convert each digit of a base r^n number to base r, and concatenate the results.

2.1.17. Multiplying n by b^m gives $b^m n = b^m(a_k b^k + a_{k-1} b^{k-1} + \cdots + a_1 b + a_0) = (a_k b^{k+m} + a_{k-1} b^{k+m-1} + \cdots + a_1 b^{m+1} + a_0 b^m + 0 \cdot b^{m-1} + \cdots + 0) = (a_k a_{k-1} \ldots a_1 a_0 00 \ldots 00)_b$, where we have placed m zeroes at the end of the base b expansion of n.

2.1.19. a. The lead digit is a one, so the number is negative. Its absolute value has a binary representation of the complement of 1001, i.e. 0110. Thus 11001 is the one's complement representation of -6.

b. 01101 is the one's complement representation of 13.

c. 10001 is the one's complement representation of -14.

d. 11111 is a one's complement representation of 0. Note that 00000 also represents 0.

2.1.21. If m is positive, then $a_{n-1} = 0$ and $a_{n-2}a_{n-3}\ldots a_0$ is the binary expansion of m. Hence, $m = \sum_{i=0}^{n-2} a_i 2^i$ as desired. If m is negative, then the one's complement expansion for m has its leading bit equal to 1. If we view the bit string $a_{n-2}a_{n-3}\ldots a_0$ as a a binary number, then it represents $(2^{n-1} - 1) - (-m)$, since finding the one's complement is equivalent to subtracting the binary number from $111\cdots 1$. That is $(2^{n-1} - 1) - (-m) = \sum_{i=0}^{n-2} a_i 2^i$. Solving for m gives us the desired identity.

2.1.23. a. Since the first digit is a 1, we know that the integer is negative and that $(1001)_2 = 9$ is the binary expansion of $2^4 - |x|$. So $|x| = 16 - 9 = 7$, and thus $x = -7$.

b. Since the first digit is a 0, we know that the integer is positive and hence $x = (1101)_2 = 13$.

c. Since the first digit is a 1, we know that the integer is negative and that $(0001)_2 = 1$ is the binary expansion of $2^4 - |x|$. So $|x| = 16 - 1 = 15$, and thus $x = -15$.

d. Since the first digit is a 1, we know that the integer is negative and that $(1111)_2 = 15$ is the binary expansion of $2^4 - |x|$. So $|x| = 16 - 15 = 1$, and thus $x = -1$.

2.1.25. If each of the digits in the two's complement representation for m is complemented and then 1 is added to the resulting binary number, the result is the two's complement representation for $-m$. To see this note that $m + (-m) + (-1) =$ (binary expansion of m) $+ (2^{n-1} +$ binary expansion for $2^{n-1} - m) + (-1) = 2^{n-1} + 2^{n-1} - 1 = 2^n - 1 = (111\ldots1)_2$. Therefore the two's complement representation of $-m - 1$ is the complement of m.

2.1.27. Since 4 bits are required for every decimal digit, $4n$ bits are required to store the number in this manner.

2.1.29. We first show that every positive integer has a Cantor expansion. To find a Cantor expansion of the positive integer n, let m be the unique positive integer such that $m! \leq n < (n+1)!$. By the division algorithm there is an integer a_m such that $n = m! \cdot a_m + r_m$ where $0 \leq a_m \leq m$ and $0 \leq r_m < m!$. We iterate, finding that $r_m = (m-1)! \cdot a_{m-1} + r_{m-1}$ where $0 \leq a_{m-1} \leq m-1$ and $0 \leq r_{m-1} < (m-1)!$. We iterate $m - 2$ more times, where we have $r_i = (i-1)! \cdot a_{i-1} + r_{i-1}$ where $0 \leq a_{i-1} \leq i-1$ and $0 \leq r_{i-1} < (i-1)!$ for $i = m+1, m, m-1, \ldots, 2$ with $r_{m+1} = n$. At the last stage we have $r_2 = 1! \cdot a_1 + 0$ where $r_2 = 0$ or 1 and $r_2 = a_1$.

Now that we have shown that every integer has a Cantor expansion, we must show that this expansion is unique. So suppose that n has two different Cantor expansions $n = a_m m! + a_{m-1}(m-1)! + \cdots + a_2 2! + a_1 1! = b_m m! + b_{m-1}(m-1)! + \cdots + b_2 2! + b_1 1!$, where a_j and b_j are integers, and $0 \leq a_j \leq j$ and $0 \leq b_j \leq j$ for $j = 1, 2, \ldots, m$. Suppose that k is the largest integer such that $a_k \neq b_k$, and without loss of generality, assume $a_k > b_k$, which implies that $a_k \geq b_k + 1$. Then $a_k k! + a_{k-1}(k-1)! + \cdots + a_1 1! = b_k k! + b_{k-1}(k-1)! + \cdots + b_1 1!$. Using the identity $\sum_{j=1}^{k} j \cdot j! = (k+1)! - 1$, proved in Exercise 16 of Section 1.3, we see that $b_k k! + b_{k-1}(k-1)! + \cdots + b_1 1! \leq b_k k! + (k-1) \cdot (k-1)! + \cdots + 1 \cdot 1! \leq b_k k! + k! - 1 = (b_k + 1)k! - 1 < a_k k!$. This is a contradiction, so the expansion is unique.

2.1.31. Call a position *good* if the number of ones in each column is even, and *bad* otherwise. Since a player can only affect one row, he or she must affect some column sums. Thus any move from a good position produces a bad position. To find a move from a bad position to a good one, construct a binary number by putting a 1 in the place of each column with odd sum, and a 0 in the place of each column with even sum. Subtracting this number of matches from the largest pile will produce a good position.

2.1.33. a. First show that the result of the operation must yield a multiple of 9. Then, it suffices to check only multiples of 9 with decreasing digits. There are only 79 of these. If we perform the operation on each of these 79 numbers and reorder the digits, we will have one of the following 23 numbers: 7551, 9954, 5553, 9990, 9981, 8820, 9810, 9620, 8532, 8550, 9720, 9972, 7731, 6543, 8730, 8640, 8721, 7443, 9963, 7632, 6552, 6642, or 6174. It will suffice to check only 9810, 7551, 9990, 8550, 9720, 8640, and 7632.

b. From the solution in part (a), construct a tree from the last seven numbers. The longest branch is six steps. Every number will reach the tree in two steps. The maximum is given by 8500 (for instance) which takes eight steps.

2.1.35. Consider $a_0 = (3043)_6$. Then $T_6((3043)_6) = (3552)_6$, $T_6((3552)_6) = (3133)_6$, $T_6((3133)_6) = (1554)_6$, $T_6((1554)_6) = (4042)_6$, $T_6((4042)_6) = (4132)_6$, and $T_6((4132)_6) = (3043)_6 = a_0$. So T_6 repeats with period 6. Therefore, it never goes to a Kaprekar's constant for the base 6. Hence, there is no Kaprekar's constant for the base 6.

2.2. Computer Operations with Integers

2.2.1. To add $(101111011)_2$ and $(1100111011)_2$ we first add 1 and 1, obtaining the rightmost bit 0 and the carry 1. Then we add the bits 1 and 1 and the carry 1, obtaining the second bit from the right in the sum 1 and the carry 1. Then we add the bits 0 and 0, and the carry 1, obtaining the third bit from the right in the sum, 1. Then we add the bits 1 and 1, obtaining the fourth bit from the right in the sum, 0, and the carry 1. Then we add the bits 1 and 1 and the carry 1, obtaining the fifth bit from the right in the sum 1, and the carry 1. Then we add the bits 1 and 1 and the carry obtaining the sixth bit from the right in the sum 1, and the carry 1. Then we add the bits 1 and 0 and the carry 1 obtaining the seventh bit from the right in the sum, 0, and the carry, 1. Then we add the bits 0 and 0 and the carry 1, obtaining the eighth bit from the right in the sum 1. Then we add the bits 1 and 1, obtaining the ninth bit from the right, 0, and the carry 1. Then we add the (leading) bit 0 and the bit 1 and the carry 1, obtaining the tenth bit in the sum, 0, and the carry, 1, which is the leading bit from the left. Hence the sum is $(10010110110)_2$.

2.2.3. We have $(1111000011)_2 - (11010111)_2 = (1011101100)_2$

2.2.5. To multiply $(11101)_2$ and $(110001)_2$ we need to add $2^0(110001)_2 = (110001)_2$, $2^2(110001)_2 = (11000100)_2$, $2^3(110001)_2 = (110001000)_2$, and $2^4(110001)_2 = (1100010000)_2$. The first bit and carry are computed from $1+0+0+0 = 1$. The second bit and carry are computed from $0+0+0+0 = 0$. The third bit and carry are computed from $0+1+0+0 = 1$. The fourth bit and carry are computed from $0+0+1+0 = 1$. The fifth bit and carry are computed from $1+0+0+1 = 10$. The sixth bit and carry are computed from (with the carry 1) $1+1+0+0+0 = 10$. The seventh bit and carry are computed from (with the carry 1) $1+0+1+0+0 = 10$. The eighth bit and carry are computed from (with the carry 1) $1+0+1+1+0=11$. The ninth bit and carry are computed from (with the carry 1) $1+0+0+1+1= 11$. The tenth bit and eleventh bit are computed from (with the carry 1) $1+0+0+0+1=10$. Hence $(11101)_2 \cdot (110001)_2 = (10110001101)_2$.

2.2.7. We have $(110011111)_2 = (11111)_2 \cdot (1101)_2 + (1100)_2$

2.2.9. We have $(1234321)_5 + (2030104)_5 = (3314430)_5$

2.2.11. We have $(1234)_5 \cdot (3002)_5 = (3023)_5 + (4312000)_5 = (4320023)_5$

2.2.13. To add $(ABAB)_{16}$ and $(BABA)_{16}$ we first add the rightmost hexadecimal digits B and A obtaining the rightmost digit of the sum, 5, and carry, 1. Then we add the hexadecimal digits in the second position from the right and the carry, namely A, B and 1, obtaining the second digit from the right in the sum, 6, and the carry, 1. Then we add the hexadecimal digits in the third position from the right, namely B, A, and 1, obtaining the digit in the third position from the right, 6, and the carry, 1. Finally, we add the hexadecimal digits in the leftmost position and the carry, namely A, B, and 1, obtaining the second hexadecimal digit from the left in the sum, 6, and the leftmost hexadecimal digit in the sum 1. Hence the sum is $(16665)_{16}$.

2.2.15. We have $(FACE)_{16} \cdot (BAD)_{16} = (B705736)_{16}$

2.2.17. We represent the integer $(18235187)_{10}$ using three words: $((018)(235)(187))_{1000}$ and the integer $(22135674)_{10}$ using three words: $((022)(135)(674))_{1000}$, where each base 1000 digit is represented by three base 10 digits in parentheses. To find the sum, difference, and product of these integers from their base 1000 representations we carry out the algorithms for such computations for base 1000. The details are omitted.

2.2.19. To add numbers using the one's complement representation, first decide whether the answer will be negative or positive. To do this is easy if both numbers have the same lead (sign) bit; otherwise conduct a bit-by-bit comparison of a positive summand's digits and the complement of the negative's. Now, add the other digits (all but the initial (sign) bit) as an ordinary binary number. If the sum is greater than 2^n we have an overflow error. If not, consider the three quantities of the two summands and the sum. If exactly zero or two of these are negative, we're done. Otherwise, we need to add $(1)_2$ to this answer.

Also, add an appropriate sign bit to the front of the number.

2.2.21. Let $a = (a_m a_{m-1} \ldots a_2 a_1)_!$ and $b = (b_m b_{m-1} \ldots b_2 b_1)_!$. Then $a+b$ is obtained by adding the digits from right to left with the following rule for producing carries. If $a_j + b_j + c_{j-1}$, where c_{j-1} is the carry from adding a_{j-1} and b_{j-1}, is greater than j, then $c_j = 1$, and the resulting jth digit is $a_j + b_j + c_{j-1} - j - 1$. Otherwise, $c_j = 0$. To subtract b from a, assuming $a > b$, we let $d_i = a_i - b_i + c_{i-1}$ and set $c_i = 0$ if $a_i - b_i + c_{i-1}$ is between 0 and j. Otherwise, $d_i = a_i - b_i + c_{i-1} + j + 1$ and set $c_i = -1$. In this manner, $a - b = (d_m d_{m-1} \ldots d_2 d_1)_!$.

2.2.23. We have $(a_n \ldots a_1 5)_{10}^2 = (10(a_n \ldots a_1)_{10} + 5)^2 = 100(a_n \ldots a_1)_{10}^2 + 100(a_n \ldots a_1)_{10} + 25 = 100(a_n \ldots a_1)_{10}((a_n \ldots a_1)_{10} + 1) + 25$. The decimal digits of this number consist of the decimal digits of $(a_n \ldots a_1)_{10}((a_n \ldots a_1)_{10} + 1)$ followed by 25 since this first product is multiplied by 100 which shifts its decimal expansion two digits.

2.3. Complexity and Integer Operations

2.3.1. a. We have $2n + 7$ is $O(n)$ since $2n + 7 \leq 9n$ for every positive integer n.

  b. Note that $n^2/3$ is not $O(n)$ for if C is a real number it follows $n^2/3 > Cn$ whenever $n > 3C$.

  c. We have 10 is $O(n)$ since $10 \leq 10n$ whenever n is a positive integer.

  d. We have $n^2 + 1 \leq 2n^2$ whenever n is a positive integer. Hence $\log(n^2 + 1) \leq \log(2n^2) = 2 \log n + \log 2 \leq 3n$ whenever n is a positive integer. It follows that $\log(n^2 + 1)$ is $O(n)$.

  e. Note that $\sqrt{n^2 + 1} \leq \sqrt{2n^2} \leq \sqrt{2} \cdot n$ whenever n is a positive integer. Hence $\sqrt{n^2 + 1}$ is $O(n)$.

  f. We have $(n^2 + 1)/(n + 1) < (2n^2/n = 2n$ whenever n is a positive integer. Hence $(n^2 + 1)/(n + 1)$ is $O(n)$.

2.3.3. First note that $(n^3 + 4n^2 \log n + 101n^2)$ is $O(n^3)$ and that $(14n \log n + 8n)$ is $O(n \log n)$ as in Example 1.36. Now applying Theorem 1.12 yields the result.

2.3.5. Use Exercise 4 and follow Example 1.34, noting that $(\log n)^3 \leq n^3$ whenever n is a positive integer.

2.3.7. Let k be an integer with $1 \leq k \leq n$. Consider the function $f(k) = (n + 1 - k)k$, whose graph is a concave-down parabola with k-intercepts at $k = 0$ and $k = n + 1$. Since $f(1) = f(n) = n$, it is clear that $f(k) \geq n$ for $k = 1, 2, 3, \ldots, n$. Now consider the product $(n!)^2 = \prod_{k=1}^n k(n + 1 - k) \geq \prod_{k=1}^n n$, by the inequality above. This last is equal to n^n. Thus we have $n^n \leq (n!)^2$. Taking logarithms of both sides yields $n \log(n) \leq 2 \log(n!)$, which shows that $n \log(n)$ is $O(\log(n!))$.

2.3.9. Suppose that f is $O(g)$ where $f(n)$ and $g(n)$ are positive integers for every integer n. Then there is an integer C such that $f(n) < Cg(n)$ for all $x \in S$. Then $f^k(n) < C^k g^k(n)$ for all $x \in S$. Hence f^k is $O(g^k)$.

2.3.11. The number of digits in the base b expansion of n is $1 + k$ where k is the largest integer such that $b^k \leq n < b^{k+1}$ since there is a digit for each of the powers of $b^0, b^1, \ldots, b^k$. Note that this inequality is equivalent to $k \leq \log_b n < k + 1$, so that $k = [\log_b n]$. Hence there are $[\log_b n] + 1$ digits in the base b expansion of n.

2.3.13. To multiply an n-digit integer by an m-digit integer in the conventional manner, one must multiply every digit of the first number by every digit of the second number. There are nm such pairs.

2.3.15. a. We use the result of Theorem 2.6. Let $m = [\log_2 n + 1]$. If we first multiply consecutive pairs of integers in the the product, we have $O(n/2)$ multiplications of integers with at most m bits. By Theorem 2.6, there is an algorithm for doing this using $O(m \log_2 m \log_2 \log_2 m)$ operations. Now we have $[n/2]$ integers of at most $2m$ bits. If we multiply pairs of these integers together, then

2.3. COMPLEXITY AND INTEGER OPERATIONS

by Theorem 2.6 again, this results in $O((n/4)(2m)\log_2 m \log_2 \log_2 m)$, where we use the fact that $\log_2 km \log_2 \log_2 km = O(\log_2 m \log_2 \log_2 m)$ for any constant k. Continuing in this manner we find that computing $n!$ takes $O(\sum_{j=1}^{m} n/(2^j) 2^{j-1} \log_2 m \log_2 \log_2 m) = O((n/2)m^2 \log_2 m \log_2 \log_2 m) = O(n \log_2^2 n \log_2 \log_2 n \log_2 \log_2 \log_2 n)$ operations.

b. We need to find three factorials, which will have the same big-O value as in part (a). We will also need to perform one subtraction (which will not affect the big-O value), one multiplication and one division. The factorials have at most $n \log n$ bits, so by Theorem 2.5, the multiplication will take at most $O((n \log n)^{1+\epsilon})$ bit operations. By Theorem 2.7, the division will take $O((n \log n)^{1+\epsilon})$, so in total the number of bit operations is $O((n \log n)^{1+\epsilon})$.

2.3.17. $(1001)_2 \cdot (1011)_2 = (2^4 + 2^2)(10)_2(10)_2 + 2^2(10 - 01)_2(11 - 10))_2 + (2^2 + 1)(01)_2(11)_2 = (10100)_2(100)_2 + (100)_2(01)_2(01)_2 + (101)_2(01)_2(11)_2 = (1010000)_2 + (100)_2 + (1111)_2 = (1100011)_2$

2.3.19. a. $ab = (10^{2n} + 10^n)A_1 B_1 + 10^n(A_1 - A_0)(B_0 - B_1) + (10^n + 1)A_0 B_0$ where A_i and B_i are defined as in identity (1.9).

b. $73 \cdot 87 = (10^2 + 10)7 \cdot 8 + 10(7 - 3)(7 - 8) + (11)3 \cdot 7 = 5600 + 560 - 40 + 210 + 21 = 6351$.

c. $4216 \cdot 2733 = (10100)42 \cdot 27 + (100)(42 - 16)(33 - 27) + (101)16 \cdot 33$. Then, $42 \cdot 27 = (10^2 + 10)4 \cdot 2 + 10(4 - 2)(7 - 2) + (11)2 \cdot 7 = 1134$, and, $26 \cdot 06 = (10^2 + 10)2 \cdot 0 + 10(2 - 6)(6 - 0) + (11)6 \cdot 6 = 156$, and $16 \cdot 33 = (10^2 + 10)1 \cdot 3 + 10(1 - 6)(3 - 3) + (11)6 \cdot 3 = 528$. Then $4216 \cdot 2733 = (10100)1134 + (100)156 + (101)528 = 11522328$.

2.3.21. That the given equation is an identity may be seen by direct calculation. The seven multiplications necessary to use this identity are: $a_{11}b_{11}$, $a_{12}b_{21}$, $(a_{11} - a_{21} - a_{22})(b_{11} - b_{12} - b_{22})$, $(a_{21} + a_{22})(b_{12} - b_{11})$, $(a_{11} + a_{12} - a_{21} - a_{22})b_{22}$, $(a_{11} - a_{21})(b_{22} - b_{12})$, $a_{22}(b_{11} - b_{21} - b_{12} + b_{22})$.

2.3.23. Let $k = [\log_2 n] + 1$. Then the number of multiplications for $2^k \times 2^k$ matrices is $O(7^k)$. But, $7^k = 2^{(\log_2 7)([\log_2 n]+1)} = O(2^{\log_2 n \log_2 7} 2^{\log_2 7}) = O(n^{\log_2 7})$. The other bit operations are absorbed into this term.

CHAPTER 3

Primes and Greatest Common Divisors

3.1. Prime Numbers

3.1.1. a. We see that 101 is prime since it is not divisible by any positive integers other than 1 or 101. To verify this it is sufficient to check that 101 is not divisible by any prime not exceeding $\sqrt{101}$. The only such primes are 2, 3, 5, and 7 and none of these divide 101.

b. We see that 103 is prime since it is not divisible by any positive integers other than 1 or 103. To verify this it is sufficient to check that 103 is not divisible by any prime not exceeding $\sqrt{103}$. The only such primes are 2, 3, 5, and 7 and none of these divide 103.

c. We see that 107 is prime since it is not divisible by any positive integers other than 1 or 107. To verify this it is sufficient to check that 107 is not divisible by any prime not exceeding $\sqrt{107}$. The only such primes are 2, 3, 5, and 7 and none of these divide 107.

d. We see that 111 is not prime since it is divisible by 3.

e. We see that 113 is prime since it is not divisible by any positive integers other than 1 or 113. To verify this it is sufficient to check that 113 is not divisible by any prime not exceeding $\sqrt{113}$. The only such primes are 2, 3, 5, and 7 and none of these divide 113.

f. We see that 121 is not prime since it is divisible by 11.

3.1.3. The primes less than 150 are 2, 3, 5, 7, 11, 13, 17, 19, 23, 29, 31, 37, 41, 43, 47, 53, 59, 61, 67, 71, 73, 79, 83, 89, 97, 101, 103, 107, 109, 113, 127, 131, 137, 139, 149

3.1.5. Suppose that $n = x^4 - y^4 = (x-y)(x+y)(x^2+y^2)$, where $x > y$. The integer n can not be prime since it divisible by $x + y$ which can not be 1 or n.

3.1.7. Using the identity given in the hint with k such that $1 < k < n$ and $k \mid n$, then $a^k - 1 \mid a^n - 1$. Since $a^n - 1$ is prime by hypothesis, $a^k - 1 = 1$. From this, we see that $a = 2$ and $k = 1$, contradicting the fact that $k > 1$. Thus we must have $a = 2$ and n is prime.

3.1.9. We need to assume $n \geq 3$ to assure that $S_n > 1$. Then by Lemma 3.1, S_n has a prime divisor p. If $p \leq n$ then $p \mid n!$, and so $p \mid n! - S_n = 1$, a contradiction. Therefore we must have $p > n$. Since we can find arbitrarily large primes, there must be infinitely many.

3.1.11. $Q_1 = 3, Q_2 = 7, Q_3 = 31, Q_4 = 211, Q_5 = 2311, Q_6 = 30031$. The smallest prime factors are 3, 7, 31, 211, 2311, and 59, respectively.

3.1.13. If n is prime, we are done. Otherwise $n/p < (\sqrt[3]{n})^2$. If n/p is prime, then we are done. Otherwise, by Theorem 3.2, n/p has a prime factor less than $\sqrt{n/p} < \sqrt[3]{n}$, a contradiction.

3.1.15. a. The arithmetic progression is $3n + 1$ and the first values are $4, 7, 10, \ldots$. The first prime is 7.

b. We list the first few numbers of the shape $5n + 4$ until we find a prime: 9, 14, 19, which is prime.

c. We list the first few numbers of the shape $11n + 16$ until we find a prime: $27, 38, 49, 60, 71$, which is prime.

3.1.17. If n is prime the statement is true for n. Otherwise, n is composite, so n is the product of two integers a and b such that $1 < a \leq b < n$. Since $n = ab$ and by the inductive hypothesis both a and b are the product of primes, we conclude that n is also the product of primes.

3.1.19. Using Exercise 18, we have, $\pi(250) = (\pi(\sqrt{250}) - 1) + 250 - ([250/2] + [250/3] + [250/5] + [250/7] + [250/11] + [250/13]) + ([250/6] + [250/10] + [250/14] + [250/22] + [250/26] + [250/15] + [250/21] + [250/33] + [250/39] + [250/35] + [250/55] + [250/65] + [250/77] + [250/91] + [250/143]) - ([250/30] + [250/42] + [250/66] + [250/70] + [250/78] + [250/105] + [250/110] + [250/130] + [250/132] + [250/154] + [250/165] + [250/195] + [250/231]) + ([250/210]) = 5 + 250 - (125 + 83 + 50 + 35 + 22 + 19) + (41 + 25 + 17 + 11 + 9 + 16 + 11 + 7 + 6 + 7 + 4 + 3 + 3 + 2 + 1) - (8 + 5 + 3 + 3 + 3 + 2 + 2 + 1 + 1 + 1 + 1 + 1 + 1) + 1 = 53$.

3.1.21. For $n = 1, 2, \ldots 10$, the values of the function are $13, 19, 29, 43, 61, 83, 109, 139, 173, 211$, each of which is prime. But $2 \cdot 11^2 + 11 = 11(2 \cdot 11 + 1) = 11 \cdot 23$, so it is not prime.

3.1.23. Assume not. Let x_0 be a positive integer. It follows that $f(x_0) = p$ where p is prime. Let k be an integer. We have $f(x_0 + kp) = a_n(x_0 + kp)^n + \cdots + a_1(x_0 + kp) + a_0$. Note that by the binomial theorem, $(x_0 + kp)^j = \sum_{i=1}^{j} \binom{j}{i} x_0^{j-i} (kp)^i$. It follows that $f(x_0 + kp) = \sum_{j=0}^{n} a_j x_0^j + Np = f(x_0) + Np$, for some integer N. Since $p \mid f(x_0)$ it follows that $p \mid (f(x_0) + Np) = f(x_0 + kp)$. Since $f(x_0 + kp)$ is supposed to be prime, it follows that $f(x_0 + kp) = p$ for all integers k. This contradicts the fact that a polynomial of degree n takes on each value no more than n times. Hence $f(y)$ is composite for at least one integer y.

3.1.25. At each stage of the procedure for generating the lucky numbers the smallest number left, say k, is designated to be a lucky number and infinitely many numbers are left after the deletion of every kth integer left. It follows that there are infinitely many steps, and at each step a new lucky number is added to the sequence. Hence there are infinitely many lucky numbers.

3.2. The Distribution of Primes

3.2.1. The smallest 5 consecutive composite integers can be found by locating the first pair of consecutive composite odd integers, 25 and 27. Hence the smallest 5 consecutive composite integers are 24, 25, 26, 27, and 28. (These are considerably smaller than the integers $(5+1)! + j = 6! + j = 720 + j$ for $j = 2, 3, 4, 5, 6$.)

3.2.3. Suppose that $p, p + 2$, and $p + 4$ were all prime. We consider three cases. First, suppose that p is of the form $3k$. Then p cannot be prime unless $k = 1$, and the prime triplet is 3, 5, and 7. Next, suppose that p is of the form $3k + 1$. Then $p + 2 = 3k + 3 = 3(k + 1)$ is not prime. We obtain no prime triplets in this case. Finally, suppose that p is of the form $3k + 2$. Then $p + 4 = 3k + 6 = 3(k + 2)$ is not prime. We obtain no prime triplet in this case either. Since the three cases are exhaustive, we have only one prime triplet of this kind, 3, 5, and 7.

3.2.5. By searching through a table we find these triples: $(7, 11, 13), (13, 17, 19), (37, 41, 43)$, and $(67, 71, 73)$.

3.2.7. a. The smallest prime between 4 and 8 is 5.

  b. The smallest prime between 6 and 12 is 7.

  c. The smallest prime between 23 and 46 is 29.

  d. The smallest prime between 47 and 94 is 53.

3.2.9. To see that the primes are indeed in the range, we print them as triples $(n^2, \text{smallest prime}, (n+1)^2)$. For $n = 11, 12, \ldots, 20$ we have $(121, 127, 144), (144, 149, 169), (169, 173, 196), (196, 197, 225), (225, 227, 256), (256, 257, 289), (289, 293, 324), (324, 331, 361), (361, 367, 400)$, and $(400, 401, 441)$.

3.2. THE DISTRIBUTION OF PRIMES

3.2.11. a. We have $7 = 3 + 2 + 2$.

b. We have $17 = 11 + 3 + 3$.

c. We have $27 = 23 + 2 + 2$.

d. We have $97 = 89 + 5 + 3$.

e. We have $101 = 97 + 2 + 2$.

f. We have $199 = 191 + 5 + 3$.

3.2.13. Suppose Goldbach's conjecture is true and let $n > 5$ be an integer. If n is even, then $n - 2$ is an even integer greater than 2 and so is the sum of two primes, p and q. Then $n = p + q + 2$, the sum of three primes. If n is odd, then $n - 3$ is an even integer greater than 2 and so is the sum of two primes p and q. Then $n = p + q + 3$, the sum of three primes.

Conversely, suppose every integer greater than 5 is the sum of three primes. Let $n > 2$ be an even integer. Then $n + 2$ is also even and is greater than 5. (It is not equal to 5 since it is even.) By hypothesis, $n + 2$ is the sum of 3 primes. If all three primes were odd, then $n + 2$ would be odd, a contradiction, so at least one of the primes is even, that is, one of the primes must be 2, so $n + 2 = p + q + 2$ for some primes p and q. Therefore $n = p + q$, the sum of two primes.

3.2.15. Let $p < n$ be prime. Using the division algorithm, we divide each of the first $p + 1$ integers in the sequence by p to get $a = q_0 p + r_0, a + k = q_1 p + r_1, \ldots, a + pk = q_p + r_p$, with $0 \leq r_i < p$ for each i. By the pigeonhole principle, at least two of the remainders must be equal, say $r_i = r_j$. We subtract the corresponding equations to get $a + ik - a - jk = q_i p + r_i - q_j p + r_j$, which reduces to $(i - j)k = (q_i - q_j)p$. Therefore $p | (i - j)k$, and since p is prime, it must divide one of the factors. But since $(i - j) < p$ we must have $p | k$.

3.2.17. From Exercise 15, we know that every prime less than four must must divide the difference, so 6 must divide the difference. Therefore the smallest possible difference is 6. This minimum is achieved with $5, 11, 17, 23$.

3.2.19. From Exercise 15, we know that every prime less than six must must divide the difference, so 30 must divide the difference. Therefore the smallest possible difference is 30. This minimum is achieved by the example given in Exercise 16.

3.2.21. If $p^\alpha - q^\beta = 1$, with p, q primes, then p or q is even, so p or q is 2. If $p = 2$, there are several cases: we have $2^\alpha - q^\beta = 1$. If α is even, say $\alpha = 2k, (2^{2k} - 1) = (2^k - 1)(2^k + 1) = q^\beta$. So $q | (2^k - 1)$ and $q | (2^k + 1)$, hence $q = 1$, a contradiction. If α is odd and β is odd, $2^\alpha = 1 + q^\beta = (1 + q)(q^{\beta-1} - q^{\beta-2} + \cdots + 1)$. So $1 + q = 2^n$ for some n. Then $2^\alpha = (2^n - 1)^\beta + 1 = 2^n(\text{odd number})$, since β is odd. So $2^{\alpha-n} = $ odd number and so $\alpha = n$. Therefore $2^\alpha = 1 + (2^\alpha - 1)^\beta$ and so $\beta = 1$ which is not allowed. If $\alpha = 2k + 1$ and $\beta = 2n$ we have $2^{2k+1} = 1 + q^{2n}$. Since q is odd, q^2 is of the form $4m + 1$, and by the binomial theorem, so is q^{2n}. Thus the right hand side of the last equation is of the form $4m + 2$, but this forces $k = 0$, a contradiction. If $q = 2$, we have $p^\alpha - 2^\beta = 1$. Whence $2^\beta = (p - 1)(p^{\alpha-1} + p^{\alpha-2} + \cdots + p + 1)$, where the last factor is the sum of α odd terms but must be a power of 2, therefore, $\alpha = 2k$ for some k. Then $2^\beta = (p^k - 1)(p^k + 1)$. These last two factors are powers of 2 which differ by 2 which forces $k = 1, \alpha = 2, \beta = 3, p = 3$, and $q = 2$ as the only solution: $3^2 - 2^3 = 1$.

3.2.23. Since $3p > 2n, p$, and $2p$ are the only multiples of p that appear as factors in $(2n)!$. So p divides $(2n)!$ exactly twice. Since $2p > n, p$ is the only multiple of p that appears as a factor in $n!$. So $p \mid n!$ exactly once. Then since $\binom{2n}{n} = 2n!/(n!n!)$, the two factors of p in the numerator are cancelled by the two in the denominator.

3.2.25. By Bertrand's conjecture, there must be a prime in each interval of the form $(2^{k-1}, 2^k)$, for $k = 2, 3, 4, \ldots$. Thus, there are at least $k - 1$ primes less than 2^k. Since the prime 2 isn't counted here, we have at least k primes less than 2^k.

3.2.27. Since $1/1$ is an integer, we may assume $n > 1$. First suppose that $m < n$. Then $1/n + 1/(n+1) + \cdots + 1/(n+m) \leq 1/n + 1/(n+1) + \cdots + 1/(2n-1) < 1/n + 1/n + \cdots + 1/n \leq n(1/n) = 1$, so the sum can not be an integer. Now suppose $m \geq n$. Then by Bertrand's postulate, there is a prime p such that $n < p < n + m$. Let p be the largest such prime. Then $n + m < 2p$, otherwise there would be a prime q with $p < q \leq 2p \leq n + m$ contradicting the choice of p. Suppose that $1/n + 1/(n+1) + \cdots + 1/p + \cdots + 1/(n+m) = a$ where a is an integer. Note that p occurs as a factor in only one denominator, since $2p > n + m$. Let $Q = \prod_{j=n}^{n+m} j$, and let $Q_i = Q/i$, for $i = n, n+1, \ldots, n+m$. If we multiply the equation by Q we get $Q_n + Q_{n+1} + \cdots + Q_p + \cdots + Q_{n+m} = Qa$. Note that every term on both sides of the equation is divisible by p except for Q_p. If we solve the equation for Q_p and factor a p out of the other side we have an equation of the form $Q_p = pN$ where N is some integer. But this implies that p divides Q_p, a contradiction. Therefore a can not be an integer.

3.2.29. Suppose n has the stated property and $n \geq p^2$ for some prime p. Since p^2 is not prime, there must a prime dividing both p^2 and n, and the only possibility for this is p itself, that is, $p|n$. Now if $n \geq 7^2$, then it is greater than $2^2, 3^2$, and 5^2 and hence divisible by $2, 3, 5$, and 7. This is the basis step for induction. Now assume n is divisible by $p_1, p_2, \ldots, p_k$. By Bonse's inequality $p_{k+1}^2 < p_1 p_1 \cdots p_k < n$, so $p_{k+1}|n$ also. This induction implies that every prime divides n, which is absurd. Therefore if n has the stated property, it must be less than $7^2 = 49$.

Now we note that the integers less that 30 sharing no common prime factor with 30 are $1, 7, 11, 13, 17, 19, 23$ and 29, all of which are prime or 1. So 30 has the property. It remains to show that the numbers from 31 to 48 do not have the property. We exhibit a counterexample in each case. For $n = 31, 33, 35, 37, 39, 41, 43, 45$ and 47 we note that $k = 8$ shares no prime factor with n, and yet is not prime. For $n = 32, 34, 38, 40, 44$ and 46, we note that $k = 9$ shares no prime factor with n and yet is not prime. For the remaining cases $n = 36, 42$, and 48, we note that $k = 25$ shares no prime factor with n and yet is not prime.

3.2.31. First suppose $n \geq 8$. Note that by Bertrand's postulate we have $p_{n-1} < p_n < 2p_{n-1}$ and $p_{n-2} < p_{n-1} < 2p_{n-2}$. Therefore, $p_n^2 < (2p_{n-1})(2p_{n-1}) < (2p_{n-1})(4p_{n-2}) = 8p_{n-1}p_{n-2} = p_{n-1}p_{n-2}p_5 \leq p_{n-1}p_{n-2}p_{n-3}$, since $n \geq 8$. All that remains is to check that the inequality it true for $n = 6$ and 7. When $n = 7$ we have $p_7^2 = 17^2 = 289 < 1001 = 13 \cdot 11 \cdot 7 = p_6 p_5 p_4$, and when $n = 6$ we have $p_6^2 = 13^2 = 169 < 385 = 11 \cdot 7 \cdot 5 = p_5 p_4 p_3$. This completes the proof. To see that the inequality does not hold for smaller n, we check that for $n = 5$, we have $p_5^2 = 11^2 = 121 > 7 \cdot 5 \cdot 3 = 105$ and when $n = 4$, we have $p_4^2 = 7^2 = 49 > 5 \cdot 3 \cdot 2 = 30$.

3.3. Greatest Common Divisors

3.3.1. a. The positive divisors of 15 are $1, 3, 5$, and 15 and the positive divisors of 25 are $1, 5$, and 25. Hence the greatest common divisor of 15 and 25 is 5.

b. Every positive integer is a divisor of 0. Hence the greatest common divisor of 0 and 111 is 111.

c. The positive divisor of -12 are $1, 2, 3, 4, 6$, and 12 and the positive divisors of 18 are $1, 2, 3, 6, 9$, and 18. Hence the greatest common divisor of -12 and 18 is 6.

d. No positive integer greater than 1 can divide 99 and 100 since any common divisor of 99 and 100 divides $100 - 99 = 1$. Hence the greatest common divisor of 99 and 100 is 1.

e. The positive divisors of 11 are 1 and 11 and the positive divisors of 121 are $1, 11$, and 121. Hence the greatest common divisor of 11 and 121 is 11.

f. A common divisor of 100 and 102 is also a divisor of $102 - 100 = 2$. Since 2 is a common divisor of 100 and 102, 2 is the greatest common divisor of these integers.

3.3.3. The greatest common divisor of a and $2a$ is also a divisor of their difference, $2a - a = a$. As a divides both a and $2a$, the greatest common divisor of a and $2a$ is a.

3.3.5. As $(a+1, a)$ is the least positive linear combination of a and $a+1$, it is clear that $(a+1, a) \leq (a+1) - a = 1$. It follows that $(a+1, a) = 1$.

3.3.7. By Theorem 3.8, $(ca, cb) = cma + cnb = |c| \cdot |ma + nb|$, where $cma + cnb$ is as small as possible. Therefore, $|ma + nb|$ is as small a positive integer as possible, i.e. equal to (a, b).

3.3.9. Let p be a prime dividing $(a^2 + b^2, a + b)$. Then $p \mid (a+b)^2 - (a^2 + b^2) = 2ab$. Now if $p \mid a$, then $p \mid b$ since $p \mid a + b$. But $(a, b) = 1$, so $p \nmid a$. Similarly, $p \nmid b$. Therefore $p \mid 2$ and so $p = 1$ or $p = 2$. If a and b have the same parity, then $2 \mid a + b$ and $2 \mid a^2 + b^2$, and so $(a^2 + b^2, a + b) = 2$. But if a and b have opposite parity, then $a + b$ and $(a^2 + b^2, a + b) = 1$.

3.3.11. Let $a = 2k$. Since $(a, b) \mid b$, and b is odd, (a, b) is odd. But $(a, b) \mid a = 2k$. Thus $(a, b) \mid k$. So $(a, b) = (k, b) = (\frac{a}{2}, b)$.

3.3.13. Let $d = (a, b)$. Then $(a/d, b/d) = 1$, so if $g \mid a/d$, then $(g, b/d) = 1$. In particular, if we let $e = (a/d, bc/d)$, then $e \mid a/d$, so $(e, b/d) = 1$, so we must have $e \mid c$. Since $e \mid a/d$, then $e \mid a$, so $e \mid (a, c)$. Conversely, if $f = (a, c)$, then $(f, b) = 1$, so $(d, f) = 1$, so $f \mid a/d$ and trivially, $f \mid bc/d$. Therefore $f \mid e$, whence $e = f$. Then $(a, b)(a, c) = de = d(a/d, bc/d) = (a, bc)$.

3.3.15. Let p, q, r be prime numbers. The set $\{pq, qr, pr\}$ is a set of three integers that are mutually relatively prime, but no two of which are relatively prime.

3.3.17. a. We have $(8, 10, 12) = 2$.

b. We have $(5, 25, 75) = 5$.

c. We have $(99, 9999, 0) = 99$.

d. We have $(6, 15, 21) = 3$.

e. We have $(-7, 28, -35) = 7$.

f. We have $(0, 0, 1001) = 1001$

3.3.19. Let $d \mid a_i, 1 \leq i \leq n$. Then clearly $dc \mid ca_i, 1 \leq i \leq n$. So $dc \mid (ca_1, ca_2, \ldots, ca_n)$. To see the other direction, note that $c \mid ca_i$ for all i, so $c \mid (ca_1, ca_2, \ldots, ca_n) = d$. Express d as $d = cd'$, where d' is as great as possible. But since $cd' \mid ca_i$, $d' \mid a_i$ and $d' \mid (a_1, a_2, \ldots, a_n)$, say $d'k = (a_1, a_2, \ldots, a_n)$. If $k > 1$, this contradicts the maximality of d', so we must have $d' = (a_1, a_2, \ldots, a_n)$.

3.3.21. Suppose that $(6k + a, 6k + b) = d$. Then $d \mid b - a$. We have $a, b \in \{-1, 1, 2, 3, 5\}$, so if $a < b$ it follows that $b - a \in \{1, 2, 3, 4, 6\}$. Hence $d \in \{1, 2, 3, 4, 6\}$. To show that $d = 1$ it is sufficient to show that neither 2 nor 3 divides $(6k + a, 6k + b)$. If $p = 2$ or $p = 3$ and $p \mid (6k + a, 6k + b)$ then $p \mid a$ and $p \mid b$. However, there are no such pairs a, b in the set $\{-1, 1, 2, 3, 5\}$.

3.3.23. We proceed with the Euclidean algorithm. $8a + 3 = 1(5a + 2) + (3a + 1)$. $5a + 2 = 1(3a + 1) + (2a + 1)$. $3a + 1 = 1(2a + 1) + (a)$. $2a + 1 = 2(a) + (1)$. Therefore $(8a + 3, 5a + 2) = 1$.

3.3.25. From Exercise 21, we know that $6k - 1, 6k + 1, 6k + 2, 6k + 3$, and $6k + 5$ are pairwise relatively prime. To represent n as the sum of two relatively prime integers greater than one, let $n = 12k + h, 0 \leq h < 12$. We now examine the twelve cases, one for each possible value of h, in the following chart:

h	n
0	$(6k-1)+(6k+1)$
1	$(6k-1)+(6k+2)$
2	$(6k-1)+(6k+3)$
3	$(6k+1)+(6k+2)$
4	$(6k+1)+(6k+3)$
5	$(6k+2)+(6k+3)$
6	$(6k+1)+(6k+5)$
7	$(6k+2)+(6k+5)$
8	$(6k+3)+(6k+5)$
9	$(12k+7)+2$
10	$(12k+7)+3$
11	$(12k+9)+2$

3.3.27. Let S be the set of all fractions $P/Q = (xa+ye)/(xb+yf)$ where x, y are relatively prime positive integers. Then every element of S lies between a/b and e/f and is in lowest terms. The first element of S to appear in a Farey series will have the smallest Q, i.e. $x = y = 1$. This fraction must be c/d by hypothesis.

3.3.29. Since $a/b < (a+c)/(b+d) < c/d$, we must have $b+d > n$, or a/b and c/d would not be consecutive, since otherwise, $(a+c)/(b+d)$ would have appeared in the Farey series of order n.

3.3.31. Since $(a/b) + (c/d) = (ad+bc)/bd$ is an integer, $bd \mid ad+bc$. Certainly, then, $bd \mid d(ad+bc) = ad^2 + cbd$. Now since $bd \mid cbd$, it must be that $bd \mid ad^2$. From this, $bdn = ad^2$ for some integer n, and it follows that $bn = ad$, or $b \mid ad$. Since $(a, b) = 1$, we must have $b \mid d$. Similarly, we can find that $d \mid b$ hence, $b = d$.

3.3.33. Consider the lattice points inside or on the triangle with vertices $(0,0), (a,0)$, and (a,b). Note that a lattice point lies on the diagonal from $(0,0)$ to (a,b) if and only if $[bx/a]$ is an integer. Let $d = (a,b)$ and $a = cd$, so that $(c,b) = 1$. Then $[bx/a]$ will be an integer exactly when x is a multiple of c, since then $d|b$ and $c|x$ so then $a = cd|bx$. But there are exactly d multiples of c less than or equal to a since $cd = a$, so there are exactly $d+1$ lattice points on the diagonal when we count $(0,0)$ also. So one way to count the lattice points in the triangle is to consider the rectangle which has $(a+1)(b+1)$ points and divide by 2. But we need to add back in half the points on the diagonal, which gives us $(a+1)(b+1)/2 + ((a,b)+1)/2$ total points in or on the triangle. Another way to count all the points is to count each column above the horizontal axis, starting with $i = 1, 2, \ldots, a-1$. The equation of the diagonal is $y = (b/a)x$, so for a given i, the number of points on or below the diagonal is $[bi/a]$. So the total number of interior points in the triangle plus the points on the diagonal is $\sum_{i=1}^{a-1}[bi/a]$. Then the right-hand boundary has b points (not counting $(a, 0)$) and the lower boundary has $a+1$ points, counting $(0,0)$. So in all, we have $\sum_{i=1}^{a-1}[bi/a] + a + b + 1$ points in or on the triangle. If we equate our two expressions and multiply through by 2 we have $(a+1)(b+1) + (a,b) + 1 = 2\sum_{i=1}^{a-1}[bi/a] + 2a + 2b + 2$ which simplifies to our expression.

3.3.35. Assume there are exactly r primes and consider the $r+1$ numbers $(r+1)! + 1$. From Lemma 3.1, each of these numbers has a prime divisor, but from Exercise 34, these numbers are pairwise relatively prime, so these prime divisors must be unique, so we must have at least $r+1$ different prime divisors, a contradiction.

3.4. The Euclidean Algorithm

3.4.1. a. We have $75 = 1 \cdot 45 + 30, 45 = 1 \cdot 30 + 15, 30 = 2 \cdot 15 + 0$, so $(45, 75) = 15$.

b. We have $222 = 2 \cdot 102 + 18, 102 = 5 \cdot 18 + 12, 18 = 1 \cdot 12 + 6, 12 = 2 \cdot 6 + 0$, so $(222, 102) = 6$.

c. We have $1414 = 2 \cdot 666 + 82, 666 = 8 \cdot 82 + 10, 82 = 8 \cdot 10 + 2, 10 = 5 \cdot 2 + 0$, so $(1414, 666) = 2$.

d. We have $44350 = 2 \cdot 20785 + 2780, 20785 = 7 \cdot 2780 + 1325, 2780 = 2 \cdot 1325 + 130, 1325 = 10 \cdot 130 + 25, 130 = 5 \cdot 25 + 5, 25 = 5 \cdot 5 + 0$, so $(44350, 2780) = 5$.

3.4.3. a. We have $q_1 = 1, q_2 = 1, q_3 = 2$, so $s_0 = 1, s_1 = 0, s_2 = s_0 - q_1 s_1 = 1, s_3 = s_1 - q_2 s_2 = -1$ and $t_0 = 0, t_1 = 1, t_2 = t_0 - q_1 t_1 = -1, t_3 = t_1 - q_2 t_2 = 2$. Thus, $(75, 45) = (-1)75 + (2)45$.

b. We have $q_1 = 2, q_2 = 5, q_3 = 1$, so $s_0 = 1, s_1 = 0, s_2 = 1, s_3 = -5, s_4 = 6$ and $t_0 = 0, t_1 = 1, t_2 = -2, t_3 = 11, t_4 = -13$. Thus $(222, 102) = (6)222 + (-13)102$.

c. We have, from Exercise 1(c), that $2 = 82 - 8 \cdot 10 == (1414 - 2 \cdot 666) - 8(666 - 8 \cdot 82) = 1414 - 10 \cdot 666 + 64(1414 - 2 \cdot 666) = -138(666) + (65)1414$.

d. We have, from Exercise 1(d), that $5 = 130 - 5 \cdot 25 = (2780 - 2 \cdot 1325) - 5(1325 - 10 \cdot 130) = (44350 - 2 \cdot 20785) - 7(20785 - 7 \cdot 2780) + 50(2780 - 2 \cdot 1325) = 44350 - 9 \cdot 20785 + 99(44350 - 2 \cdot 20785) - 100(20785 - 7 \cdot 2780) = 100 \cdot 44350 - 307 \cdot 20785 - 7(44350 - 2 \cdot 20785) = -1707(20785) + 800(44350)$.

3.4.5. a. We have $(6, 10, 15)((6, 10), 15) = (2, 15) = 1$.

b. We have $(70, 98, 105) = (70, (98, 105)) = (70, (98, 105 - 98)) = (70, (98, 7)) = (70, 7) = 7$.

c. We have $(280, 330, 405, 490) = (10(28, 33), 5(81, 98)) = (10, 5) = 5$.

3.4.7. a. Since $(6, 10) = 2 = 2 \cdot 6 - 10$, we have $1 = (6, 10, 15) = (2, 15) = 8 \cdot 2 - 15 = 8(2 \cdot 6 - 10) - 15 = 16 \cdot 6 - 8 \cdot 10 - 15$.

b. Since $(70, 98) = 14 = 3 \cdot 70 - 2 \cdot 98$, we have $7 = (70, 98, 105) = (14, 105) = 105 - 7 \cdot (14) = 105 - 7(3 \cdot 70 - 2 \cdot 98) = 105 - 21 \cdot 70 + 14 \cdot 98$.

c. Since $(280, 330) = 10 = 17 \cdot 330 - 20 \cdot 280$, and $(405, 490) = 5 = -75 \cdot 405 + 62 \cdot 490$, we have $(280, 330, 405, 490) = 5 = 0 \cdot 280 + 0 \cdot 330 - 75 \cdot 405 + 62 \cdot 490$.

3.4.9. Applying the reductions in the algorithm we find that $(2106, 8318) = 2(1053, 4159) = 2(3106, 1053) = 2(1553, 1053) = 2(500, 1053) = 2(250, 1053) = 2(125, 1053) = 2(125, 928) = 2(125, 464) = 2(125, 232) = 2(125, 116) = 2(125, 58) = 2(125, 29) = 2(96, 29) = 2(48, 29) = 2(24, 29) = 2(12, 29) = 2(6, 29) = 2(3, 29) = 2(3, 26) = 2(3, 13) = 2(3, 10) = 2(3, 5) = 2(3, 2) = 2(3, 1) = 2(2, 1) = 2(1, 1) = 2$.

3.4.11. The algorithm stops after $2n - 2$ steps. To prove this we use mathematical induction. When $n = 2$, $a = 1$ and $b = 2$. The first step leaves $a = 1$ and $b = 1$, and the second step will find the g.c.d.. Thus, the basis step holds. For the inductive hypothesis, we assume that the algorithm uses $2n - 2$ steps to find the g.c.d.. of $(2^n - (-1)^n)/3$ and $(2(2^{n-1} - (-1)^{n-1})/3$. To find the g.c.d. of $(2^{n+1} - (-1)^{n+1})/3$ and $(2(2^n - (-1)^n))/3$, the first step reduces this to the g.c.d. of $(2^{n+1} - (-1)^{n+1})/3$ and $(2^n - (-1)^n)/3$. The next step, as neither of these numbers is even, gives us $(2^n - (-1)^n)/3$ and $(1/3)(2^{n+1} - (-1)^{n+1} - 2^n + (-1)^n) = (1/3)(2^n + 2(-1)^n) = (2/3)(2^{n-1} - (-1)^{n-1})$. By the inductive hypothesis, the algorithm will take $2n - 2$ more steps, for a total of $2n = 2(n - 1) - 2$ steps.

3.4.13. Suppose we have the balanced ternary expansions for integers $a \geq b$. If both expansions end in zero, then both are divisible by 3, and we can divide this factor of 3 out by deleting the trailing zeros (a shift) in which case $(a, b) = 3(a/3, b/3)$. If exactly one expansion ends in zero, then we can divide the factor of 3 out by shifting, and we have $(a, b) = (a/3, b)$, say. If both expansions end in 1 or in -1, we can subtract the larger from the smaller to get $(a, b) = (a - b, b)$, say, and then the expansion for $a - b$ ends in zero. Finally, if one expansion ends in 1 and the other in -1, then we can add the two to get $(a + b, b)$, where the expansion of $a + b$ now ends in zero. Since $a + b$ is no larger than $2a$ and since we can now divide $a + b$ by three, the larger term is reduced by a factor of at least $2/3$ after two steps. Therefore this algorithm will terminate in a finite number of steps, when we finally have $a = b = 1$.

3.4.15. Lemma: If c and d are integers and $c = dq \pm r$ where q and r are integers, then $(c, d) = (d, r)$.

Proof of Lemma: If an integer e divides both c and d, then since $r = \pm(c - dq)$, Theorem 1.8 shows that $e \mid r$. If $e \mid d$ and $e \mid r$, then since $c = dq + r$, from Theorem 1.8 we see that $e \mid c$. Since the common divisors of c and d are the same as the common divisors of d and r, we see that $(c, d) = (d, r)$.

Proof of proposition: Let $r_0 = a$ and $r_1 = b$ be positive integers with $a \geq b$. By successively applying the least-remainder division algorithm, we find that

$$r_0 = r_1 q_1 + e_2 r_2, \quad \frac{-r_1}{2} < e_2 r_2 \leq \frac{r_1}{2}$$

$$\vdots$$

$$r_{n-2} = r_{n-1} q_{n-1} + e_n r_n, \quad \frac{-r_{n-1}}{2} < e_n r_n \leq \frac{r_{n-1}}{2}$$

$$r_{n-1} = r_n q_n.$$

We eventually obtain a remainder of zero since the sequence of remainders $a = r_0 > r_1 > r_2 > \cdots \geq 0$ cannot contain more than a terms. By the Lemma we see that $(a,b) = (r_0, r_1) = (r_1, r_2) = \cdots = (r_{n-2}, r_{n-1}) = (r_{n-1}, r_n) = (r_n, 0) = r_n$. Hence $(a,b) = r_n$, the last nonzero remainder.

3.4.17. Let $v_2 = v_3 = 2$, and for $i \geq 4$, $v_i = 2v_{i-1} + v_{i-2}$. Thus the least remainder algorithm will proceed with $e_i = 1$ and $q_i = 2$ for all i. To prove this we use induction. It clearly requires one division in the least-remainder division algorithm to find the g.c.d. of v_2 and v_3. This completes the basis step. For the induction hypothesis, we assume that it takes n steps to find the g.c.d. of v_{n+1} and v_{n+2}. To find the g.c.d. of v_{n+2} and v_{n+3}, the first step will be: $v_{n+3} = 2v_{n+2} + v_{n+1}$ by the definition of our v_i's. From this point, the algorithm will look identical to that for v_{n+1} and v_{n+2}. By our induction hypothesis, this will require n more steps. Hence, the total number of steps is $n+1$.

3.4.19. Performing the Euclidean algorithm with $r_0 = m$ and $r_1 = n$, we find that $r_0 = r_1 q_1 + r_2, 0 \leq r_2 < r_1, r_1 = r_2 q_2 + r_3, 0 \leq r_3 < r_2, \ldots, r_{k-3} = r_{k-2} q_{k-2} + r_{k-1}, 0 \leq r_{k-1} < r_{k-2}$, and $r_{k-2} = r_{k-1} q_{k-1}$. We have $(m,n) = r_{k-1}$. We will use these steps to find the greatest common divisor $a^m - 1$ and $a^n - 1$. First, we show that if u and v are positive integers, then the least positive residue of $a^u - 1$ modulo $a^v - 1$ is $a^r - 1$ where r is the least positive residue of u modulo v. To see this, note that $u = vq + r$ where r is the least positive residue of u modulo v. It follows that $a^u - 1 = a^{vq+r} - 1 = (a^v - 1)(a^{v(q-1)+r} + \cdots + a^{v+r} + a^r) + (a^r - 1)$. This shows that the remainder is $a^r - 1$ when $a^u - 1$ is divided by $a^v - 1$. Now let $R_0 = a^m - 1$ and $R_1 = a^n - 1$. When we perform the Euclidean algorithm starting with R_0 and R_1 we obtain $R_0 = R_1 Q_1 + R_2$, where $R_2 = a^{r_2} - 1$, $R_1 = R_2 Q_2 + R_3$ where $R_3 = a^{r_3} - 1, \ldots, R_{k-3} = R_{k-2} Q_{k-2} + R_{k-1}$ where $R_{k-1} = a^{r_{k-1}} - 1$. Hence the last nonzero remainder, $R_{k-1} = a^{r_{k-1}} - 1 = a^{(m,n)} - 1$ is the greatest common divisor of $a^m - 1$ and $a^n - 1$.

3.4.21. Note that $(x, y) = (x - ty, y)$, as any divisor of x and y is also a divisor of $x - ty$. So, every move in the game of Euclid preserves the g.c.d. of the two numbers. Since $(a, 0) = a$, if the game beginning with $\{a, b\}$ terminates, then it must do so at $\{(a,b), 0)\}$. Since the sum of the two numbers is always decreasing and positive, the game must terminate.

3.4.23. Choose the integer m so that d has no more than m bits and that q has $2m$ bits, appending extra zeros to the front of q if necessary. Then $m = O(\log_2 q) = O(\log_2 d)$. Then from Theorems 2.7 and 2.5 we know that there is an algorithm for dividing q by d in $O(m^2) = O(\log_2 q \log_2 d)$ bit operations. Now let n be the number of steps needed in the Euclidean algorithm to find the greatest common divisor of a and b. Then by Theorem 3.12, $n = O(\log_2 a)$. Let q_i and r_i be as in the proof of Theorem 3.12. Then the total number of bit operations for divisions in the Euclidean algorithm is $\sum_{i=1}^{n} O(\log_2 q_i \log_2 r_i) = \sum_{i=1}^{n} O(\log_2 q_i \log_2 b) = O\left(\log_2 b \sum_{i=1}^{n} \log_2 q_i\right) = O\left(\log_2 b \log_2 \prod_{i=1}^{n} q_i\right)$. By dropping the remainder in each step of the Euclidean algorithm, we have the system of inequalities $r_i \geq r_{i+1} q_{i+1}$, for $i = 0, 1, \ldots, n-1$. Multiplying these inequalities together yields $\prod_{i=0}^{n-1} r_i \geq \prod_{i=1}^{n} r_i q_i$. Cancelling common factors reduces this to $a = r_0 \geq r_n \prod_{i=1}^{n} q_i$. Therefore, from above we have that the total number of bit operations is $O\left(\log_2 b \log_2 \prod_{i=1}^{n} q_i\right) = O(\log_2 b \log_2 a) = O((\log_2 a)^2)$.

3.4.25. We apply the Q_i's one at a time. When we multiply $\begin{pmatrix} q_n & 1 \\ 1 & 0 \end{pmatrix} \begin{pmatrix} r_n \\ 0 \end{pmatrix} = \begin{pmatrix} q_n r_n \\ r_n \end{pmatrix} = \begin{pmatrix} r_{n-1} \\ r_n \end{pmatrix}$, the top component is the last equation in the series of equations in the proof of Lemma 3.3. When we multiply this result on the left by the next matrix we get $\begin{pmatrix} q_{n-1} & 1 \\ 1 & 0 \end{pmatrix} \begin{pmatrix} r_{n-1} \\ r_n \end{pmatrix} = \begin{pmatrix} q_{n-1} r_{n-1} + r_n \\ r_{n-1} \end{pmatrix} = \begin{pmatrix} r_{n-2} \\ r_{n-1} \end{pmatrix}$, which is the matrix version of the last two equations the proof of Lemma 3.3. In general, at the ith step we have $\begin{pmatrix} q_{n-i} & 1 \\ 1 & 0 \end{pmatrix} \begin{pmatrix} r_{n-i-1} \\ r_{n-i} \end{pmatrix} = \begin{pmatrix} q_{n-i}r_{n-i-1} + r_{n-i} \\ r_{n-i-1} \end{pmatrix} = \begin{pmatrix} r_{n-i-2} \\ r_{n-i-1} \end{pmatrix}$, so that we inductively work our way up the equations in the proof of Lemma 3.3, until finally we have $\begin{pmatrix} r_0 \\ r_1 \end{pmatrix} = \begin{pmatrix} a \\ b \end{pmatrix}$.

3.5. The Fundamental Theorem of Arithmetic

3.5.1. a. We have $36 = 6^2 = 2^2 \cdot 3^2$.

b. We have $39 = 3 \cdot 13$.

c. We have $100 = 10^2 = 2^2 \cdot 5^2$.

d. We have $289 = 17^2$.

e. We have $222 = 2 \cdot 111 = 2 \cdot 3 \cdot 37$.

f. We have $256 = 2^8$.

g. We have $515 = 5 \cdot 103$.

h. We have $989 = 23 \cdot 43$.

i. We have $5040 = 10 \cdot 504 = 2 \cdot 5 \cdot 4 \cdot 126 = 2^4 \cdot 3^2 \cdot 5 \cdot 7$.

j. We have $8000 = 8 \cdot 10^3 = 2^6 \cdot 5^3$.

k. We have $9555 = 3 \cdot 5 \cdot 7^2 \cdot 13$.

l. We have $9999 = 9 \cdot 1111 = 3^2 \cdot 11 \cdot 101$.

3.5.3. We have $4849845 = 3 \cdot 5 \cdot 7 \cdot 11 \cdot 13 \cdot 17 \cdot 19$.

3.5.5. a. We have $196608 = 2^{16} 3$.

b. We have $7290000 = 729 \cdot 10^4 = 2^4 3^6 5^4$.

c. If a prime divides 20!, then it must divide one of the factors from 1 to 20. Thus the prime factors are exactly those less than or equal to 20.

d. We have $\binom{50}{25} = (26 \cdot 27 \cdot 28 \cdot 29 \cdot 30 \cdot 31 \cdot 32 \cdot 33 \cdot 34 \cdot 35 \cdot 36 \cdot 37 \cdot 38 \cdot 39 \cdot 40 \cdot 41 \cdot 42 \cdot 43 \cdot 44 \cdot 45 \cdot 46 \cdot 47 \cdot 48 \cdot 49 \cdot 50)/(2 \cdot 3 \cdot 4 \cdot 5 \cdot 6 \cdot 7 \cdot 8 \cdot 9 \cdot 10 \cdot 11 \cdot 12 \cdot 13 \cdot 14 \cdot 15 \cdot 16 \cdot 17 \cdot 18 \cdot 19 \cdot 20 \cdot 21 \cdot 22 \cdot 23 \cdot 24 \cdot 25) = 2^3 3^2 7^2 13 \cdot 29 \cdot 31 \cdot 37 \cdot 41 \cdot 43 \cdot 47$.

3.5.7. The integers with exactly three positive divisors are those of the form p^2 where p is prime. The integers with exactly four positive divisors are those of the form pq or p^3 where p and q are distinct primes. These results can be proved considering the cases where the integer is a power of a prime, the product of powers of two primes, and the product of powers of more that two primes.

3.5.9. Let $n = p_1^{2a_1} p_2^{2a_2} \cdots p_k^{2a_k} q_1^{2b_1+3} q_2^{2b_2+3} \cdots q_l^{2b_l+3}$ be the factorization of a powerful number. Then $n = (p_1^{a_1} p_2^{a_2} \cdots p_k^{a_k} q_1^{b_1} q_2^{b_2} \cdots q_l^{b_l})^2 (q_1 q_2 \cdots q_l)^3$ is a product of a square and a cube.

3.5.11. Suppose that $p^a \| m$ and $p^b \| n$. Then $m = p^a Q$ and $n = p^b R$ where both Q and R are products of primes other than p. Hence $mn = (p^a Q)(p^b R) = p^{a+b} QR$. It follows that $p^{a+b} \| mn$ since p does not divide QR.

3.5.13. Suppose that $p^a \| m$ and $p^b \| n$ with $a \neq b$. Then $m = p^a Q$ and $n = p^b R$ where both Q and R are products of primes other than p. Suppose, without loss of generality, that $a = \min(a,b)$. Then $m + n = p^a Q + p^b R = p^{\min(a,b)}(Q + p^{b-a} R)$. Then $p \nmid (Q + p^{b-a} R)$ because $p \nmid Q$ but $p \mid p^{b-a} R$. It follows that $p^{\min(a,b)} \| (m+n)$.

3.5.15. We know that in the prime power factorization of 20! the number 2 occurs $[20/2] + [20/4] + [20/8] + [20/16] = 10 + 5 + 2 + 1 = 18$ times, 3 occurs $[20/3] + [20/9] = 6 + 2 = 8$ times, 5 occurs $[20/5] = 4$ times, 7 occurs $[20/7] = 2$ times, 11 occurs $[20/11] = 1$ time, 13 occurs $[20/13] = 1$ time, 17 occurs $[20/17] = 1$ time, and 19 occurs $[20/19] = 1$ time. Hence $20! = 2^{18} \cdot 3^8 \cdot 5^4 \cdot 7^2 \cdot 11 \cdot 13 \cdot 17 \cdot 19$.

3.5.17. Suppose $n!$ ends with exactly 74 zeroes. Then $5^{74} \cdot 2^{74} = 10^{74} \mid n!$. Since there are more multiples of 2 than 5 in $1, 2, \ldots, n$, we need only concern ourselves with the fact that $5^{74} \mid n!$. Thus, via Exercise 12, we need to find an n such that $74 = [n/5] + [n/25] + \cdots$. By direct calculation, $74 = [300/5] + [300/25] + [300/125]$. It follows that $300!, 301!, 302!, 303!$, and $304!$ end with exactly 74 zeroes.

3.5.19. We compute $\alpha\beta = (ac - 5bd) + (ad + bc)\sqrt{-5}$. Thus $N(\alpha\beta) = (ac - 5bd)^2 + 5(ad + bc)^2 = a^2c^2 - 10acbd + 25b^2d^2 + 5a^2d^2 + 10adbc + 5b^2c^2 = a^2(c^2 + 5d^2) + 5b^2(5d^2 + c^2) = (a^2 + 5b^2)(c^2 + 5d^2) = N(\alpha)N(\beta)$.

3.5.21. Suppose $3 = \alpha\beta$. Then by Exercise 19, $9 = N(3) = N(\alpha)N(\beta)$. Then $N(\alpha) = 1, 3$ or 9. Let $\alpha = a + b\sqrt{-5}$. Then we must have $a^2 + 5b^2 = 1, 3$, or 9. So either $b = 0$ and $a = \pm 1$ or ± 3, or $b = \pm 1$ and $a = \pm 2$. Since $a = \pm 1, b = 0$ is excluded, and since $a = \pm 3$ forces $\beta = \pm 1$, we must have $b = \pm 1$. That is, $\alpha = \pm 2 \pm \sqrt{-5}$. But then $N(\alpha) = 9$, and hence $N(\beta) = 1$, which forces $\beta = \pm 1$.

3.5.23. Note that $21 = 3 \cdot 7 = (1 + 2\sqrt{-5})(1 - 2\sqrt{-5})$. We know 3 is prime from Exercise 21. Similarly if we seek $\alpha = a + b\sqrt{-5}$ such that $N(\alpha) = a^2 + 5b^2 = 7$, we find there are no solutions. For $|b| = 0$ implies $a^2 = 7$, $|b| = 1$ implies $a^2 = 2$ and $|b| > 1$ implies $a^2 < 0$, and in each case there is no such a. Hence if $\alpha\beta = 7$, then $N(\alpha\beta) = N(\alpha)N(\beta) = N(7) = 49$. So one of $N(\alpha)$ and $N(\beta)$ must be equal to 49 and the other equal to 1. Hence 7 is also prime. We have shown that there are no numbers of the form $a + b\sqrt{-5}$ with norm 3 or 7. So in a similar fashion to the argument above, if $\alpha\beta = 1 \pm 2\sqrt{-5}$, then $N(\alpha\beta) = N(\alpha)N(\beta) = N(1 \pm 2\sqrt{-5}) = 21$. And there are no numbers with norm 3 or 7, so one of α and β has norm 21 and the other has norm 1. Hence $1 \pm 2\sqrt{-5}$ is also prime.

3.5.25. The product of $4k + 1$ and $4l + 1$ is $(4k+1)(4l+1) = 16kl + 4k + 4l + 1 = 4(4kl + k + l) + 1 = 4m + 1$ where $m = 4kl + k + l$. Hence the product of two integers of the form $4k + 1$ is also of this form.

3.5.27. We proceed by mathematical induction on the elements of H. The first Hilbert number greater than 1, 5, is a Hilbert prime because it is an integer prime. This completes the basis step. For the inductive step, we assume that all numbers in H less than or equal to n can be factored into Hilbert primes. The next greatest number in H is $n + 4$. If $n + 4$ is a Hilbert prime, then we are done. Otherwise, $n = hk$, where h and k are less than n and in H. By the inductive hypothesis, h and k can be factored into Hilbert primes. Thus, $n + 4$ can be written as the product of Hilbert primes.

3.5.29. Suppose that n is divisible by all primes not exceeding $\sqrt{n}$. Let M be the least common multiple of the integers m with $1 \leq m \leq \sqrt{n}$. Then for every prime p with $p \leq \sqrt{n}$, $p^k \mid M$ but p^{k+1} does not divide M where p^k is the largest power of p not exceeding $\sqrt{n}$. Then $M = p_1^{k_1} \cdots p_t^{k_t}$ where the powers of the prime p_i is the largest power of this prime not exceeding $\sqrt{n}$. Since $\sqrt{n} < p_i^{k_i+1}$ for $i = 1, 2, \ldots t$ we have $(\sqrt{n})^t < p_1^{k_1+1} \cdots p_t^{k_t+1}$. But note that $p_1^{k_1+1} \cdots p_t^{k_t+1} = (p_1^{k_1} \cdots p_t^{k_t}) \cdot (p_1 \cdots p_k) \leq M \cdot p_1 \cdots p_t \leq M^2$. It follows that $(\sqrt{n})^t < M^2$. Since $M \mid n$ it follows that $M \leq n$, so $(\sqrt{n})^t < n^2$. It follows that $t < 4$. If t is the number of primes less than $\sqrt{n}$ and there are four or fewer primes less than $\sqrt{n}$ and 7 is the fourth prime, it follows that $\sqrt{n} \leq 7$, so $n < 49$. Examining the integers less than 49 shows that the only integers satisfying the conditions are $n = 1, 2, 3, 4, 6, 8, 12$, and 24.

3.5.31. a. We have $[7, 11] = 77$.

    b. We have $[12, 18] = 36$.

c. We have $[25, 30] = 150$.

   d. We have $[101, 333] = 33633$.

   e. We have $[1331, 5005] = 605605$.

   f. We have $[5040, 7700] = 277200$.

3.5.33. a. We have $(2^2 3^3 5^5 7^7, 2^7 3^5 5^3 7^2) = 2^2 3^3 5^3 7^2$; $[2^2 3^3 5^5 7^7, 2^7 3^5 5^3 7^2] = 2^7 3^5 5^5 7^7$.

   b. We have $(2 \cdot 3 \cdot 5 \cdot 7 \cdot 11 \cdot 13, 17 \cdot 19 \cdot 23 \cdot 29) = 1$; $[2 \cdot 3 \cdot 5 \cdot 7 \cdot 11 \cdot 13, 17 \cdot 19 \cdot 23 \cdot 29] = 2 \cdot 3 \cdot 5 \cdot 7 \cdot 11 \cdot 13 \cdot 17 \cdot 19 \cdot 23 \cdot 29$.

   c. We have $(2^3 5^7 11^{13}, 2 \cdot 3 \cdot 5 \cdot 7 \cdot 11 \cdot 13) = 2 \cdot 5 \cdot 11$; $[2^3 5^7 11^{13}, 2 \cdot 3 \cdot 5 \cdot 7 \cdot 11 \cdot 13] = 2^3 \cdot 3 \cdot 5^7 \cdot 7 \cdot 11^{13} \cdot 13$.

   d. We have $(47^{11} 79^{111} 101^{1001}, 41^{11} 83^{111} 101^{1000}) = 101^{1000}$; $[47^{11} 79^{111} 101^{1001}, 41^{11} 83^{111} 101^{1000}] = 41^{11} 47^{11} 79^{111} 83^{111} 101^{1001}$.

3.5.35. Suppose that both 13-year and 17-year cicadas emerge in a location in 1900. The 13-year cicada will emerge again in years $1900 + 13k$ where k is a positive integer. The 17-year cicadas will emerge again in years $1900 + 17k$ where k is a positive integer. Both 13-year and 17-year cicadas will emerge again in years $1900 + [13, 17]k = 1900 + 221k$ where k is a positive integer. Hence they both will emerge again in the year 2121.

3.5.37. Let $a = p_1^{r_1} p_2^{r_2} \cdots p_k^{r_k}$ and $b = p_1^{s_1} p_2^{s_2} \cdots p_k^{s_k}$, where p_i is a prime and r_i and s_i are nonnegative. $(a, b) = p_1^{\min(r_1, s_1)} \cdots p_k^{\min(r_k, s_k)}$ and $[a, b] = p_1^{\max(r_1, s_1)} \cdots p_k^{\max(r_k, s_k)}$. So $[a, b] = (a, b) p_1^{\max(r_1, s_1) - \min(r_1, s_1)} \cdots p_k^{\max(r_k, s_k) - \min(r_k, s_k)}$. Since $\max(r_i, s_i) - \min(r_i, s_i)$ is clearly nonnegative, we now see that $(a, b) \mid [a, b]$.

3.5.39. If $[a, b] \mid c$, then since $a \mid [a, b]$, $a \mid c$. Similarly, $b \mid c$. Conversely, suppose that $a = p_1^{a_1} p_2^{a_2} \cdots p_n^{a_n}$ and $b = p_1^{b_1} p_2^{b_2} \cdots p_n^{b_n}$ and $c = p_1^{c_1} p_2^{c_2} \cdots p_n^{c_n}$. If $a \mid c$ and $b \mid c$, then $\max(a_i, b_i) \leq c_i$ for $i = 1, 2, \ldots, n$. Hence, $[a, b] \mid c$.

3.5.41. Assume that $p \mid a^n = \pm \mid a \mid \cdot \mid a \mid \cdots \mid a \mid$. Then by Lemma 3.5, $p \mid\mid a \mid$ and so $p \mid a$.

3.5.43. a. Suppose that $(a, b) = 1$ and $p \mid (a^n, b^n)$ where p is a prime. It follows that $p \mid a^n$ and $p \mid b^n$. By Exercise 41, $p \mid a$ and $p \mid b$. But then $p \mid (a, b) = 1$, which is a contradiction.

   b. Suppose that a does not divide b, but $a^n \mid b^n$. Then there is some prime power, say p^r that divides a but does not divide b (else $a \mid b$ by the Fundamental Theorem of Arithmetic). Thus, $a = p^r Q$, where Q is an integer. Now, $a^n = (p^r Q)^n = p^{rn} Q^n$, so $p^{rn} \mid a^n \mid b^n$. Then $b^n = m p^{rn}$, from which it follows that each of the n b's must by symmetry contain r p's. But this is a contradiction.

3.5.45. Suppose that $x = \sqrt{2} + \sqrt{3}$. Then $x^2 = 2 + 2\sqrt{2}\sqrt{3} + 3 = 5 + 2\sqrt{6}$. Hence $x^2 - 5 = 2\sqrt{6}$. It follows that $x^4 - 10x^2 + 25 = 24$. Consequently, $x^4 - 10x^2 + 1 = 0$. By Theorem 3.17 it follows that $\sqrt{2} + \sqrt{3}$ is irrational, since it is not an integer (we can see this since $3 < \sqrt{2} + \sqrt{3} < 4$).

3.5.47. Suppose that $m/n = \log_p b$. This implies that $p^{\frac{m}{n}} = b$, from which it follows that $p^m = b^n$. Since b is not a power of p, there must be another prime, say q, such that $q \mid b$. But then $q \mid b \mid b^n = p^m = p \cdot p \cdots p$. By Lemma 2.4, $q \mid p$, which is impossible since p is a prime number.

3.5.49. Let p be a prime that divides a or b. Then p divides $a + b$ and $[a, b]$. Hence p divides both sides of the equation. Define s, t by $p^s \mid\mid a$, $p^t \mid\mid b$, say that $a = xp^s$ and $b = yp^t$. Without loss of generality, suppose $s \leq t$. Then $a + b = p^s(x + p^{t-s})$, so $p^s \mid\mid a + b$. Also, $p^{\max(s,t)} \mid\mid [a, b]$. But $\max(s, t) = t$, so $p^t \mid\mid [a, b]$. Therefore $p^{\min(s,t)} \mid\mid (a + b, [a, b])$. But $\min(s, t) = s$, so the same power of p divides both sides of the equation. Therefore the two sides must be equal.

3.5.51. Let $a = p_1^{r_1} p_2^{r_2} \cdots p_k^{r_k}, b = p_1^{s_1} p_2^{s_2} \cdots p_k^{s_k}$, and $c = p_1^{t_1} p_2^{t_2} \cdots p_k^{t_k}$, with p_i prime and r_i, s_i, and t_i nonnegative. Observe that $\min(x, \max(y, z)) = \max(\min(x, y), \min(x, z))$. We also know that $[a, b] = p_1^{\max(r_1,s_1)} p_2^{\max(r_2,s_2)} \cdots p_k^{\max(r_k,s_k)}$, and so $([a, b], c) = p_1^{\min(t_1,\max(r_1,s_1))} p_2^{\min(t_2,\max(r_2,s_2))} \cdots p_k^{\min(t_k,\max(r_k,s_k))}$. We also know that $(a, c) = p_1^{\min(r_1,t_1)} p_2^{\min(r_2,t_2)} \cdots p_k^{\min(r_k,t_k)}$ and $(b, c) = p_1^{\min(s_1,t_1)} p_2^{\min(s_2,t_2)} \cdots p_k^{\min(s_k,t_k)}$. Then, $[(a, c), (b, c)] = p_1^{\max(\min(r_1,t_1),\min(s_1,t_1))} p_2^{\max(\min(r_2,t_2),\min(s_2,t_2))} \cdots p_k^{\max(\min(r_k,t_k),\min(s_k,t_k))}$. Therefore, $([a, b], c) = [(a, c), (b, c)]$. In a similar manner, noting that $\min(\max(x, z), \max(y, z)) = \max(\min(x, y), z)$, we find that $[(a, b), c] = ([a, c], [b, c])$.

3.5.53. Let $c = [a_1, \ldots, a_n]$, $d = [[a_1, \ldots, a_{n-1}], a_n]$, and $e = [a_1, \ldots, a_{n-1}]$. If $c \mid m$, then all a_i's divide m, hence $e \mid m$ and $a_n \mid m$, so $d \mid m$. Conversely, if $d \mid m$, then $e \mid m$ and $a_n \mid m$, so all a_i's divide m, thus $c \mid m$. Since c and d divide all the same numbers, they must be equal.

3.5.55. a. There are six cases, all handled the same way. So without loss of generality, suppose that $a \leq b \leq c$. Then $\max(a, b, c) = c, \min(a, b) = a, \min(a, c) = a, \min(b, c) = b$, and $\min(a, b, c) = a$. Hence $c = \max(a, b, c) = a + b + c - \min(a, b) - \min(a, c) - \min(b, c) + \min(a, b, c) = a + b + c - a - a - b + a$.

b. The power of a prime p that occurs in the prime factorization of $[a, b, c]$ is $\max(a, b, c)$ where a, b, and c are the powers of this prime in the factorizations of a, b, and c, respectively. Also $a + b + c$ is the power of p in abc, $\min(a, b)$ is the power of p in (a, b), $\min(a, c)$ is the power of p in (a, c), $\min(b, c)$ is the power of p in (b, c), and $\min(a, b, c)$ is the power of p in (a, b, c). It follows that $a + b + c - \min(a, b) - \min(a, c) - \min(b, c)$ is the power of p in $abc(a, b, c)/((a, b)(a, c)(b, c))$. Hence $[a, b, c] = abc(a, b, c)/((a, b)(a, c)(b, c))$.

3.5.57. Let $a = p_1^{r_1} p_2^{r_2} \cdots p_k^{r_k}, b = p_1^{s_1} p_2^{s_2} \cdots p_k^{s_k}$, and $c = p_1^{t_1} p_2^{t_2} \cdots p_k^{t_k}$, with p_i prime and r_i, s_i, and t_i nonnegative. Then $p_i^{r_i+s_i+t_i} \parallel abc$, but $p_i^{\min(r_i,s_i,t_i)} \parallel (a, b, c)$ and $p_i^{r_i+s_i+t_i-\min(r_i,s_i,t_i)} \parallel [ab, ac, ab]$, and $p_i^{\min(r_i,s_i,t_i)} \cdot p_i^{r_i+s_i+t_i-\min(r_i,s_i,t_i)} = p_i^{r_i+s_i+t_i}$.

3.5.59. Let $a = p_1^{r_1} p_2^{r_2} \cdots p_k^{r_k}, b = p_1^{s_1} p_2^{s_2} \cdots p_k^{s_k}$, and $c = p_1^{t_1} p_2^{t_2} \cdots p_k^{t_k}$, with p_i prime and r_i, s_i, and t_i nonnegative. Then, using that $(a, b, c) = p_1^{\min(r_1,s_1,t_1)} p_2^{\min(r_2,s_2,t_2)} \cdots p_k^{\min(r_k,s_k,t_k)}$, and $[a, b, c] = p_1^{\max(r_1,s_1,t_1)} p_2^{\max(r_2,s_2,t_2)} \cdots p_k^{\max(r_k,s_k,t_k)}$, we can write the prime factorization of $([a, b], [a, c], [b, c])$ and $[(a, b), (a, c), (b, c)]$. For instance, consider the case where $k = 1$. Then $([a, b], [a, c], [b, c]) = (p_1^{\max(r_1,s_1)}, p_1^{\max(r_1,t_1)}, p_1^{\max(s_1,t_1)}) = p_1^{\min(\max(r_1,s_1),\max(r_1,t_1),\max(s_1,t_1))}$. Similarly, $[(a, b), (a, c), (b, c)] = p_1^{\max(\min(r_1,s_1),\min(r_1,t_1),\min(s_1,t_1))}$. Clearly, these two are equal (examine the six orderings $r_1 \geq s_1 \geq t_1, \ldots$).

3.5.61. First note that there are arbitrarily long sequences of composites in the integers. For example, $(n + 2)! + 2, (n + 2)! + 3, \ldots, (n + 2)! + (n + 2)$ is a sequence of n consecutive composites. To find a sequence of n composites in the sequence $a, a + b, a + 2b, \ldots$, look at the integers in $a, a + b, a + 2b, \ldots$ with absolute values between $(nb + 2)! + 2$ and $(nb + 2)! + (nb + 2)$. There are clearly n or $n + 1$ such integers, and all are composite.

3.5.63. We have $8137 = 79 \cdot 103$. Since the price of the camera is an integer and less than 99 dollars, it follows that the discounted price of a camera is 79 dollars. Hence they sold 103 cameras at 79 dollars each.

3.5.65. Since $139499 = 199 \cdot 701$, the price must have been \$199 and so the number of electronic organizers sold was 701.

3.5.67. Let $a = \prod_{i=1}^{s} p_i^{\alpha_i}$ and $b = \prod_{i=1}^{t} p_i^{\beta_i}$. The condition $(a, b) = 1$ is equivalent to $\min(\alpha_i, \beta_i) = 0$ for all i and the condition $ab = c^n$ is equivalent to $n \mid (\alpha_i + \beta_i)$ for all i. Hence $n \mid \alpha_i$ and $\beta_i = 0$ or $n \mid \beta_i$ and $\alpha_i = 0$. Let d be the product of $p_i^{\alpha_i/n}$ over all i of the first kind, and let e be the product of $p_i^{\beta_i/n}$ over all i of the second kind. Then $d^n = a$ and $e^n = b$.

3.5.69. Suppose the contrary and that $a \leq n$ is in the set. Then $2a$ cannot be in the set. Thus, if there are k elements in the set not exceeding n then, there are k integers between $n + 1$ and $2n$ which cannot be in

3.5. THE FUNDAMENTAL THEOREM OF ARITHMETIC

the set. So there are at most $k + (n-k) = n$ elements in the set.

3.5.71. he fundamental theorem of arithmetic implies that m and n have the same prime divisors. So suppose that m and n have prime-power factorizations $m = p_1^{a_1} p_2^{a_2} \cdots p_k^{a_k}$ and $n = p_1^{b_1} p_2^{b_2} \cdots p_k^{b_k}$. From the equation $m^n = n^m$ it follows that $a_i n = b_i m$ for $i = 1, 2, \ldots, k$. We first assume that $n > m$. Then $a_i < b_i$ for $i = 1, 2, \ldots, k$. Hence n is divisible by m, so $n = dm$ for some integer d. This implies that $m^{dm} = (dm)^m$. Taking the mth roots of both sides gives $m^d = dm$, which implies that $m^{d-1} = d$. Since $n > m$ we know that $d > 1$, so $m > 1$. However $2^{2-1} = 2$ and when $d > 2$ it follows that $m^{d-1} > d$. When $d > 2$ and $m \geq 2$ we have $m^{d-1} \geq 2^{d-1} > d$ since $2^{3-1} > 3$ and when $d = 2$ and $m > 2$ we have $m^{d-1} = m > 2 = d$. Hence the only solution with $n > m$ has $m = 2$ and $n = 2d = 2 \cdot 2 = 4$. Consequently all solutions are given by $m = 2$ and $n = 4$, $m = 4$ and $n = 2$, or $m = n$.

3.5.73. By Lemma 3.1, S must have a prime divisor, and by our assumption, it must be one of the p_i, $i = 1, 2, \ldots r$. For $j \neq i$, $p_i | Q_j$, since it is one of the factors. So p_i must divide $S - \sum_{j \neq i} Q_j = Q_i = p_1 \cdots p_{i-1} p_{i+1} \cdots p_r$, but by the Fundamental Theorem of Arithmetic, p_i must be equal to one of these last factors, a contradiction, therefore S must have a prime factor different from the list we have. Since no finite list can contain all the primes, there must be infinitely many primes.

3.5.75. Let p be the largest prime less than or equal to n. If $2p$ were less than or equal to n then Bertrand's postulate would guarantee another prime q such that $p < q < 2p \leq n$ contradicting the choice of p. Therefore, we know that $n < 2p$. Therefore, in the product $n! = 1 \cdot 2 \cdot 3 \cdots n$, there appears only one multiple of p, namely p itself, and so in the prime factorization of n, p appears with exponent 1.

3.5.77. a. Uniqueness follows from the Fundamental Theorem. If a prime p_i doesn't appear in the prime factorization, then we include it in the product with an exponent of 0. Since $e_i \geq 0$, we have $p_1^{e_1} = p_1^{e_1} p_2^0 \cdots p_r^0 \leq p_1^{e_1} p_2^{e_2} \cdots p_r^{e_r} = m$.

b. Since $p_1^{e_i} < p_i^{e_i} \leq m \leq Q = p_r^n$, we take logs of both sides to get $e_i \log p_1 \leq n \log p_r$. Solving for e_i gives the first inequality. If $1 \leq m \leq Q$, then m has a prime-power factorization of the form given in part (a), so the r-tuples of exponents count the number of integers in the range $1 \leq m \leq Q$.

c. To bound the number of r-tuples, by part (b) there are at most $Cn + 1$ choices for each e_i, therefore there are at most $(Cn + 1)^r$ r-tuples, which by part (b) gives us $p_r^n \leq (Cn + 1)^r = (n(C + 1/n))^r \leq n^r (C+1)^r$.

d. Taking logs of both sides of the inequality in part (c) and solving for n yields $n \leq (r \log n + \log(C + 1))/\log p_r$, but since n grows much faster than $\log n$, the left side must be larger than the right for large values of n. This contradiction shows there must be infinitely many primes.

3.5.79. Since 40 has lots of small factors in its prime factorization, we expect it to have a small Smarandache value. Since it's divisible by 5, the smallest possible value will be 5, and since 40 does indeed divide 5!, we have $S(40) = 5$. Since 41 and 43 are primes, after Exercise 80 we have S(41)=41, S(43)=43.

3.5.81. From Exercise 83, we have $a(2) = 2$, $a(3) = 3$, $a(5) = 5$, $a(7) = 7$, and $a(11) = 11$. The smallest value of m such that $S(m) = 1$ is $m = 1$, so $a(1) = 1$. $S(4) = 4$, but not for any smaller argument, so $a(4) = 4$. To find $a(6)$ we consider the smallest number which would require two factors of 3 in the factorial, and that number would be 9, so $a(6) = 9$. To find $a(8)$, we consider the smallest number which would require 5 factors of 2 in the factorial (one factor from 2, two factors from 4, one factor from 6 and the additional factor from 8.) And that number would be 32, so $a(8) = 32$. Similarly $a(9) = 27$, since 27 is the smallest number requiring the 3 factors of 3 (one from 3, one from 6 and the additional one from 9.) Similarly $a(10) = 25$, since 25 is the smallest number needing both factors of 5. In sum, the sequence is $a(n) = 1, 2, 3, 4, 5, 9, 7, 32, 27, 25, 11$.

3.5.83. From Exercise 80, we have $S(p) = p$ whenever p is prime. If $m < p$ and $m | S(p)! = p!$ then $m|(p-1)!$, so $S(p)$ must be the first time that $S(n)$ takes on the value p. Therefore of all the inverses of p, p is the least.

3.5.85. Let n be a positive integer and suppose n is square-free. Then no prime can appear to a power greater than one in the prime-power factorization of n. So $n = p_1 p_2 \cdots p_r$ for some distinct primes p_i. Then $rad(n) = p_1 p_2 \cdots p_r = n$. Conversely, if n is not square-free, then some square $d^2 | n$ and some prime factor p_1 of d appears to an even power in the prime-power factorization of n. So $n = p_1^{2a} p_2^{b_2} \cdots p_r^{b_r}$. Then $rad(n) = p_1 p_2 \cdots p_r \neq n$.

3.5.87. Since every prime occurring in the prime-power factorization of mn occurs in either the factorization of m or n, every factor in $rad(mn)$ occurs at least once in the product $rad(m)rad(n)$, which gives us the inequality. If $m = p_1^{a_1} \cdots p_r^{a_r}$ and $n = q_1^{b_1} \cdots q_s^{b_s}$ are relatively prime, then we have $rad(mn) = p_1 \cdots p_r q_1 \cdots q_s = rad(m)rad(n)$.

3.5.89. First note that if $p \mid \binom{2n}{n}$, then $p \leq 2n$. This is true because every factor of the numerator of $\binom{2n}{n} = \frac{(2n)!}{(n!)^2}$ is less than or equal to $2n$. Let $\binom{2n}{n} = p_1^{r_1} p_2^{r_2} \cdots p_k^{r_k}$ be the factorization of $\binom{2n}{n}$ into distinct primes. By the definition of π, $k \leq \pi(2n)$. By Exercise 72, $p_i^{r_i} \leq 2n$. It now follows that $\binom{2n}{n} = p_1^{r_1} p_2^{r_2} \cdots p_k^{r_k} \leq (2n)(2n) \cdots (2n) \leq (2n)^{\pi(2n)}$.

3.5.91. Note that $\binom{2n}{n} \leq \sum_{a=0}^{2n} \binom{2n}{a} = (1+1)^{2n} = 2^{2n}$. Then from Exercise 74, $n^{\pi(2n)-\pi(n)} < \binom{2n}{n} \leq 2^{2n}$. Taking logarithms gives $(\pi(2n) - \pi(n)) \log n < \log(2^{2n}) = n \log 4$. Now divide by $\log n$.

3.5.93. Note that $2^n = \prod_{a=1}^n 2 \leq \prod_{a=1}^n (n+a)/a = \binom{2n}{n}$. Then by Exercise 73, $2^n \leq (2n)^{\pi(2n)}$. Taking logs gives $\pi(2n) \geq n \log 2/\log 2n$. Hence, for a real number x, we have $\pi(x) \geq [x/2] \log 2/ \log [x] > c_1 x/\log x$. For the other half, Exercise 65 gives $\pi(x) - \pi(x/2) < ax/\log x$, where a is a constant. Then $\log x/2^m \pi(x/2^m) - \log x/2^{m+1} \pi(x/2^{m+1}) < ax/2^m$ for any positive integer m. Then, $\log x \pi(x) = \sum_{m=0}^v (\log x/2^m \pi(x/2^m) - \log x/2^{m+1} \pi(x/2^{m+1})) < ax \sum_{m=0}^v 1/2^m < c_2 x$, where v is the largest integer such that $2^{v+1} \leq x$. Then $\pi(x) < c_2 x/\log x$.

3.6. Factorization Methods and the Fermat Numbers

3.6.1. a. We see that 2 does not divide 33776925. Next we see that 3 does divide 33775925, with $33776925 = 3 \cdot 11258975$. Note that 3 does not divide 11258975. Next note that 5 does divide 11258975 with $11258975 = 5 \cdot 2251795$. We see that 5 also divides 2251795, with $2251795 = 5 \cdot 450359$. Next we see that 5 does not divide 450359. Next we note that 7 does divide 450359 with $450359 = 7 \cdot 64337$. Again dividing by 7 we see that $64337 = 7 \cdot 9191$. Dividing by 7 another time shows that $9191 = 7 \cdot 1313$. Next we note that 7 does not divide 1313. We see that 11 does not divide 1313. Dividing by 13 gives $1313 = 13 \cdot 101$. Since $\sqrt{101} < 13$, we conclude that 101 is prime. Hence the prime factorization is $33776925 = 3 \cdot 5^2 \cdot 7^3 \cdot 13 \cdot 101$.

b. We first note that neither $2, 3, 5$, nor 7 divides 210733237. Next we see that $210733237 = 11 \cdot 19157567$. Dividing by 11 again gives $19157567 = 11 \cdot 1741597$, and dividing by 11 yet again shows that $1741597 = 11 \cdot 158327$. We see that 11 does not divide 158327. Dividing by 13 shows that $158327 = 13 \cdot 12179$. Note that 12179 is not divisible by 13 nor by 17. We see that it is divisible by 19 with $12179 = 19 \cdot 641$. We see that 641 is not divisible by 19 or 23. Since 23 is the largest prime not exceeding $\sqrt{641}$ it follows that 641 is prime. It follows that the prime factorization is $210733237 = 11^3 \cdot 13 \cdot 19 \cdot 641$.

c. We first note that neither $2, 3, 5, 7$, nor 11 divides 1359170111. Next we see that $1359170111 = 13 \cdot 104551547$, and that 13 does not divide 104551547. Dividing by 17 gives $104551547 = 17 \cdot 6150091$, but 17 does not divide 6150091. Next we see that $6150091 = 19 \cdot 323689$, but 19 does not divide 323689. We see that neither $23, 29, 31, 37, 41$, nor 43 divides 323689, but $323689 = 47 \cdot 6887$. We see that 47 does not divide 6887. Neither $53, 59, 61$, nor 67 divides 6887, but $6887 = 71 \cdot 97$. Since 97 is prime, we conclude that $1359170111 = 13 \cdot 17 \cdot 19 \cdot 47 \cdot 71 \cdot 97$.

3.6.3. a. Since $11 < \sqrt{143} < 12$, we begin by noting that $12^2 - 143 = 1$ is a perfect square. So, $143 = 12^2 - 1 = (12+1)(12-1) = 13 \cdot 11$.

b. Since $47 < \sqrt{2279} < 48$, we begin by noting that $48^2 - 2279 = 25 = 5^2$ is a perfect square. So, $2279 = 48^2 - 5^2 = (48 + 5)(48 - 5) = 53 \cdot 43$.

c. Since $6 < \sqrt{43} < 7$, we begin by looking for a perfect square in the sequence $7^2 - 43 = 6, 8^2 - 43 = 21, 9^2 - 43 = 38, 10^2 - 43 = 57, 11^2 - 43 = 78, \ldots$. The smallest such perfect square is $22^2 - 43 = 21^2$. From this, it follows that $43 = (22 + 21)(22 - 21) = 43 \cdot 1$, which shows that 43 is prime.

d. Since $106 < \sqrt{11413} < 107$, we begin by looking for a perfect square in the sequence $107^2 - 11413 = 36 = 6^2, \ldots$. Thus, $11413 = 107^2 - 6^2 = (107 + 6)(107 - 6) = 113 \cdot 101$.

3.6.5. Note that $(50+n)^2 = 2500 + 100n + n^2$ and $(50-n)^2 = 2500 - 100n + n^2$. The first equation shows that the possible final two digits of squares can be found by examining the squares of the integers $0, 1, \ldots, 49$, and the second equation shows that these final two digits can be found by examining the squares of the integers $0, 1, \ldots, 25$. We find that $0^2 = 0, 1^2 = 1, 2^2 = 4, 3^2 = 9, 4^2 = 16, 5^2 = 25, 6^2 = 36, 7^2 = 49, 8^2 = 64, 9^2 = 81, 10^2 = 100, 11^2 = 121, 12^2 = 144, 13^2 = 169, 14^2 = 196, 15^2 = 225, 16^2 = 256, 17^2 = 289, 18^2 = 324, 19^2 = 361, 20^2 = 400, 21^2 = 441, 22^2 = 484, 23^2 = 529, 24^2 = 576$, and $25^2 = 625$. It follows that the last two digits of a square are $00, e1, e4, 25, o6$, and $e9$ where e represents an even digit and o represents an odd digit.

3.6.7. Suppose that $x^2 - n$ is a perfect square with $x > (n + p^2)/2p$, say a^2. Now, $a^2 = x^2 - n > ((n + p^2)/2p)^2 - n = ((n - p^2)/2p)^2$. It follows that $a > (n - p^2)/2p$. From these inequalities for x and a, we see that $x + a > n/p$, or $n < p(x + a)$. Also, $a^2 = x^2 - n$ tells us that $(x-a)(x+a) = n$. Now, $(x-a)(x+a) = n < p(x+a)$. Cancelling, we find that $x - a < p$. But since $x - a$ is a divisor of n less than p, the smallest prime divisor of n, $x - a = 1$. In this case, $x = (n-1)/2$.

3.6.9. From the identity in Exercise 8, it is clear that if $n = n_1$ is a multiple of $2k + 1$, then so is n_k, since it is the sum of two multiples of $2k + 1$. If $(2k+1) \mid n_k$, then $(2k+1) \mid r_k$ and it follows from $r_k < 2k + 1$ that $r_k = 0$. Thus, $n_k = (2k+1)q_k$. Continuing, we see that $n = n + 2n_k - 2(2k+1)q_k = (2k+1)n + 2(n_k - kn) - 2(2k+1)q_k$. It follows from Exercise 8 that $n = (2k+1)n - 2(2k+1)\sum_{i=1}^{k-1} q_i - 2(2k+1)q_k = (2k+1)n - 2(2k+1)\sum_{i=1}^{k} q_i$. Using Exercise 8 again, we conclude that $n = (2k+1)(n - 2\sum_{i=1}^{k} q_i) = (2k+1)m_{k+1}$.

3.6.11. To see that u is even, note that $a - c$ is the difference of odd numbers and that $b - d$ is the difference of even numbers. Thus $a - c$ and $b - d$ are even, and u must be as well. That $(r, s) = 1$ follows trivially from Theorem 2.1 (i). To continue, $a^2 + b^2 = c^2 + d^2$ implies that $(a+c)(a-c) = (d-b)(d+b)$. Dividing both sides of this equation by u, we find that $r(a+c) = s(d+b)$. From this, it is clear that $s \mid r(a+c)$. But since $(r, s) = 1$, $s \mid a + c$.

3.6.13. To factor n, observe that $[(\frac{u}{2})^2 + (\frac{v}{2})^2](r^2 + s^2) = (1/4)(r^2u^2 + r^2v^2 + s^2u^2 + s^2v^2)$. Substituting $a - c, d - b, a + c$, and $d + b$ for ru, su, sv, and rv respectively, will allow everything to be simplified down to n. As u and v are both even, both of the factors are integers.

3.6.15. We have $2^{4n+2} + 1 = 4(2^n)^4 + 1 = (2 \cdot 2^{2n} + 2 \cdot 2^n + 1)(2 \cdot 2^{2n} - 2 \cdot 2^n + 1)$. Using this identity we have the factorization: $2^{18} + 1 = 4(2^4)^4 + 1 = (2 \cdot 2^8 + 2 \cdot 2^4 + 1)(2 \cdot 2^8 - 2 \cdot 2^4 + 1) = (2^9 + 2^5 + 1)(2^9 - 2^5 + 1) = 545 \cdot 481$.

3.6.17. We can prove that the last digit in the decimal expansion of F_n is 7 for $n \geq 2$ by proving that the last digit in the decimal expansion of 2^{2^n} is 6 for $n \geq 2$. This can be done using mathematical induction. We have $2^{2^2} = 16$ so the result is true for $n = 2$. Now assume that the last decimal digit of 2^{2^n} is 6, that is $2^{2^n} \equiv 6 \pmod{10}$. It follows that $2^{2^{n+1}} = (2^{2^n})^{2^{n+1}-2^n} \equiv 6^{2^{n+1}-2^n} \equiv 6 \pmod{10}$. This completes the proof.

3.6.19. Since every prime factor of $F_5 = 2^{2^5} + 1 = 4294967297$ is of the form $2^7k + 1 = 128k + 1$, attempt to factor F_5 by trial division by primes of this form. We find that $128 \cdot 1 + 1 = 129$ is not prime, $128 \cdot 2 + 1 = 257$ is prime but does not divide 4294967297, $128 \cdot 3 + 1 = 385$ is not prime, $128 \cdot 4 + 1 = 513$ is not prime, and $128 \cdot 5 + 1 = 641$ is prime and does divide 4294967297 with $4294967297 = 641 \cdot 6700417$. Any factor of 6700417 is also a factor of 4294967297. We attempt to factor 6700417 by trial division by primes of the form $128k + 1$ beginning with 641. We first note that 641 does not divide 6700417. Among the other integers of the form $128k + 1$ less than $\sqrt{6700417}$, namely the integers 769, 897, 1025, 1153, 1281, 1409, 1537, 1665, 1793, 1921, 2049, 2177, 2305, 2433, and 2561, only 769, 1153, and 1409 are prime, and none of them divide 6700417. Hence 6700417 is prime and the prime factorization of F_5 is $641 \cdot 6700417$.

3.6.21. The number of decimal digits of F_n is $[\log_{10} F_n] + 1 = [\log_2 F_n / \log_2 10] + 1$ by the change of base formula for logarithms. But this is approximately $\log_2 2^{2^n} / \log_2 10 + 1 = 2^n / \log_2 10 + 1$.

3.6.23. Suppose $n^a - 2^m = 1$ for some integer n. Then $2^m = (n-1)(n^{a-1} + n^{a-2} + \cdots + n + 1)$, where the last factor is the sum of a odd terms but must be a power of 2, therefore, $a = 2k$ for some k. Then $2^m = (n^k - 1)(n^k + 1)$. These last two factors are powers of 2 which differ by 2 which forces $k = 1, a = 2, m = 3$, and $n = 3$ as the only solution.

3.7. Linear Diophantine Equations

3.7.1. a. Using the Euclidean algorithm we find that $2 \cdot 3 + 5 \cdot (-1) = 1$. Multiplying both sides by 11 gives $2 \cdot 33 + 5 \cdot (-11) = 11$. Hence $x = 33, y = -11$ is a solution. All solutions are given by $x = 33 - 5t, y = -11 + 2t$ where t is an integer.

  b. Using the Euclidean algorithm we find that $17 \cdot (-3) + 13 \cdot 4 = 1$. Multiplying both sides by 100 gives $17 \cdot (-300) + 13 \cdot 400 = 100$. Hence $x = -300, y = 400$ is a solution. All solutions are given by $x = -300 + 13t, y = 400 - 17t$, where t is an integer.

  c. Using the Euclidean algorithm we see that $21 \cdot 1 + 14 \cdot (-1) = 7$. Multiplying both sides by 21 gives $21 \cdot 21 + 14 \cdot (-21) = 147$. Hence $x = 21, y = -21$ is a solution. All solutions are given by $x = 21 - 2t, y = -21 + 3t$ where t is an integer.

  d. Since $(60,18)=3$ and 97 is not divisible by 3, it follows that there are no solutions in integer of $60x + 18y = 97$.

  e. Using the Euclidean algorithm it follows that $1402 \cdot 889 + 1969 \cdot (-633) = 1$. Hence $x = 889, y = -633$ is a solution. All solutions are given by $x = 889 - 1969t, y = -633 + 1402t$ where t is an integer.

3.7.3. Let x be the number of U.S. dollars and y be the number of Canadian dollars the businessman exchanges. Then $122x + 112y = 15286$. Since $(122, 112) \mid 15286$, there exist solutions with integer x and y. Using the Euclidean algorithm we find that $112(12) - 122(11) = 2$. It follows that $122(-84073) + 112(91716) = 15286$. Consequently all solutions of the linear diophantine equation are given by $x = -84073 + 56t, y = 91716 - 61t$. But our situation requires that both x and y be positive. We can see that x is positive when $t > 1501$, and y is positive when $t < 1504$. It follows that the only positive solutions which occur when $t = 1502$ and $t = 1503$, namely $x = 39, y = 94$ and $x = 95, y = 33$, respectively.

3.7.5. Let e be the number of euros and p be the number of pounds. Then $111e + 169p = 11798$. Since $(111, 169) = 1$, there exist solutions. Using the Euclidean algorithm, we find that $111(-102) + 169(67) = 1$, so that multiplying by 11798 gives us $111(-1203396) + 169(790466) = 11798$, so all solutions are given by $e = -1203396 + 169t, p = 790466 - 111t$. Since e is positive we must have $169t > 1203396$, which implies $t > 7120$. Since p is positive we must have $111t < 790466$, which implies that $t \leq 7121$, so we must have $t = 7121$. Therefore $e = 53$ and $p = 35$.

3.7.7. Let x be the number of apples and y the number of oranges. We have $25x + 18y = 839$. Using the Euclidean algorithm we find that $-5 \cdot 25 + 7 \cdot 18 = 1$. It follows that $25(-5 \cdot 839) + 18(7 \cdot 839) = 25(-4195) + 18 \cdot 5873 = 839$. Consequently all solutions of the linear diophantine equation are given by $x = -4195 + 18t, y = 5873 - 25t$ where t is an integer. For x and y to both be nonnegative, we must have $4195/18 \leq t \leq 5873/25$. Since t must be an integer, this requires that $t = 234$. This give the unique nonnegative solution $x = -4195 + 18 \cdot 234 = 17, y = 5873 - 25 \cdot 234 = 23$.

3.7. LINEAR DIOPHANTINE EQUATIONS

3.7.9. a. Suppose that x 14-cent stamps and y 21-cent stamps are combined to form $\$\,3.50$. Then $14x + 21y = 350$. Since $(14, 21) = 7$ and $7 \mid 350$ it follows that there are solutions in integers to this diophantine equation. We can find these by first noting that $7 = -1 \cdot 14 + 1 \cdot 21$, so $350 = 50 \cdot 7 = -50 \cdot 14 + 50 \cdot 21$. This implies that all solutions in integers are given by $x = -50 + (21/7)t = -50 + 3t$ and $y = 50 - (14/7)t = 50 - 2t$ where t is an integer. For x to be positive we must have $t \geq 17$ and for y to be positive we must have $t \leq 25$. This gives the solutions, for $17 \leq t \leq 25$, $x = 1, y = 16; x = 4, y = 14; x = 7, y = 12; x = 10, y = 10; x = 13, y = 8; x = 16, y = 6; x = 19, y = 4; x = 22, y = 2$; and $x = 25, y = 0$.

b. Let x be the number of 14-cent stamps and y be the number of 21-cent stamps. Then $14x + 21y = 400$. However, $(14, 21) = 7$ but 7 does not divide 400. Hence there are no solutions in integers and it is impossible to use 14-cent and 21-cent stamps to form postage of $\$\,4.00$.

c. We have 18 solutions: $(0, 37), (3, 35), \ldots, (54, 1)$.

3.7.11. a. Since $(2, 3) = 1$, we can take z to be any integer t and solve the diophantine equation $2x + 3y = 5 - 4t$, which leads to the solution $x = -5 + 3s - 2t, y = 5 - 2s, z = t$.

b. Since $(7, 21, 35) = 7 \nmid 8$, there are no solutions.

c. Since $(101, 102) = 1$, we can take z to be any integer t and solve the diophantine equation $101x + 102y = 1 - 103z$, which leads to the solution $x = -1 + 102s + t, y = 1 - 101s - 2t, z = t$.

3.7.13. Let x be the number of pennies, y the number of dimes, and z the number of quarters. Then $x + 10y + 25z = 99$. Since $x, y,$ and z are all nonnegative, it follows that $z = 0, 1, 2,$ or 3. First suppose that $z = 0$. Then $x + 10y = 99$. We find the nonnegative solutions to this by letting y range form 0 to 9. We see that $x = 9, y = 9; x = 19, y = 8; x = 29, y = 7; x = 39, y = 6; x = 49, y = 5; x = 59, y = 4; x = 69, y = 3; x = 79, y = 2; x = 89, y = 1$; and $x = 99, y = 0$ are the solutions for $z = 0$. Now let $z = 1$. Then $x + 10y = 74$. The nonnegative solutions to this are determined by letting y range from 0 to 7. We see that $x = 4, y = 7; x = 14, y = 6; x = 24, y = 5; x = 34, y = 4; x = 44, y = 3; x = 54, y = 2; x = 64, y = 1$; and $x = 74, y = 0$ are the solutions with $z = 1$. Now let $z = 2$. Then $x + 10y = 49$. The nonnegative solutions to this are determined by letting y range from 0 to 4. We see that $x = 9, y = 4; x = 19, y = 3; x = 29, y = 2; x = 39, y = 1$; and $x = 49, y = 0$ are the solutions with $z = 2$. Finally, let $z = 3$. Then $x + 10y = 24$. The nonnegative solutions to this are determined by letting y range from 0 to 2. We see that $x = 4, y = 2; x = 14, y = 1$; and $x = 24, y = 0$ are the solutions with $z = 3$. We have exhausted all nonnegative solutions of our equation.

3.7.15. a. We subtract the first equation from the second to get the diophantine equation $7y + 49y = 56$, which has solutions $y = 8 - 7t, z = t$. Substituting these expressions into the first equation gives us $x = 92 + 6t, y = 8 - 7t, z = t$.

b. We subtract the first equation from the second to get the diophantine equation $5y + 20z = 21$. Since $(5, 20) = 5 \nmid 21$, there is no solution.

c. We subtract the first equation from the other two to get the system $y + 2z + 3w = 200$, and $3y + 8z + 15w = 900$. We subtract 3 times this first equation from the second to get $2z + 6w = 300$, which has solutions $z = 150 - 3t, w = t$. Substituting these expressions into $y + 2z + 3w = 200$ gives us $y = -100 + 3t$, and substituting all three expressions into the first equation gives us $x = 50 - t$.

3.7.17. Let x be the number of first-class tickets sold, y be the number of second-class tickets sold, and z be the number of stand-by tickets sold. Then we have the system of diophantine equations $140x + 110y + 78z = 6548$, $x + y + z = 69$. Substituting $z = 69 - x - y$ into the first equation yields $62x + 32y = 1166$, which has solutions $x = 9 + 16t, y = 19 - 31t$. Then $z = 41 + 15t$. The only value of t that leaves all three quantities positive is $t = 0$, so the only solution is $x = 9, y = 19, z = 41$.

3.7.19. The quadrilateral with vertices $(b, 0), (0, a), (b - 1, -1)$, and $(-1, a - 1)$, has area $a + b$. Pick's Theorem, from elementary geometry, states that the area of a simple polygon whose vertices are lattice points

(points with integer coordinates) is given by $\frac{1}{2}x + y - 1$, where x is the number of lattice points on the boundary and y is the number of lattice points inside the polygon. Since $(a, b) = 1$, $x = 4$, and therefore, by Pick's Theorem, the quadrilateral contains $a + b - 1$ lattice points. Every point corresponds to a different value of n in the range $ab - a - b < n < ab$. Therefore every n in the range must get hit, so the equation is solvable.

3.7.21. See the solution to Exercise 19. The line $ax + by = ab - a - b$ bisects the rectangle with vertices $(-1, a-1), (-1, -1), (b-1, a-1)$, and $(b-1, -1)$ but contains no lattice points. Hence, half the interior points are below the line and half are above. The half below correspond to $n < ab - a - b$ and there are $(a-1)(b-1)/2$ of them.

3.7.23. Let x, y and z be the number of cocks, hens and chickens respectively. The problem leads to the system of diophantine equations $x + y + z = 100, 5x + 3y + z/3 = 100$. Substituting $z = 100 - x - y$ into the second equation and clearing fractions yields $14x + 8y = 200$, which has solutions $x = 4t, y = 25 - 7t$. It follows that $z = 75 + 3t$. The only values for t which make all three of these numbers nonnegative are $t = 0, 1, 2$, and 3. Thus the solutions to the problem are $(x, y, z) = (0, 25, 75); (4, 18, 78); (8, 11, 81); (12, 4, 84.)$

CHAPTER 4

Congruences

4.1. Introduction to Congruences

4.1.1. a. We have $2 \mid (13 - 1) = 12$, so $13 \equiv 1 \pmod{2}$.

b. We have $5 \mid (22 - 7) = 15$, so $22 \equiv 7 \pmod{5}$.

c. We have $13 \mid (91 - 0) = 91$, so $91 \equiv 0 \pmod{13}$.

d. We have $7 \mid (69 - 62) = 7$, so $69 \equiv 62 \pmod{7}$.

e. We have $3 \mid (-2 - 1) = -3$, so $-1 \equiv 1 \pmod{3}$.

f. We have $11 \mid (-3 - 30) = -33$, so $-3 \equiv 30 \pmod{11}$.

g. We have $40 \mid (111 - (-9)) = 120$, so $111 \equiv -9 \pmod{40}$.

h. We have $37 \mid (666 - 0) = 666$, so that $666 \equiv 0 \pmod{37}$.

4.1.3. a. Since the positive divisors of $27 - 5 = 22$ are $1, 2, 11,$ and 22 it follows that $27 \equiv 5 \pmod{m}$ if and only if $m = 1, m = 2, m = 11,$ or $m = 22$.

b. Since the positive divisors of $1000 - 1 = 999$ are $1, 3, 9, 27, 37, 111, 333,$ and 999, it follows that $1000 \equiv 1 \pmod{m}$ if and only if m is one of these eight integers.

c. Since the only positive divisors of $1331 - 0 = 1331$ are $1, 11, 121,$ and 1331 it follows that $1331 \equiv 0 \pmod{m}$ if and only if m is one of these four integers.

4.1.5. Suppose that a is odd. Then $a = 2k + 1$ for some integer k. Then $a^2 = (2k + 1)^2 = 4k^2 + 4k + 1 = 4k(k + 1) + 1$. If k is even, then $k = 2l$ where l is an integer. Then $a^2 = 8l(2l + 1) + 1$. Hence $a^2 \equiv 1 \pmod{8}$. If k is odd, then $k = 2l + 1$ when l is an integer. Then $a^2 = 4(2l + 1)(2l + 2) + 1 = 8(2l + 1)(l + 1) + 1$. Hence $a^2 \equiv 1 \pmod{8}$. It follows that $a^2 \equiv 1 \pmod{8}$ whenever a is odd.

4.1.7. a. Since $n! \equiv 0 \pmod{2}$ if $n \geq 2$, we have $1! + 2! + 3! + \cdots + 100! \equiv 1 \pmod{2}$.

b. We have $n! \equiv 0 \pmod{7}$ whenever $n \geq 7$. Since $1! \equiv 1 \pmod{7}, 2! \equiv 2 \pmod{7}, 3! \equiv 6 \pmod{7}, 4! = 24 \equiv 3 \pmod{7}, 5! = 120 \equiv 1 \pmod{7}$ and $6! = 720 \equiv 6 \pmod{7}$, we have $1! + 2! + 3! + \cdots + 100! \equiv 1! + 2! + 3! + 4! + 5! + 6! \equiv 1 + 2 + 6 + 3 + 1 + 6 \equiv 5 \pmod{7}$.

c. Since $n! \equiv 0 \pmod{12}$ whenever $n \geq 4$, it follows that $1! + 2! + 3! + \cdots + 100! \equiv 1 + 2 + 6 \equiv 9 \pmod{12}$.

d. Since $n! \equiv 0 \pmod{25}$ whenever $n \geq 10$, it follows that $1! + 2! + 3! + \cdots + 100! \equiv 1! + 2! + 3! + 4! + 5! + 6! + 7! + 8! + 9! \equiv 1 + 2 + 6 + 24 + 20 + 20 + 15 + 20 + 5 \equiv 13 \pmod{25}$.

4.1.9. Since $a \equiv b \pmod{m}$, there exists an integer k such that $a = b + km$. Thus, $ac = (b + km)c = bc + k(mc)$. By Theorem 4.1, $ac \equiv bc \pmod{mc}$.

4.1.11. a. We proceed by induction on n. It is clearly true for $n = 1$. For the inductive step we assume that $\sum_{j=1}^{n} a_j \equiv \sum_{j=1}^{n} b_j \pmod{m}$ and that $a_{n+1} \equiv b_{n+1} \pmod{m}$. Now $\sum_{j=1}^{n+1} a_j = (\sum_{j=1}^{n} a_j) + a_{n+1} \equiv (\sum_{j=1}^{n} b_j) + b_{n+1} = \sum_{j=1}^{n+1} b_j \pmod{m}$ by Theorem 4.5(i). This completes the proof.

b. We use induction on n. For $n = 1$, the identity clearly holds. This completes the basis step. For the inductive step we assume that $\prod_{j=1}^{n} a_j \equiv \prod_{j=1}^{n} b_j \pmod{m}$ and $a_{n+1} \equiv b_{n+1} \pmod{m}$. Then $\prod_{j=1}^{n+1} a_j = a_{n+1}(\prod_{j=1}^{n} a_j) \equiv b_{n+1}(\prod_{j=1}^{n} b_j) = \prod_{j=1}^{n+1} b_j \pmod{m}$ by Theorem 4.5(iii). This completes the proof.

4.1.13.

$-$	0	1	2	3	4	5
0	0	5	4	3	2	1
1	1	0	5	4	3	2
2	2	1	0	5	4	3
3	3	2	1	0	5	4
4	4	3	2	1	0	5
5	5	4	3	2	1	0

4.1.15. a. Since $11 + 29 = 40 \equiv 4 \pmod{12}$, the (12-hour) clock reads 4 o'clock 29 hours after reading 11 o'clock.

b. Since $2 + 100 = 102 \equiv 6 \pmod{12}$, the (12-hour) clock reads 6 o'clock 100 hours after it reads 2 o'clock.

c. Since $6 - 50 = -44 \equiv 4 \pmod{12}$, the (12-hour) clock reads 4 o'clock 50 hours before it reads 6 o'clock.

4.1.17. If $a^2 \equiv b^2 \pmod{p}$ then $p \mid (a^2 - b^2) = (a+b)(a-b)$. Since p is prime, either $p \mid (a+b)$ or $p \mid (a-b)$. Hence either $a \equiv b \pmod{p}$ or $a \equiv -b \pmod{p}$.

4.1.19. Note that $1 + 2 + 3 + \cdots + (n+1) = (n-1)n/2$. If n is odd, then $(n-1)$ is even, so $(n-1)n/2$ is an integer. Hence $n \mid (1 + 2 + 3 + \cdots + (n-1))$ if n is odd, and $1 + 2 + 3 + \cdots + (n-1) \equiv 0 \pmod{n}$. If n is even, then $n = 2k$ where k is an integer. Then $(n-1)n/2 = (n-1)k$. We can easily see that n does not divide $(n-1)k$ since $(n, n-1) = 1$ and $k < n$. It follows that $1 + 2 + \cdots + (n-1)$ is not congruent to 0 modulo n if n is even.

4.1.21. $1^2 + 2^2 + \cdots + (n-1)^2 \equiv 0 \pmod{n}$ if and only if n is relatively prime to 6. If $(n, 6) = 1$, then $1^2 + 2^2 + \cdots + (n-1)^2 = n(n-1)(2n-1)/6 \equiv 0 \cdot (n-1)(2n-1)/6 \equiv 0 \pmod{n}$, using Exercise 7 from Section 1.2 and Theorem 4.3(iii). This works because $1^2 + 2^2 + \cdots + (n-1)^2$ is an integer and $(n, 6) = 1$ implies that $(n-1)(2n-1)/6$ is an integer, and so we are dealing with integer-only arithmetic. If however, $2 \mid n$ so that $2 \mid (n, 6)$ and if $1^2 + 2^2 + \cdots + (n-1)^2 = n(n-1)(2n-1)/6 \equiv 0 \pmod{n}$, then $nk = n(n-1)(2n-1)/6$ for some integer k by Theorem 4.1. It follows that $6k = (n-1)(2n-1)$. But $6k$ is even, and $(n-1)(2n-1)$ is odd since both $n-1$ and $2n-1$ are odd. If $3 \mid n$, and $n(n-1)(2n-1)/6 \equiv 0 \pmod{n}$, then $nk = n(n-1)(2n-1)/6$ by Theorem 4.1. Hence, $6k = (n-1)(2n-1)$. But if we look at this equality modulo 3, we see that $0 \equiv 6k = (n-1)(2n-1) \equiv (-1)(-1) = 1 \pmod{3}$. Again, a contradiction.

4.1.23. If $n = 1$, then $5 = 5^1 \equiv 1 + 4(1) \pmod{16}$, so the basis step holds. For the inductive step, we assume that $5^n \equiv 1 + 4n \pmod{16}$. Now $5^{n+1} \equiv 5^n 5 \equiv (1 + 4n)5 \pmod{16}$ by Theorem 4.3(iii). Further, $(1 + 4n)5 = 5 + 20n \equiv 5 + 4n \pmod{16}$. Finally $5 + 4n = 1 + 4(n+1)$. So, $5^{n+1} \equiv 1 + 4(n+1) \pmod{16}$. This completes the proof.

4.1.25. Note that if $x \equiv 0 \pmod{4}$ then $x^2 \equiv 0 \pmod{4}$, if $x \equiv 1 \pmod{4}$ then $x^2 \equiv 1 \pmod{4}$, if $x \equiv 2 \pmod{4}$ then $x^2 \equiv 4 \equiv 0 \pmod{4}$, and if $x \equiv 3 \pmod{4}$ then $x^2 \equiv 9 \equiv 1 \pmod{4}$. Hence $x^2 \equiv 0$ or $1 \pmod{4}$ whenever x is an integer. It follows that $x^2 + y^2 \equiv 0, 1$ or $2 \pmod{4}$ whenever x and y are integers. We see that n is not the sum of two squares when $n \equiv 3 \pmod{4}$.

4.1.27. By Theorem 4.1, for some integer a, $ap^k = x^2 - x = x(x-1)$. By the Fundamental Theorem of Arithmetic, p^k is a factor of $x(x-1)$. Since p cannot divide both x and $x-1$, we know that $p^k \mid x$ or $p^k \mid x-1$.

Thus, $x \equiv 0$ or $x \equiv 1 \pmod{p^k}$.

4.1.29. First note that there are m_1 possibilities for a_1, m_2 possibilities for a_2, and in general m_i possibilities for a_i. Thus there are $m_1 m_2 \cdots m_k$ expressions of the form $M_1 a_1 + M_2 a_2 + \cdots M_k a_k$ where $a_1, a_2, \ldots, a_k$ run through complete systems of residues modulo $m_1, m_2, \ldots, m_k$, respectively. Since this is exactly the size of a complete system of residues modulo M, the result will follow if we can show distinctness of each of these expressions modulo M. Suppose, by way of contradiction, that $M_1 a_1 + M_2 a_2 + \cdots + M_k a_k \equiv M_1 a'_1 + M_2 a'_2 + \cdots + M_k a'_k \pmod{M}$. Then $M_1 a_1 \equiv M_1 a'_1 \pmod{m_1}$, since m_1 divides each of $M_2, M_3, \ldots, M_k$, and further $a_1 \equiv a'_1 \pmod{m_1}$ since $(M_1, m_1) = 1$. Similarly $a_i \equiv a'_i \pmod{m_i}$. Thus a'_i is in the same congruence class modulo m_i as a_i for all i. The result now follows.

4.1.31. a. Let $\sqrt{n} = a + r$, where a is an integer, and $0 \le r < 1$. We now consider two cases, when $0 \le r < \frac{1}{2}$ and when $\frac{1}{2} \le r < 1$. For the first case, $T = [\sqrt{n} + \frac{1}{2}] = a$, and so $t = T^2 - n = -(2ar + r^2)$. Thus $|t| = 2ar + r^2 < 2a(\frac{1}{2}) + (\frac{1}{2})^2 = a + \frac{1}{4}$. Since both T and n are integers, t is also an integer. It follows that $|t| \le [a + \frac{1}{4}] = a = T$. For the second case, when $\frac{1}{2} \le r < 1$, we find that $T = [\sqrt{n} + \frac{1}{2}] = a + 1$ and $t = 2a(1 - r) + (1 - r^2)$. Since $\frac{1}{2} \le r < 1$, $0 < (1 - r) \le \frac{1}{2}$ and $0 < 1 - r^2 < 1$. It follows that $t \le 2a(\frac{1}{2}) + (1 - r^2)$. Because t is an integer, we can say that $t \le [a + (1 - r^2)] = a < T$.

b. By the division algorithm, we see that if we divide x by T we get $x = aT + b$, where $0 \le b < T$. If a were negative, then $x = aT + b \le (-1)T + b < 0$; but we assumed x to be nonnegative. This shows that $0 \le a$. Suppose now that $a > T$. Then $x = aT + b \ge (T+1)T = T^2 + T \ge (\sqrt{n} - \frac{1}{2})^2 + (\sqrt{n} - \frac{1}{2}) = n - \frac{1}{4}$ and, as x and n are integers, $x \ge n$. This is a contradiction, which shows that $a \le T$. Similarly, $0 \le c \le T$ and $0 \le d < T$.

c. $xy = (aT + b)(cT + d) = acT^2 + (ad + bc)T + bd \equiv ac(T^2 - n) + zT + bd \equiv act + zT + bd \pmod{n}$.

d. Use part (c), substituting $eT + f$ for ac.

e. The first half is identical to part (b); the second half follows by substituting $gT + h$ for $z + et$ and noting that $T^2 \equiv t \pmod{n}$.

f. Certainly, ft and gt can be computed since all three numbers are less than T, which is less than $\sqrt{n} + 1$. So $(f + g)t$ is less than $2n < w$. Similarly, we can compute $j + bd$ without exceeding the word size. And, finally, using the same arguments, we can compute $hT + k$ without exceeding the word size.

4.1.33. a. We have $3^{10} \equiv (3^2)^5 \equiv 9^5 \equiv (-2)^5 \equiv -32 \equiv 1 \pmod{11}$.

b. We have $2^{12} \equiv (2^4)^3 \equiv 16^3 \equiv 3^3 \equiv 27 \equiv 1 \pmod{13}$.

c. We have $5^{16} \equiv (5^2)^8 \equiv 25^8 \equiv 8^8 \equiv (8^2)^4 \equiv 64^4 \equiv (-1)^4 \equiv 1 \pmod{17}$.

d. We have $3^{22} \equiv (3^3)^7 \cdot 3 \equiv 27^7 \cdot 3 \equiv 4^7 \cdot 3 \equiv (4^3)^2 \cdot 4 \cdot 3 \equiv 64^2 \cdot 12 \equiv (-5)^2 \cdot 12 \equiv 2 \cdot 12 \equiv 24 \equiv 1 \pmod{23}$.

e. The theorem is that $a^{p-1} \equiv 1 \pmod{p}$ whenever p is prime and p does not divide a. This is Fermat's little theorem which will be proved in Chapter 5.

4.1.35. Since $f_{n-2} + f_{n-1} \equiv f_n \pmod{m}$, if two consecutive numbers recur in the same order, then the sequence must be repeating both as n increases and as it decreases. But there are only m residues, and so m^2 ordered sequence of two residues. As the sequence is infinite, some two elements of the sequence must recur by the pigeonhole principle. Thus the sequence of least positive residues of the Fibonacci numbers repeats. It follows that if m divides some Fibonacci number, that is, if $f_n \equiv 0 \pmod{m}$, then m divides infinitely many Fibonacci numbers. To see that m does divide some Fibonacci number, note that the sequence must contain a 0, namely $f_0 \equiv 0 \pmod{m}$.

4.1.37. Let a and b be positive integers less than m. Then they have $O(\log m)$ digits (bits). Therefore by Theorem 2.4, we can multiply them using $O(\log^2 m)$ operations. Division by m takes $O(\log^2 m)$ operations by Theorem 2.7. Then, in all we have $O(\log^2 m)$ operations.

4.1.39. Let N_i be the number of coconuts the ith man leaves for the next man and $N_0 = N$. At each stage, the ith man finds N_{i-1} coconuts, gives k coconuts to the monkeys, takes $(1/n)(N_{i-1} - k)$ coconuts for himself and leaves the rest for the next man. This yields the recursive formula $N_i = (N_{i-1} - k)(n-1)/n$. For convenience, let $w = (n-1)/n$. If we iterate this formula a few times we get $N_1 = (N_0 - k)w$, $N_2 = (N_1 - k)w = ((N_0 - k)w - k)w = N_0 w^2 - kw^2 - kw$, $N_3 = N_0 w^3 - kw^3 - kw^2 - kw, \ldots$. The general pattern $N_i = N_0 w^i - kw^i - kw^{i-1} - \cdots - kw = N_0 w^i - kw(w^i - 1)/(w-1)$ may be proved by induction. When the men rise in the morning they find $N_n = N_0 w^n - kw(w^n - 1)/(w-1)$ coconuts, and we must have $N_n \equiv k \pmod{n}$, that is, $N_n = N_0 w^n - kw(w^n - 1)/(w-1) = k + tn$ for some integer t. Substituting $w = (n-1)/n$ back in for w, solving for N_0, and simplifying yields $N = N_0 = n^{n+1}(t+k)/(n-1)^n - kn + k$. For N to be an integer, since $(n, n-1) = 1$, we must have $(t+k)/(n-1)^n$ an integer. Since we seek the smallest positive value for N, we take $t + k = (n-1)^n$, so $t = (n-1)^n - k$. Substituting this value back into the formula for N yields $N = n^{n+1} - kn + k$.

4.1.41. a. Let $f_1(x) = \sum_{i=0}^{m} a_i x^i$, $f_2(x) = \sum_{i=1}^{m} b_i x^i$, $g_1(x) = \sum_{i=1}^{m} c_i x^i$, and $g_2(x) = \sum_{i=1}^{m} d_i x^i$ where the leading coefficients may be zero to keep the limits of summation the same for all polynomials. Then $a_i \equiv c_i \pmod{n}$ and $b_i \equiv d_i \pmod{n}$, for $i = 0, 1, \ldots, m$. Therefore by Theorem 4.5 part (i), $a_i + b_i \equiv c_i + d_i \pmod{n}$ for $i = 0, 1, \ldots, m$. Since $(f_1 + f_2)(x) = \sum_{i=1}^{m}(a_i + b_i)x^i$ and $(g_1 + g_2)(x) = \sum_{i=1}^{m}(c_i + d_i)x^i$, this shows the sums of the polynomials are congruent modulo n.

b. With the same set up as in part (a), the coefficient on x^k in $(f_1 f_2)(x)$ is given by $a_0 b_k + a_1 b_{k-1} + \cdots + a_k b_0$, and the corresponding coefficient in $(g_1 g_2)(x)$ is given by $c_0 d_k + c_1 d_{k-1} + \cdots + c_k d_0$. Since each $a_i \equiv c_i \pmod{n}$ and $b_i \equiv d_i \pmod{n}$, by Theorem 4.5, the two expressions are congruent modulo n, and so, therefore, are the polynomials.

4.1.43. The basis step for induction on k is Exercise 42. Assume that $f(x) \equiv h(x) \pmod{p}$ and $f(x) = (x - a_1) \cdots (x - a_{k-1}) h(x)$, where $h(x)$ is a polynomial with integer coefficients. Substituting a_k for x in this congruence gives us $0 \equiv (a_k - a_1) \cdots (a_k - a_1) h(a_k) \pmod{p}$. None of the factors $a_k - a_i$ can be congruent to zero modulo p, so we must have $h(a_k) \equiv 0 \pmod{p}$. Applying Exercise 42 to $h(x)$ and a_k gives us $h(x) \equiv (x - a_k)g(x) \pmod{p}$ and substituting this in the congruence for $f(x)$ yields $f(x) \equiv (x - a_1) \cdots (x - a_k)g(x) \pmod{p}$, which completes the induction step.

4.2. Linear Congruences

4.2.1. a. Since $(2,7) = 1 \mid 5$, Theorem 4.10 tells us that there is one class of solutions. We solve the diophantine equation $2x + 7y = 5$, to get $x \equiv 6 \pmod{7}$.

b. Since $(3,9) = 3 \mid 6$, Theorem 4.10 tells us that there are three classes of solutions. We solve the diophantine equation $3x + 9y = 6$, to get $x = 2 + 3t$. All solutions are thus congruent to 2, 5, or 8 modulo 9.

c. Since $(19, 40) = 1 \mid 30$, Theorem 4.10 tells us that there is one class of solutions. We solve the diophantine equation $19x + 40y = 30$, to get $x \equiv 10 \pmod{40}$.

d. All solutions are given by $x \equiv 20 \pmod{25}$.

e. All solutions are given by $x \equiv 111 \pmod{999}$.

f. Since $(980, 1600) = 20 \mid 1500$, Theorem 4.10 tells us that there are twenty classes of solutions. All solutions are given by $x \equiv 75 + 80k \pmod{1600}$ where k is an integer such that $0 \leq k \leq 19$.

4.2.3. Since $(28927591, 6789783) = 9163 \mid 2474010$, Theorem 4.10 tells us that there are 9163 classes of solutions. Reducing the congruence by dividing each side of the equation and the modulus by 9163, we look at the congruence $741 \equiv 270 \pmod{3157}$. The single class of solutions of this congruence is congruent to 1074. Thus, the 9163 solutions to the original congruence are given by $x \equiv 1074 + 3157k \pmod{28927591}$ where k is an integer such that $0 \leq k \leq 9162$.

4.2.5. This is equivalent to saying that $11x \equiv 17 \pmod{24}$. This has one solution modulo 24, by Theorem 4.10, $x \equiv 19 \pmod{24}$. So the satellite orbits the Earth every 19 hours.

4.2.7. We know by Theorem 4.10 that $154x \equiv c \pmod{1001}$ has solutions if and only if $(1001, 154) = 77 \mid c$. Also, by Theorem 4.10, we know that when there are solutions, there are exactly 77 of them.

4.2.9. a. To find an inverse of 4 modulo 17 we must solve the congruence $4x \equiv 1 \pmod{17}$. Form the Euclidean algorithm we find that $1 \cdot 17 - 4 \cdot 4 = 1$. Hence $x = -4 \equiv 13 \pmod{17}$ is a solution, so that 13 is an inverse of 4 modulo 17.

b. To find an inverse of 5 modulo 17 we must solve the congruence $5x \equiv 1 \pmod{17}$. From the Euclidean algorithm we find that $-2 \cdot 17 + 7 \cdot 5 = 1$. Hence $x = 7 \pmod{17}$ is a solution, so that 7 is an inverse of 5 modulo 17.

c. To find an inverse of 7 modulo 17 we must solve the congruence $7x \equiv 1 \pmod{17}$. From the Euclidean algorithm we find that $-2 \cdot 17 + 5 \cdot 7 = 1$. Hence $x = 5 \pmod{17}$ is a solution, so that 5 is an inverse of 7 modulo 17.

d. To find an inverse of 16 modulo 17 we must solve the congruence $16x \equiv 1 \pmod{17}$. Since $16 \equiv -1 \pmod{17}$, this implies that $-x \equiv 1 \pmod{17}$, or that $x \equiv -1 \equiv 16 \pmod{17}$. Hence $x = 16$ is an inverse of 16 modulo 17.

4.2.11. a. The integers a with inverses modulo 30 are exactly those that are relatively prime to 30. Therefore, only 1, 7, 11, 13, 17, 19, 23, and 29 have inverses modulo 30.

b. Note that 1 and 29 are their own inverses. Solving the congruence $7x \equiv 1 \pmod{30}$ yields $x = 13$, so 7 and 13 are inverses of each other. And so are $-7 \equiv 23$ and $-13 \equiv 17$. Solving $11x \equiv 1 \pmod{30}$ yields $x = 11$, so 11 is its own inverse, and so is $-11 \equiv 19$.

4.2.13. If $ax + by \equiv c \pmod{m}$, then there exists an integer k such that $ax + by - mk = c$. Since $d \mid ax + by - mk, d \mid c$. Thus there are no solutions when $d \nmid c$. Now, assume that $d \mid c$ and let $a = da', b = db', c = dc'$, and $m = dm'$, so that $(a', b', m') = 1$. Then we can divide the original congruence by d to get (*) $a'x + b'y \equiv c' \pmod{m'}$, or $a'x \equiv c' - b'y \pmod{m'}$, which has solutions if and only if $g = (a', m') \mid c - b'y$, which is equivalent to $b'y \equiv c' \pmod{g}$ having solutions. Since $(a', b', m') = 1$, and $(a', m') = g$, we must have $(b', g) = 1$ and so the last congruence has only one incongruent solution y_0 modulo g. But the m'/g solutions, $y_0, y_0 + g, y_0 + 2g, \ldots, y_0 + (m'/g + 1)g$ are incongruent modulo m'. Each of these yields g incongruent values of x in the congruence (*). Therefore, there are $g(m'/g) = m'$ incongruent solutions to (*).

Now let (x_1, y_1) be one solution of the original congruence. Then the d values $x_1, x_1 + m', x_1 + 2m', \ldots, x_1 + (d-1)m'$ are congruent modulo m' but incongruent modulo m. Likewise, the d values $y_1, y_1 + m', y_1 + 2m', \ldots, y_1 + (d-1)m'$ are congruent modulo m' but incongruent modulo m. So for each solution of (*), we can generate d^2 solutions of the original congruence. Since there are m' solutions to (*), we have $d^2 m' = dm$ solutions to the original congruence.

4.2.15. Suppose that $x^2 \equiv 1 \pmod{p^k}$ where p is an odd prime and k is a positive integer. Then $x^2 - 1 \equiv (x+1)(x-1) \equiv 0 \pmod{p^k}$. Hence $p^k \mid (x+1)(x-1)$. Since $(x+1) - (x-1) = 2$ and p is an odd prime, we know that p divides at most one of $(x-1)$ and $(x+1)$. It follows that either $p^k \mid (x+1)$ or $p^k \mid (x-1)$, so that $p \equiv \pm 1 \pmod{p^k}$.

4.2.17. To find the inverse of a modulo m, we must solve the Diophantine equation $ax + my = 1$, which can be done using the Euclidean algorithm. Using Corollary 2.5.1, we can find the greatest common divisor

4.3. The Chinese Remainder Theorem

4.3.1. The integers x that leave a remainder of one when divided by 2 or 3 are those integers x that are solutions of $x \equiv 1 \pmod{2}$ and $x \equiv 1 \pmod{3}$. The solutions of these two simultaneous congruences are those integers x such that $x \equiv 1 \pmod{6}$. These integers are the integers leaving a remainder of 1 when divided by either 2 or 3.

4.3.3. We want a solution to the congruences $x \equiv 2 \pmod{3}, x \equiv 2 \pmod{5}$, and $x \equiv 0 \pmod{4}$. Using the iterative method described in the text (because our moduli aren't relatively prime!), $x = 4k$, and so $k \equiv 2 \pmod{5}$. Thus $k = 3 + 5j \equiv 2 \pmod{3}$. Finally, $j = 1 + 3m$. So $x = 4k = 4(3 + 5j) = 12 + 20(1 + 3m) = 32 + 60m$. The smallest possible such number is 32.

4.3.5. We have $m_1 = 2, m_2 = 3, m_3 = 5, m_4 = 7$, and $m_5 = 11$. Also $M_1 = 1115, M_2 = 770, M_3 = 462, M_4 = 330$, and $M_5 = 210$. By the Chinese remainder theorem, $x = M_1 y_1 + 2 M_2 y_2 + 3 M_3 y_3 + 4 M_4 y_4 + 5 M_5 y_5$, where $M_i y_i \equiv \pmod{m_i}$. We find that solutions are $y_1 = 1, y_2 = 2, y_3 = 3, y_4 = 1$, and $y_5 = 1$. So, $x \equiv 1523 \pmod{2310}$.

4.3.7. Let b be the number of bananas. Then $b \equiv 6 \pmod{11}$ and $b \equiv 0 \pmod{17}$. This implies that $b \equiv 6 \cdot 17 \cdot 2 + 0 \cdot 11 \cdot 14 \equiv 204 \equiv 17 \pmod{187}$. We also know that $b > 11 \cdot 7 + 6 = 83$ since the equal piles contain at least 7 bananas each. It follows that the least number of bananas in the pile is 204.

4.3.9. The situation we have here is $0 \leq x \leq 1200, x \equiv 3 \pmod{5}, x \equiv 3 \pmod{6}, x \equiv 1 \pmod{7}$, and $x \equiv \pmod{11}$. Using the iterative method described in the text, $x = 11x_0, 11x_0 \equiv 1 \pmod{7}, x_0 = 2 + 7x_1, x = 11x_0 = 22 + 77x_1, 22 + 77x_1 \equiv 3 \pmod{6}, x_1 = 1 + 6x_2, x = 99 + 462x_2, 99 + 462x_2 \equiv 3 \pmod{5}, x_2 = 2 + 5x_3, x = 1023 + 2310x_3$. The only solution satisfying $0 \leq x \leq 1200$ is $x = 1023$. It follows that 1023 troops remained.

4.3.11. We solve the system $x \equiv 0 \pmod{11}, x \equiv 1 \pmod{2}, x \equiv 1 \pmod{3}, x \equiv 1 \pmod{5}, x \equiv 1 \pmod{7}$, to find that $x \equiv 2101 \pmod{2310}$.

4.3.13. We can construct a sequence of k consecutive integers each divisible by a square as follows. Consider the system of congruences $x \equiv 0 \pmod{p_1^2}, x \equiv -1 \pmod{p_2^2}, x \equiv -4 \pmod{p_3^2}, \ldots, x \equiv -k+1 \pmod{p_k^2}$, where p_k is the kth prime. By the Chinese remainder theorem there is a solution to this simultaneous system of congruence since the moduli are relatively prime. It follows that there is a positive integer N that satisfies each of these congruences. Each of the k integers $n, N+1, \ldots, N+k-1$ is divisible by a square since p_j^2 divides $N + j - 1$ for $j = 1, 2, \ldots, k$.

4.3.15. Suppose that x is a solution to the system of congruences. Then $x \equiv a_1 \pmod{m_1}$, so that $x = a_1 + km_1$ for some integer k. We substitute this into the second congruence to get $a_1 + km_1 \equiv a_1 \pmod{m_2}$ or $km_1 \equiv (a_2 - a_1) \pmod{m_2}$, which has a solution in k if and only if $(m_1, m_2) \mid (a_1, a_2)$. Now assume such a solution k_0 exists. Then all incongruent solutions are given by $k = k_0 + m_2 t/(m_1, m_2)$, where t is an integer. Then $x = a_1 + km_1 = a_1 + \left(k_0 + \dfrac{m_2 t}{(m_1, m_2)}\right) m_1 = a_1 + k_0 m_1 + \dfrac{m_1 m_2}{(m_1, m_2)} t$. Note that $m_1 m_2/(m_1, m_2) = [m_1, m_2]$ so that if we set $x_1 = a_1 + k_0 m_1$, we have $x = x_1 + [m_1, m_2] t \equiv x_1 \pmod{[m_1, m_2]}$, and so the solution is unique modulo $[m_1, m_2]$.

4.3.17. a. Using Exercise 15, there is one solution modulo $[60, 350] = 2100$ because $(60, 350) = 10 \mid (80 - 10)$. Because $x \equiv 10 \pmod{60}$, we know that $x = 10 + 60k$, where k is an integer. Continuing onward, $x = 10 + 60k \equiv 80 \pmod{350}$, so $60k \equiv 70 \pmod{350}$ and so $k \equiv 7 \pmod{350}$; thus $k = 7 + (350/(350, 60))j$, where j is an integer. In conclusion, $x = 10 + 60k = 10 + 60(7 + 35j) = 430 + 2100j$.

b. Using Exercise 15, there is one solution modulo $[910,1001]=2100$ because $(910,1001) = 91 \mid (93-2)$. Because $x \equiv 2 \pmod{910}$, we know that $x = 2 + 910k$, where k is an integer. Continuing onward, $x = 2 + 910k \equiv 93 \pmod{1001}$, so $910k \equiv 91 \pmod{1001}$ and so $k \equiv 10 \pmod{1001}$; thus $k = 10 + (1001/(1001,910))j$, where j is an integer. In conclusion, $x = 2 + 910k = 2 + 910(10 + 11j) = 9102 + 10010j$.

4.3.19. The basis step $r = 2$ is given by Exercise 15. Suppose that the system of the first k congruences has a unique solution A modulo $M = [m_1, \ldots, m_k]$ and $(m_i, m_j) | a_j - a_i$ for $1 \le i < j \le k$. Consider the system $x \equiv A \pmod{M}, x \equiv a_{r+1} \pmod{m_{r+1}}$. First suppose it has a solution B modulo $[[m_1, m_2, \ldots, m_k], m_{k+1}]$. Then by Exercise 15, $([m_1, m_2, \ldots, m_k], m_{r+1}) | B - a_{k+1}$. Since $m_i | [m_1, m_2, \ldots, m_k]$ for $1 \le i \le k$, we have $(m_i, m_{k+1}) | B - a_{kr+1}$. That is, there exists an integer n such that $(m_i, m_{k+1})n = B - a_{k+1}$. If we reduce this equation modulo m_i, for $1 \le i \le k$ we have $(0, m_{k+1})n \equiv m_{k+1} = a_i - a_{k+1} \pmod{m_i}$. If we reduce modulo m_{k+1} we have $(m_i, 0)n = m_i n = 0 \pmod{m_{k+1}}$. In either case we have that $(m_i, m_j) | a_j - a_i$ for $1 \le i < j \le k+1$. Conversely, suppose we have the conditions $(m_i, m_j) | a_j - a_i$ for $1 \le i < j \le k+1$. Then as we have just shown, $([m_1, m_2, \ldots, m_k], m_{k+1}) | A - a_{k+1}$. Therefore, by exercise 15, there is a unique solution B to the first $k+1$ congruences. This completes the induction step.

4.3.21. This is equivalent to the system: $x \equiv 1 \pmod{2}, x \equiv 1 \pmod{3}, x \equiv 1 \pmod{5}, x \equiv 1 \pmod{7}, x \equiv 0 \pmod{11}$. So, using the iterative method described in the text, $x = 11k_1 \equiv 7 \pmod 7$, and we see that $k_1 = 2 + 7k_2$. Now, $x = 11(2 + 7k_2) \equiv 1 \pmod{5}$ and $k_2 = 2 + 5k_3$. Now, $x = 176 + 385k_3 \equiv 1 \pmod{3}$ and $k_3 = 2 + 3k_4$. Now, $x = 946 + 1155k_4 \equiv 1 \pmod{2}$ and $k_4 = 1 + 2k_5$. So $x = 2101 + 2310k_5$. The smallest such number is 2101.

4.3.23. Let x be the number of grams of rice each farmer took to market. The problem yields the system of congruences $x \equiv 32 \pmod{83}, x \equiv 70 \pmod{110}, x \equiv 30 \pmod{135}$. In order to apply the Chinese remainder theorem, we replace the modulus 110 by 22. The solution is then given by $x = 24600$. This solution remains consistent modulo 110. Thus the original amount of rice was $3 \cdot 24600 = 73800$.

4.3.25. Suppose that x is a base 10 automorph with four digits. Then $x^2 \equiv x \pmod{10^4}$ since the last four digits of x and x^2 must agree. It follows that $x^2 - x = x(x-1) \equiv 0 \pmod{10^4}$. This is equivalent to the two congruences $x(x-1) \equiv 0 \pmod{2^4}$ and $x(x-1) \equiv 0 \pmod{5^4}$. We can conclude that either $x \equiv 0 \pmod{2^4}$ or $x \equiv 1 \pmod{2^4}$, since 2^4 must divide either x or $x-1$ since x and $x-1$ have no common factors. Similarly, either $x \equiv 0 \pmod{2^4}$ or $x \equiv 1 \pmod{5^4}$. It follows that x satisfies one of four simultaneous congruences: $x \equiv 0 \pmod{2^4}$ and $x \equiv 0 \pmod{5^4}$; $x \equiv 0 \pmod{2^4}$ and $x \equiv 1 \pmod{5^4}$; $x \equiv 1 \pmod{2^4}$ and $x \equiv 0 \pmod{5^4}$; or $x \equiv 1 \pmod{2^4}$ and $x \equiv \pmod{5^4}$. Using the Chinese remainder theorem for each of these sets of congruences gives $x \equiv 0 \pmod{10000}, x \equiv 625 \pmod{10000}, x \equiv 9376 \pmod{10000}$, and $x \equiv 1 \pmod{10000}$. The base 10 automorphs with four digits, allowing initial digits of 0 are 0000,0001,0625, and 9376.

4.3.27. We need to solve the system $x \equiv 23 + 2 \pmod{4 \cdot 23}, x \equiv 28 + 1 \pmod{4 \cdot 28}, x \equiv 33 \pmod{4 \cdot 33}$, where we have added 2 and 1 to make the system solvable under the conditions of Exercise 19. The solution to this system is $x \equiv 4257 \pmod{85008}$.

4.3.29. We need to solve the system $x \equiv 0 \pmod{4 \cdot 23}, x \equiv 0 \pmod{4 \cdot 28}, x \equiv 0 \pmod{4 \cdot 33}$. The solution to this system is $x \equiv 0 \pmod{85008}$. Every 85008 quarter-days, starting at 0.

4.3.31. If the set of distinct congruences cover the integers modulo the least common multiple of the moduli, then that set will cover all integers. Examine the integers modulo 210, the l.c.m. of the moduli in this set of congruences. The first four congruences take care of all numbers containing a prime divisor of 2, 3, 5, or 7. The remaining numbers can be examined one at a time, and each can be seen to satisfy one (or more) of the congruences.

4.3.33. Let x be the length in inches of the dining room. Then $x \equiv 3 \pmod 5, x \equiv 3 \pmod 7$, and $x \equiv 3 \pmod 9$. Since 5,7,and 9 are pairwise relatively prime, the Chinese remainder theorem tells us that there is a unique solution to this system of congruences modulo $5 \cdot 7 \cdot 9 = 315$. This solution is immediately seen to be $x \equiv 3 \pmod{315}$. Since x is a length it is positive. Hence possible values for x are 3,318,633,948, and so on. Since x is the length of a room in inches, the possibility that $x = 3$ is absurd, and it is most likely that $x = 318$, so that the room is 26 feet and 6 inches long. This is a big dining room. Of course, it is possible that the dining room is 633 inches, or 52 feet and 9 inches long. However, unless the house is huge this, and larger possible answers, are extremely unlikely.

4.3.35. Examining $x^2 + 18x - 823 \equiv 0 \pmod{1800}$ modulo 8, we see that $x^2 + 18x - 823 \equiv x^2 + 2x + 1 = (x+1)^2 \equiv 0 \pmod{8}$ has solutions $x \equiv 3 \pmod{8}$ and $x \equiv 7 \pmod{8}$. Examining $x^2 + 18x - 823 \equiv 0 \pmod{1800}$ modulo 9, we see that $x^2 + 18x - 823 \equiv x^2 + 5 \equiv 0$ has solutions $x \equiv 2 \pmod 9$ and $x \equiv 7 \pmod 9$. Examining $x^2 + 18x - 823 \equiv 0 \pmod{1800}$ modulo 25, we see that $x^2 + 18x - 823 \equiv x^2 + 18x + 77 = (x+11)(x+7) \equiv 0 \pmod{25}$. This has solutions $x \equiv 18 \pmod{25}$, and $x \equiv 14 \pmod{14}$. Thus there are $2^3 = 8$ systems to examine. We may find, by the iterative method discussed in the text, that the solutions are given by $x = 225a_1 + 1000a_2 + 576a_3 + 1800k$, where k is an integer and a_1 is 3 or 7, a_2 is 2 or 7, and a_3 is 14 or 18.

4.4. Solving Polynomial Congruences

4.4.1. a. By testing each of the integers $0, 1, \ldots, 6$, we see that $1^2 + 4(1) + 2 \equiv 0 \pmod 7$ and $2^2 + 4(2) + 2 \equiv 0 \pmod 7$. So the solutions are the integers $x \equiv 1$ or $2 \pmod 7$.

b. Let $f(x) = x^2 + 4x + 2$. Then $f'(x) = 2x + 4$. Since $f'(1) \equiv 6 \not\equiv 0 \pmod 7$, we can apply case (i) of Hensel's lemma. The solutions $x \equiv 1 \pmod 7$ lift uniquely to solutions $x \equiv 1 + 7t \pmod{49}$, where $t \equiv -\overline{f'(1)}f(1)/7 \equiv -\overline 6 \cdot 7/7 \equiv 1 \pmod 7$. So $x \equiv 8 \pmod{49}$. Similarly, since $f'(2) \equiv 1 \not\equiv 0 \pmod 7$, the solutions $x \equiv 2 \pmod 7$ lift uniquely to $x \equiv 2 + 7t \pmod{49}$, where $t = -\overline 8 f(2)/7 \equiv 5 \pmod 7$. So $x \equiv 2 + 7(5) \equiv 37 \pmod{49}$. The solutions are the integers $x \equiv 8$ or $37 \pmod{39}$.

c. Since $f'(8) \equiv 6 \pmod 7$, the solutions $x \equiv 8 \pmod{49}$ lift uniquely to solutions $x \equiv 8 + 49t$ where $t \equiv -\overline 6 f(8)/49 \equiv 2 \pmod 7$. So $x \equiv 8 + 49(2) \equiv 106 \pmod{343}$. Similarly, since $f'(37) \equiv 1 \pmod 7$, the solutions $x \equiv 37 \pmod{49}$ lift uniquely to solutions $x \equiv 37 + 49t$ where $t \equiv -\overline{f'(37)}f(37)/49 \equiv 4 \pmod 7$. So $x \equiv 37 + 49(4) \equiv 233 \pmod{343}$. The solutions are the integers $x \equiv 106$ or $233 \pmod{343}$.

4.4.3. Let $f(x) = x^2 + x + 47$. By inspection, the solutions to $f(x) \equiv x^2 + x + 5 \equiv 0 \pmod 7$ are $r \equiv 1$ or $5 \pmod 7$. Since $f'(1) \equiv 3 \pmod 7$, we know, by Corollary 4.14.1, that $r \equiv 1$ lifts successively to unique solutions modulo each power of 7. Note that $\overline{f'(1)} = \overline 3 \equiv 5 \pmod 7$. Then, with notation as in Corollary 4.14.1, $r_2 = 1 - f(1) \cdot 5 \equiv 1 - 49 \cdot 5 \equiv 1 \pmod{49}$, and $r_3 = 1 - 49 \cdot 5 \equiv 99 \pmod{343}$, and finally, $r_4 = 99 - f(99) \cdot 5 \equiv 785 \pmod{2401}$. Similarly, since $f'(5) \equiv 4 \pmod 7$, we know, by Corollary 4.14.1, that $r \equiv 5$ lifts successively to unique solutions modulo each power of 7. Note that $\overline{f'(5)} = \overline 4 \equiv 2 \pmod 7$. Then, with notation as in Corollary 4.14.1, $r_2 = 5 - f(5) \cdot 2 \equiv 47 \pmod{49}$, and $r_3 = 47 - f(47) \cdot 2 \equiv 243 \pmod{343}$, and finally, $r_4 = 243 - f(243) \cdot 2 \equiv 1615 \pmod{2401}$. Therefore the solutions are $x \equiv 785$ or $1615 \pmod{2401}$.

4.4.5. Let $f(x) = 13x^7 - 42x - 649$ and observe that $1323 = 3^3 7^2$. We start by solving the congruence modulo 3 and lifting to modulo 27. First $f(x) \equiv x^7 - 1 \equiv 0 \pmod 3$, which has only the solution $r \equiv 1 \pmod 3$. Since $f'(1) = 13 \cdot 7 \cdot 1^6 - 42 \equiv 1 \pmod 3$, this solution lifts to unique solutions modulo 9 and 27. Following Corollary 4.14.1, we have $r_2 = 1 + f(1)\overline{f'(1)} = 1 - (13 - 42 - 649)(1) \equiv 4 \pmod 9$, and $r_3 = 4 - f(4)(1) \equiv 22 \pmod{27}$. Next we solve the congruence modulo 7 and lift to 49. Then $f(x) \equiv -x^7 + 2 \equiv 0 \pmod 7$ has only the solution $r \equiv 2 \pmod 7$. Note that $f'(2) = 5782 \equiv 0 \pmod 7$ and that $f(2) = 931 \equiv 0 \pmod 7$, so $r = 2$ lifts to 7 solutions modulo 49, namely 2, 9, 16, 23, 30, 37, and 44. Finally, we pair the solution for 27 with each of the solutions for 49 to produce solutions for 1323. Solving the system $x \equiv 22 \pmod{27}, x \equiv 2 \pmod{49}$ yields $x \equiv 1129 \pmod{1323}$. Solving the system $x \equiv 22 \pmod{27}, x \equiv 9 \pmod{49}$ yields $x \equiv 940 \pmod{1323}$. Solving the system $x \equiv 22 \pmod{27}, x \equiv 16 \pmod{49}$ yields $x \equiv 751 \pmod{1323}$. Solving the system $x \equiv 22 \pmod{27}, x \equiv 23 \pmod{49}$ yields $x \equiv 562 \pmod{1323}$. Solving the system $x \equiv 22 \pmod{27}, x \equiv 30 \pmod{49}$ yields $x \equiv 373 \pmod{1323}$. Solving the system $x \equiv 22 \pmod{27}, x \equiv 37 \pmod{49}$ yields $x \equiv 184 \pmod{1323}$. Solving the system $x \equiv 22 \pmod{27}, x \equiv 44 \pmod{49}$ yields $x \equiv 1318 \pmod{1323}$. So the incongruent solutions are 184,

373, 562, 751, 940, 1129, and 1318.

4.4.7. Let $f(x) = x^4 + 2x + 36$ and note that $4375 = 5^4 7$. By inspection, the only solution to $f(x) \equiv x^4 + 2x + 1 \equiv 0 \pmod{5}$ is $r \equiv -1 \pmod 5$. Since $f'(-1) = 4(-1)^3 + 2 \equiv 3 \not\equiv 0 \pmod 5$, we know that $r = -1$ lifts uniquely to solutions modulo 5^k. Applying Corollary 4.14.1, we have $r_2 = (-1) - f(-1) \cdot \overline{3} \equiv -1 - 35 \cdot 2 \equiv 4 \pmod{25}$, and $r_3 = 4 - f(4)2 \equiv 29 \pmod{125}$ and $r_4 = 29 - f(29)2 \equiv 279 \pmod{625}$. Again, by inspection, we solve $f(x) \equiv x^4 + 2x + 1 \equiv 0 \pmod 7$ and obtain the two solutions $x \equiv 2$ or $-1 \pmod 7$. Finally we solve the two systems $x \equiv 279 \pmod{625}, x \equiv 2 \pmod 7$ and $x \equiv 279 \pmod{625}, x \equiv -1 \pmod 7$ to get the two solutions 3404 and 279 $\pmod{4375}$, respectively.

4.4.9. Let $f(x) = 5x^3 + x^2 + x + 1$. By inspection, the solution of the congruence $f(x) \equiv 0 \pmod 2$ is $x \equiv 1 \pmod 2$. Note that $f'(x) = 15x^2 + 2x + 1$, so $f'(1) \equiv 0 \pmod 2$. Since $f(1) = 8 \equiv 0 \pmod 4$, we know that $x = 1$ lifts to two solutions $x \equiv 1$ or $3 \pmod 4$. Since $f(3) \equiv 4 \pmod 8$, but $f'(3) \equiv 0 \pmod 2$, we know that 3 does not lift to solutions modulo 8. However, since $f'(1) \equiv 0 \pmod 2$ and $f(1) \equiv 0 \pmod 8$, we know that 1 lifts to the two solutions 1 and 5 $\pmod 8$. Since $f(1) \equiv 8 \not\equiv 0 \pmod{16}$, we know that 1 does not lift further. Since $f(5) \equiv 0 \pmod{16}$, we know that 5 lifts to solutions 5 and 13 $\pmod{16}$. Since $f(5) \equiv 16 \not\equiv 0 \pmod{32}$, we know that 5 does not lift further. Since $f(13) \equiv 0 \pmod{32}$, we know that 13 lifts to solutions 13 and 29 $\pmod{32}$. Since $f(13) \equiv 32 \not\equiv 0 \pmod{64}$, we know that 13 does not lift further. Since $f(29) \equiv 0 \pmod{32}$, we know that 29 lifts to solutions 29 and 61 $\pmod{64}$. So there are only two incongruent solutions.

4.4.11. Since $(a, p) = 1$, we know that a has an inverse b modulo p. Let $f(x) = ax - 1$. Then $x \equiv b \pmod p$ is the unique solution to $f(x) \equiv 0 \pmod p$. Since $f'(x) = a \not\equiv 0 \pmod p$, we know that $r \equiv b$ lifts uniquely to solutions modulo p^k for all natural numbers k. By Corollary 4.14.1, we have that $r_k = r_{k-1} - f(r_{k-1})\overline{f'(b)} = r_{k-1} - (ar_{k-1} - 1)\overline{a} = r_{k-1} - (ar_{k-1} - 1)b = r_{k-1}(1 - ab) + b$. This gives a recursive formula for lifting b to a solution modulo p^k for any k.

4.4.13. By inspection the only solution of $f(x) = x^2 + x + 223 \equiv 0 \pmod 3$ is $x \equiv 1 \pmod 3$. Since $f'(1) = 2 \cdot 1 + 1 = 3 \equiv 0 \pmod 3$ and $f(1) = 225 \equiv 0 \pmod 9$, we have by Theorem 4.14 that $1, 4,$ and 7 are the only solutions modulo 9. Since $f(1) = 225 \equiv 9 \pmod{27}$, this solutions doesn't lift. Since $f(4) = 243 \equiv 0 \pmod{27}$, this solution lifts to three solutions $4, 13,$ and 22 $\pmod{27}$. Since $f(7) = 279 \equiv 9 \pmod 7$, this solution doesn't lift. So the only solutions modulo 27 are 4, 13, and 22. Next $f(4) \equiv f(13) \equiv f(22) \equiv 0 \pmod{81}$, so each of these solutions lifts to three solutions modulo 81, namely 4, 31, 58, 13, 40, 67, 22, 49, and 76. Of these, $f(13) \equiv f(40) \equiv f(67) \equiv 163 \not\equiv 0 \pmod{3^5}$ and so these do not lift to solutions. But $f(4) \equiv f(31) \equiv f(58) \equiv f(22) \equiv f(49) \equiv f(76) \equiv 0 \pmod{3^5}$. Therefore, each of these 6 solutions lifts to three solutions modulo 3^5, namely $x \equiv$ 166, 112, 238, 193, 85, 130, 58, 103, 211, 31, 76, 157, 184, 49, 4, 139, 220 or 22 $\pmod{3^5}$. It is easy to check that each of these solutions satisfies the hypotheses of Exercise 12 with $p = 3$, $k = 5$ and $j = 2$. E.g., $f(166) \equiv 0 \pmod{3^5}$ and $3^2 \parallel f'(166) = 333 = 3^2 \cdot 37$. Therefore each of these solutions lifts uniquely to solutions modulo 3^n for $n \geq 5$. So there are exactly 18 solutions modulo 3^n for $n \geq 5$.

4.5. Systems of Linear Congruences

4.5.1. a. Multiplying the first congruence by 2 gives $2x + 4y \equiv 2 \pmod 5$. Subtracting the second congruence $2x + y \equiv 1 \pmod 5$ from this gives $3y \equiv 1 \pmod 5$. Since 2 is the inverse of 3 modulo 5 we have $y \equiv 2 \pmod 5$. Inserting this into the congruence $x + 2y \equiv 1 \pmod 5$ gives $x + 4 \equiv 1 \pmod 5$. Hence $x \equiv -3 \equiv 2 \pmod 5$. The unique solution modulo 5 is $x \equiv 2 \pmod 5$ and $x \equiv 2 \pmod 5$.

b. Multiplying the first congruence by 3 gives $3x + 9y \equiv 3 \pmod 5$. Subtracting the second congruence $3x + 4y \equiv 2 \pmod 5$ from this gives $5y \equiv 1 \pmod 5$ which is impossible. Hence this system has no solutions.

c. Multiplying the second congruence by 2 gives $4x + 6y \equiv 2 \pmod 5$. Subtracting the first congruence from this gives $5y \equiv 0 \pmod 5$. The solutions to this are all values of y, that is, $y \equiv 0, 1, 2, 3,$ or 4 $\pmod 5$. This implies that $4x \equiv 2, 1, 0, 4,$ or 3 $\pmod 5$, respectively, or that $x \equiv 3, 4, 0, 1$ or 2 $\pmod 5$, respectively. The solutions are $x \equiv 3 \pmod 5, y \equiv 0 \pmod 5; x \equiv 4 \pmod 5, y \equiv 1$

(mod 5); $x \equiv 0$ (mod 5), $y \equiv 2$ (mod 5); $x \equiv 1$ (mod 5), $y \equiv 3$ (mod 5); and $x \equiv 2$ (mod 5), $y \equiv 4$ (mod 5).

4.5.3. If we use one congruence to eliminate a variable from the other congruence, we are left with linear congruence of the form $ax \equiv b$ (mod p). If $(a, p) = 1$, then this congruence has a unique solution, but if $p \mid a$, we have $0 \equiv b$ (mod p), which has 0 solutions if b is not 0 modulo p and p solutions if b is 0 modulo p. So there are $0, 1$, or p solutions for this variable. Similarly, There are $0, 1$, or p solutions for the other variable. Multiplying all the possible combinations gives us $0, 1, p$, or p^2 solutions for the system.

4.5.5. The basis step, where $k = 1$, is clear by assumption. For the inductive hypothesis assume that $\mathbf{A} \equiv \mathbf{B}$ (mod m) and $\mathbf{A}^k \equiv \mathbf{B}^k$ (mod m). Then, $\mathbf{A} \cdot \mathbf{A}^k \equiv \mathbf{A} \cdot \mathbf{B}^k$ (mod m) by Theorem 4.16. Further, $\mathbf{A}^{k+1} = \mathbf{A} \cdot \mathbf{A}^k \equiv \mathbf{A} \cdot \mathbf{B}^k \equiv \mathbf{B} \cdot \mathbf{B}^k = \mathbf{B}^{k+1}$ (mod m) by simple substitution. This completes the inductive proof.

4.5.7. Note that $1 = \det(\mathbf{I}) = \det(\mathbf{A}^2) = (\det(\mathbf{A}))^2$ (mod m). So, $(\det(\mathbf{A}))^2 - 1 = (\det(\mathbf{A}) + 1)(\det(\mathbf{A}) - 1) \equiv 0$ (mod m). It follows that $\det(\mathbf{A}) = \pm 1$.

4.5.9. a. Let $\mathbf{A}$ be the matrix. We have $\det \mathbf{A} = -2$ which has inverse 3 modulo 7. Then
$$\overline{\mathbf{A}} = 3 \cdot \operatorname{adj} \mathbf{A} = 3 \begin{pmatrix} -1 & -1 & 1 \\ -1 & 1 & -1 \\ 1 & -1 & -1 \end{pmatrix} = \begin{pmatrix} 4 & 4 & 3 \\ 4 & 3 & 4 \\ 3 & 4 & 4 \end{pmatrix}.$$

b. Let $\mathbf{A}$ be the matrix. We have $\det \mathbf{A} = 3$ which has inverse 5 modulo 7. Then
$$\overline{\mathbf{A}} = 5 \cdot \operatorname{adj} \mathbf{A} = 5 \begin{pmatrix} -1 & 0 & 4 \\ -1 & 3 & -2 \\ 2 & -2 & 0 \end{pmatrix} = \begin{pmatrix} 2 & 0 & 6 \\ 2 & 1 & 4 \\ 3 & 4 & 0 \end{pmatrix}.$$

c. Let $\mathbf{A}$ be the matrix. We have $\det \mathbf{A} = 4$ which has inverse 2 modulo 7. Then
$$\overline{\mathbf{A}} = 2 \cdot \operatorname{adj} \mathbf{A} = 2 \begin{pmatrix} -1 & -1 & -1 & 2 \\ -1 & -1 & 2 & -1 \\ -1 & 2 & -1 & -1 \\ 2 & -1 & -1 & -1 \end{pmatrix} = \begin{pmatrix} 5 & 5 & 5 & 4 \\ 5 & 5 & 4 & 5 \\ 5 & 4 & 5 & 5 \\ 4 & 5 & 5 & 5 \end{pmatrix}.$$

4.5.11. a. Multiplying the first congruence by 2 gives $2x + 2y + 2z \equiv 2$ (mod 5). Subtracting this from the second congruence gives $2y + z \equiv 4$ (mod 5). There are five possible values for z modulo 5, and since $(2, 5) = 1$, each of these leads to a unique value of y modulo 5, and substituting these values of y and z modulo 5 into the first congruence we obtain a unique value of x modulo 5. Hence there are exactly 5 incongruent solutions modulo 5. There are $x \equiv 4$ (mod 5), $y \equiv 2$ (mod 5), $z \equiv 0$ (mod 5); $x \equiv 1$ (mod 5), $y \equiv 4$ (mod 5), $z \equiv 1$ (mod 5); $x \equiv 3$ (mod 5), $y \equiv 1$ (mod 5), $z \equiv 2$ (mod 5); $x \equiv 0$ (mod 5), $y \equiv 3$ (mod 5), $z \equiv 3$ (mod 5); and $x \equiv 2$ (mod 5), $y \equiv 0$ (mod 5), $z \equiv 4$ (mod 5).

b. Subtracting the last congruence from the first gives $3y \equiv 3$ (mod 5), so $y \equiv 4$ (mod 5). Let z take on the values 0, 1, 2, 3, and 4 and solve the last congruence for x to get 3, 0, 2, 4, and 1, respectively. This represents the 5 incongruent solutions.

c. Since the coefficient matrix for x and y is $\begin{pmatrix} 1 & 1 \\ 2 & 4 \end{pmatrix}$, which had determinant $2 \not\equiv 0$ (mod 5), we can find a unique solution in x and y for any of the 5 possible values for z. Therefore, there are 5 incongruent solutions.

d. Since the determinant of the coefficient matrix is $4 \not\equiv 0$ (mod 5) there is a unique solution to the system.

4.5.13. In Gaussian elimination, the chief operation is to subtract a multiple of one equation or row from another, in order to put a 0 in a desirable place. Given that an entry a must be changed to 0 by subtracting a multiple of b, we proceed as follows: Let $\bar{b}$ be the inverse for $b \pmod{k}$. Then $a - (a\bar{b})b = 0$, and elimination proceeds as for real numbers. If $\bar{b}$ doesn't exist, and one cannot swap rows to get an invertible b, then the system is underdetermined.

4.5.15. Consider summing the ith row. Let $k = xn + y$, where $0 \le y < n$. Then x and y must satisfy the Diophantine equation $i \equiv a + cy + ex \pmod{n}$, if k is in the ith row. Then $x - ct$ and $y + et$ is also a solution for any integer t. By Exercise 14, there must be n positive solutions which yield n numbers k between 0 and n^2. Let $s, s+1, \ldots, s+n-1$ be the values for t that give these solutions. Then the sum of the ith row is $\sum_{r=0}^{n-1}(n(x - c(s+r)) + y + e(s+r)) = n(n+1)$, which is independent of i.

4.6. Factoring Using the Pollard Rho Method

4.6.1. a. We compute $x_1 = 2^2 + 1 = 5$ and $x_2 = 5^2 + 1 = 26$. Then $(26 - 5, 133) = (21, 133) = 7$, so we have $133 = 7 \cdot 19$.

b. $x_1 = 5, x_2 = 26, x_3 = 677, x_4 = 565, x_5 = 574, x_6 = 124, x_7 = 1109, x_8 = 456, x_9 = 1051, x_{10} = 21, x_{11} = 442, x_{12} = 369, x_{13} = 616$, and $x_{14} = 166$. Then $(x_{2i} - x_i, 1189) = 1$ for $i = 1, 2, \ldots, 6$, but $(x_{14} - x_7, 1189) = 41$, and we have $1189 = 29 \cdot 41$.

c. We need to compute up to $x_7 = 1273$ and $x_{14} = 535$. Then we have $(535 - 1273, 1927) = 41$, and so $1927 = 41 \cdot 47$.

d. We need to compute up to $x_4 = 2994$ and $x_8 = 6973$. Then we have $(6973 - 2994, 8131) = 173$, and so $8131 = 47 \cdot 173$.

e. We need to compute up to $x_7 = 24380$ and $x_{14} = 12066$. Then we have $(12066 - 24380, 36287) = 131$, and so $36287 = 131 \cdot 277$.

f. We need to compute up to $x_8 = 18842$ and $x_{16} = 7329$. Then we have $(7329 - 18842, 48227) = 29$, and so $48227 = 29 \cdot 1663$.

4.6.3. Numbers generated by linear functions where $a > 1$ will not be random in the sense that $x_{2s} - x_k = ax_{2s-1} + b - (ax_{s-1} + b) = a(x_{2s-1} - x_{s-1})$ is a multiple of a for all s. If $a = 1$, then $x_{2s} - x_s = x_0 + sb$. In this case, if $x_0 \ne 0$, then we will not notice if a factor of b that is not a factor of x_0 is a divisor of n.

CHAPTER 5

Applications of Congruences

5.1. Divisibility Tests

5.1.1. a. Since $2 \mid 4, 4 \mid 84, 8 \mid 984, 16 \mid 1984, 64 \mid 201984, 128 \mid 201984, 256 \mid 201984$, but 512 does not divide 201984, it follows that $256 = 2^8$ is the highest power of 2 that divides 201984.

b. Since $2 \mid 8, 4 \mid 8, 8 \mid 408, 16 \mid 3408$, but 32 does not divide 23408, it follows that $16 = 2^4$ is the highest power of 2 that divides 1423408.

c. Since $2 \mid 4, 4 \mid 44, 8 \mid 744, 16 \mid 5744, 32 \mid 75744, 64 \mid 375744, 128 \mid 9375744, 256 \mid 89375744, 512 \mid 89375744, 1024 \mid 89375744$, but 2048 does not divide 89375744, it follows that $1024 = 2^{10}$ is the highest power of 2 that divides 8937544.

d. Since $2 \mid 6$ but 4 does not divide 46, it follows that $2 = 2^1$ is the highest power of 2 that divides 41578912246.

5.1.3. a. The sum of the digits of 18381 is $1 + 8 + 3 + 8 + 1 = 21$. Since this sum is divisible by 3 but not by 9, 18381 is divisible by 3, but not by 9.

b. The sum of the digits of 65412351 is $6 + 5 + 4 + 1 + 2 + 3 + 5 + 1 = 27$. Since this sum is divisible by 3 and by 9 it follows that 65412351 is divisible by both 3 and 9.

c. The sum of the digits of 987654321 is $9 + 8 + 7 + 6 + 5 + 4 + 3 + 2 + 1 = 45$. since this sum is divisible by 3 and by 9 it follows that 987654321 is divisible by both 3 and 9.

d. The sum of the digits of 78918239735 is $7 + 8 + 9 + 1 + 8 + 2 + 3 + 9 + 7 + 3 + 5 = 62$. Since this sum is not divisible by 3, 78918239735 is divisible by neither 3 nor 9.

5.1.5. By Theorem 5.1, the power of 2 dividing a number is equal to the number of zeros at the end of its binary expression. **a.** $2^1 = 2$ **b.** $2^0 = 1$ **c.** $2^6 = 64$ **d.** $2^0 = 1$

5.1.7. a. Using Theorem 5.2, we need only examine the sum of the digits. We have $1 + 2 + 1 + 0 + 1 + 2 + 2 = 9$. As 2 does not divide 9, 2 does not divide $(1210122)_3$.

b. Since 2 does not divide $2 + 1 + 1 + 1 + 0 + 2 + 1 + 0 + 1 = 9$, 2 does not divide $(211102101)_3$.

c. Since 2 divides $1 + 1 + 1 + 2 + 2 + 0 + 1 + 1 + 1 + 2 = 12$, then $2 \mid (1112201112)_3$.

d. Since 2 divides $1 + 0 + 1 + 2 + 2 + 2 + 2 + 2 + 0 + 1 + 1 + 1 + 0 + 1 = 16$, then $2 \mid (10122222011101)_3$.

5.1.9. a. As both 3 and 5 divide $16 - 1$, Theorem 5.2 tells us that we need only examine the sum of the base 16 digits.
$3 + E + A + 2 + 3 + 5 = 3 + 14 + 10 + 2 + 3 + 5 = 37$. As neither 3 nor 5 divides 7, neither 3 nor 5 divides $(3EA235)_{16}$.

b. Since $A + B + C + D + E + F = 10 + 11 + 12 + 13 + 14 + 15 = 75$ is divisible by 3 and 5, we see that both 3 and 5 divide $(ABCDEF)_{16}$.

c. Since neither 3 nor 5 divides $15 + 1 + 1 + 7 + 9 + 2 + 1 + 1 + 7 + 3 = 47$, neither 3 nor 5 divides $(F117921173)_{16}$.

d. Since 5 divides $1 + 0 + 10 + 11 + 9 + 8 + 7 + 3 + 0 + 1 + 15 = 65$, but 3 does not, we have that 5 divides $(10AB987301F)_{16}$, but 3 does not.

5.1.11. The sum of the digits of a repunit with n 1's in its decimal expansion is n. This repunit is divisible by 3 if and only if n is divisible by 3 and is divisible by 9 if and only if n is divisible by 9.

5.1.13. The alternating sum of blocks of three digits of an n-digit repunit is 0 if $n \equiv 0 \pmod 6$, 1 if $n \equiv 1 \pmod 6$, 11 if $n \equiv 2 \pmod 6$, 111 if $n \equiv 3 \pmod 6$, 110 if $n \equiv 4 \pmod 6$, 100 if $n \equiv 5 \pmod 6$. Hence a repunit with n decimal digits is divisible by 1001 if and only if $n \equiv 0 \pmod 6$. Since 7 divides this alternating sum if and only if $n \equiv 0 \pmod 6$, these are exactly the values of n for which this requnit is divisible by 7. Exactly the same reasoning and conclusion holds for divisibility by 13.

5.1.15. Let d be a divisor of $b-1$. By Theorem 5.2, a number is divisible by d if and only if the sum of its digits is a multiple of d. Since the sum of the digits of a repunit is equal to the number of digits it has, a repunit is divisible by d if and only if it has a multiple of d digits.

5.1.17. A palindromic integer with $2k$ digits has the form $(a_k a_{k-1} \ldots a_1 a_1 a_2 \ldots a_k)_{10}$. Using the test for divisibility by 11 developed in this section, we find that $a_k - a_{k-1} + \cdots \pm a_1 \mp a_1 \pm a_2 \mp \cdots - a_k = 0 \equiv 0 \pmod{11}$ and so $(a_k a_{k-1} \ldots a_1 a_1 a_2 \ldots a_k)_{10}$ is divisible by 11.

5.1.19. Let $a_k a_{k-1} \ldots a_1 a_0$ be the decimal representation of an integer. Then $a_k a_{k-1} \ldots a_1 a_0 = a_0 a_1 a_2 + 10^3 a_3 a_4 a_5 + 10^3(10^3 a_6 a_7 a_8) + \cdots$. So, $a_k a_{k-1} \ldots a_1 a_0 \equiv a_0 a_1 a_2 + a_3 a_4 a_5 + a_6 a_7 a_8 + \cdots \pmod{37}$. Thus $a_k a_{k-1} \ldots a_1 a_0$ is divisible by 37 if and only if $a_0 a_1 a_2 + a_3 a_4 a_5 + a_6 a_7 a_8 + \cdots$ is also. Hence, 443692 is divisible by 37 if and only if $443 + 692 = 1135$ is. And 1134 is divisible by 37 if and only if $1 + 135 = 136$ is. But 136 is not, and so 37 does not divide 443692. Further, 11092785 is divisible by 37 if and only if $11 + 092 + 785 = 888$ is. We know that $888 = 24 \cdot 37$, so 11092785 is a multiple of 37.

5.1.21. a. Applying Exercise 20, we have $(1)_2 - (01)_2 + (11)_2 - (01)_2 + (10)_2 = (100)_2 = 4$, Since 4 is not divisible by $5 = 2^2 + 1$, neither is $(101110110)_2$.

b. Applying Exercise 20, we have $-(12)_3 + (10)_3 - (01)_3 + (22)_3 = (12)_3 = 5$. Since $2 \nmid 12$ but $5 \mid 12$, only 5 divides $(12100122)_3$.

c. Applying Exercise 20, we have $(3)_8 - (64)_8 + (70)_8 - (12)_8 + (44)_8 = (41)_8 = 33$ which is divisible by neither 5 nor 13. Hence neither divides the number.

d. Applying Exercise 20, we have $5 - 83 + 70 - 41 + 32 - 02 + 19 = 0$ which is divisible by 101, and therefore, so is $(5837041320219)_{10}$.

5.1.23. First, note that $89878 \equiv 8 + 9 + 8 + 7 + 8 \equiv 4 \pmod 9$, $58965 \equiv 5 + 8 + 9 + 6 + 5 \equiv 6 \pmod 9$, and $5299?56270 \equiv 5 + 2 + 9 + 9 + ? + 5 + 6 + 2 + 7 + 0 \equiv ? \pmod 9$. So, $89878 \cdot 58965 \equiv 4 \cdot 6 \equiv 6 \equiv ? \pmod 9$. Thus, as the question mark represents a single decimal digit congruent to 6, the question mark represents the digit 6.

5.1.25. Casting out nines is not infallible. To see this, note that $19 \equiv 2 \cdot 5 \pmod 9$, but $19 \neq 2 \cdot 5$. The cause of this problem is that $0 \equiv 9 \pmod 9$, and so any 0 may be replace by a 9, or vice versa, and the congruence $c \equiv ab \pmod 9$ will still hold, whereas in general, the equality $c = ab$ will not hold.

5.2. The Perpetual Calendar

5.2.1. Happy Birthday!

5.2.3. For this problem, we let $k = 13, C = 20, Y = 20$, and $W = 5$. Now, $[2.6m - 0.2] \equiv W - k + 2C - Y - [\frac{Y}{4}] - [\frac{C}{4}] \equiv 2 \pmod 7$. And since $[2.6m - 0.2] \equiv 2 \pmod 7$ with $0 < m \leq 10$ only for $m = 1$ and 9, we see that March and November have Friday the 13th. But we have only checked the months after (and including March). To check January and February, let $W = 5, k = 13, C = 20$, and $Y = 19$. Now, $[2.6m - 0.2] \equiv W - k + 2C - Y - [\frac{Y}{4}] - [\frac{C}{4}] \equiv 4 \pmod 7$. But $[2.6 \cdot 11 - 0.2] \equiv 0 \pmod 7$ and $[2.6 \cdot 12 - 0.2] \equiv 3 \pmod 7$, so neither January nor February have Friday the 13th. So the 13th will fall on Friday only twice in the year 2020.

5.2.5. For each 4000 years, we need to subtract one day from the total number of days before reducing modulo 7. Therefore, we subtract $[N/4000] = [C/40]$ from the right hand side of the formula, giving $W \equiv k + [2.6m - 0.2] - 2C + Y + [Y/4] + [C/4] - [C/40] \pmod 7$.

5.2.7. If B is the number of the day of the week you were born, $0 \leq B < 7$, and M is the month and K is the day, then we need to solve the congruence $B \equiv K + [2.6M - 0.2] - 2C + Y + [Y/4] + [C/4] \pmod 7$ for C and Y. There are two cases. If $C = 19$, then the congruence reduces to $B \equiv K + [2.6M - 0.2] + Y + [Y/4] + 1 \pmod 7$. If $C = 20$ then the congruence reduces to $B \equiv K + [2.6M - 0.2] + Y + [Y/4] \pmod 7$. In both cases, there are 4 subcases depending on the residue of Y modulo 4. Restrict Y to only those years between your birth and your 100th birthday.

5.2.9. This is the sequence of years divisible by 100, but not by 400, so the next term is 2500. These are all the century years that are not leap years.

5.2.11. If the 13th falls on the same day of the week on two consecutive months, then the number of days in the first month must be congruent to 0 modulo 7, and the only such month is February during non-leap year. If February 13th is a Friday, then January 1st is $31 + 13 - 1 \equiv 1 \pmod 7$ week days earlier, that is, Thursday.

5.2.13. In the perpetual calendar formula we let $W = 5$ and $k = 13$ to get $5 \equiv 13 + [2.6m - 0.2] - 2C + Y + [Y/4] + [C/4] \pmod 7$. Then $[2.6m - 0.2] \equiv 6 + 2C - Y - [Y/4] - [C/4] \pmod 7$. We note that as the month varies from March to December, the expression $[2.6m - 0.2]$ takes on every residue class modulo 7. So regardless of the year, there is always an m which makes the left side of the last congruence congruent to the right side.

5.2.15. The months with 31 days are March, May, July, August, October, December and January, which is considered in the previous year. The corresponding numbers for these months are 1, 3, 5, 6, 8, 10, and 12. Given Y and C, we let $k = 31$ in the perpetual calendar formula and get $W \equiv 31 + [2.6m - 0.2] - 2C + Y + [Y/4] + [C/4] \equiv 3 + [2.6m - 0.2] - 2C + Y + [Y/4] + [C/4] \pmod 7$. To see which days of the week the 31st will fall on, we let m take on the values 1, 3, 5, 6, 8, 10 and reduce. Finally, we decrease the year by one (which may require decreasing the century by one) and let m take on the value 12 and reduce modulo 7. The collection of values of W tells us the days of the week on which the 31st will fall.

5.3. Round-Robin Tournaments

5.3.1. a. Teams i and j are paired in round k if and only if $i + j \equiv k \pmod 7$ with team i drawing a bye if $2i \equiv k \pmod 7$. The result is shown in the following table.

| | Team | | | | | | |
Round	1	2	3	4	5	6	7
1	7	6	5	bye	3h	2h	1h
2	bye	7h	6h	5h	4	3	2
3	2h	1	7h	6h	bye	4	3
4	3h	bye	1	7	6	5h	4h
5	4	3	2h	1h	7h	bye	5
6	5h	4h	bye	2	1	7	6h
7	6	5	4	3h	2h	1h	bye

b. Teams i and j are paired in round k if and only if $i + j \equiv k \pmod 7$. Team i draws a Team 8 if $2i \equiv k \pmod 7$. The result is shown in the following table.

| | Team | | | | | | | |
Round	1	2	3	4	5	6	7	8
1	7	6	5	8	3	2	1	4
2	8	7	6	5	4	3	2	1
3	2	1	7	6	8	4	3	5
4	3	8	1	7	6	5	4	2
5	4	3	2	1	7	8	5	6
6	5	4	8	2	1	7	6	3
7	6	5	4	3	2	1	8	7

c. Teams i and j are paired in round k if and only if $i + j \equiv k \pmod 9$. Team i draws a bye if $2i \equiv k \pmod 9$. The result is shown in the following table.

| | Team | | | | | | | | |
Round	1	2	3	4	5	6	7	8	9
1	9h	8h	7h	6h	bye	4	3	2	1
2	bye	9	8	7	6	5h	4h	3h	2h
3	2	1h	9h	8h	7h	bye	5	4	3
4	3h	bye	1	9	8	7	6h	5h	4h
5	4	3	2h	1h	9h	8h	bye	6	5
6	5h	4h	bye	2	1	9	8	7h	6h
7	6	5	4	3h	2h	1h	9	bye	7h
8	7h	6h	5h	bye	3	2	1	9	8h
9	8	7	6	5	4h	3h	2h	1h	bye

d. Teams i and j are paired in round k if and only if $i + j \equiv k \pmod 9$. Team i draws Team 10 if $2i \equiv k \pmod 9$. The result is shown in the following table.

	Team									
Round	1	2	3	4	5	6	7	8	9	10
1	9	8	7	6	10	4	3	2	1	5
2	10	9	8	7	6	5	4	3	2	1
3	2	1	9	8	7	10	5	4	3	6
4	3	10	1	9	8	7	6	5	4	2
5	4	3	2	1	9	8	10	6	5	7
6	5	4	10	2	1	9	8	7	6	3
7	6	5	4	3	2	1	9	10	7	8
8	7	6	5	10	3	2	1	9	8	4
9	8	7	6	5	4	3	2	1	10	9

5.3.3. a. For round 1, teams i and j are paired if $i + j \equiv 1 \pmod{5}$. Teams 1 and 5 are paired, and since $1 + 5 = 6$ is even, team 5 is the home team. Teams 2 and 4 are paired, and since $2 + 4 = 6$ is even, team 4 is the home team. Finally, in round 1 team 3 draws a bye.

For round 2, teams i and j are paired if $i + j = 2 \pmod{5}$. Team 1 draws a bye. Teams 2 and 5 are paired, and since $2 + 5 = 7$ is odd, team 2 is the home team. Teams 3 and 4 are paired, and since $3 + 4 = 7$ is odd, team 3 is the home team.

For round 3, teams i and j are paired if $i + j = 3 \pmod{5}$. Teams 1 and 2 are paired, and since $1 + 2 = 3$ is odd, team 1 is the home team. Teams 3 and 5 are paired, and since $3 + 5 = 8$ is even, team 5 is the home team. Team 4 draws a bye.

For round 4, teams i and j are paired if $i + j = 4 \pmod{5}$. Teams 1 and 3 are paired, and since $1 + 3 = 4$ is even, team 3 is the home team. Team 2 draws a bye. Teams 4 and 5 are paired, and since $4 + 5 = 9$ is odd, team 4 is the home team.

For round 5, teams i and j are paired if $i + j = 5 \pmod{5}$. Teams 1 and 4 are paired, and since $1 + 4 = 5$ is odd, team 1 is the home team. Teams 2 and 3 are paired, and since $2 + 3 = 5$ is odd, team 2 is the home team. Team 4 draws a bye.

We see that each team plays 2 home and 2 away games.

b. In the table in Exercise 1 part (a), the teams who play at home are marked with an "h."

c. In the table in Exercise 1 part (c), the teams who play at home are marked with an "h."

5.4. Hashing Functions

5.4.1. Let k be the six-digit number on the license plate of a car. We can assign this car the space numbered $h(k) \equiv k \pmod{101}$ where the spaces are numbered $0, 1, 2, \ldots, 100$. When a car is assigned the same space as another car we can assign it to the space $h(k) + g(k)$ where $g(k) \equiv k + 1 \pmod{99}$ and $0 < g(k) \leq 98$. When this space is occupied we next try $h(k) + 2g(k)$, then $h(k) + 3g(k)$, and so on. All spaces are examined since $(g(k), 101) = 1$.

5.4.3. a. It is clear that m memory locations will be probed as $j = 0, 1, 2, \ldots, m - 1$. To see that they are all distinct, and hence every memory location is probed, assume that $h_i(K) \equiv h_j(K) \pmod{m}$. Then $h(K) + iq \equiv h(K) + jq \pmod{m}$. From this it follows that $iq \equiv jq \pmod{m}$, and as $(q, m) = 1$, $i \equiv j \pmod{m}$ by corollary 3.4.1. And so $i = j$ since i and j are both less than m.

b. It is clear that m memory locations will be probed as $j = 0, 1, 2, \ldots, m - 1$. To see that they are all distinct, and hence every memory location is probed, assume that $h_i(K) \equiv h_j(K) \pmod{m}$. Then $h(K) + iq \equiv h(K) + jq \pmod{m}$. From this it follows that $iq \equiv jq \pmod{m}$, and as $(q, m) = 1$, $i \equiv j \pmod{m}$ by corollary 3.4.1. And so $i = j$ since i and j are both less than m.

5.4.5. We have $k_{11} = 137612044 \equiv 558 \pmod{4969}$ so that the files of the student with this social security number are assigned to location $h(k_{11}) = 558$. We find that $k_{12} = 505576452 \equiv 578 \pmod{4969}$, but location $h(k_{12}) = 578$ is taken, so we continue with the probing sequence $h_1(k_{12}) = h(k_{12}) + g(k_{12})$, where $g(k_{12}) \equiv 505576452 + 1 \equiv 424 \pmod{4967}$, so that $g(k_{12}) = 424$. We have $h_1(k_{12}) \equiv 578 + 424 = 1002 \pmod{4969}$. Since location 1002 is not occupied, we assign the files of the student with this social security number to location 1002. We find that $k_{13} = 157170996 \equiv 1526 \pmod{4969}$ but location 1526 is taken. We find that $g(k_{13}) \equiv 157170996 + 1 \equiv 216$. We probe locations $h_1(k_{13}) = h(k_{13}) + g(k_{13}) = 1742$ and $h_2(k_{13}) = h(k_{13}) + 2g(k_{13}) = 1958$, but they are taken. Finally, we probe once more and find that we can place the files of this student in location $h_3(k_{13}) = h(k_{13}) + 3g(k_{13}) = 2174$. Finally, we see that $k_{14} = 131220418 \equiv 4 \pmod{4969}$ so that we can place the files of this last student in location 4 which is not already taken.

5.5. Check Digits

5.5.1. a. Since $1+1+1+1+1+1 \equiv 0 \pmod{2}$, the check bit is 0.

b. Since $0+0+0+0+0+0 \equiv 0 \pmod{2}$, the check bit is 0.

c. Since $1+0+1+0+1+0 \equiv 1 \pmod{2}$, the check bit is 1.

d. Since $1+0+0+0+0+0 \equiv 1 \pmod{2}$, the check bit is 1.

e. Since $1+1+1+1+1+1+1+1 \equiv 0 \pmod{2}$, the check bit is 0.

f. Since $1+1+0+0+1+0+1+1 \equiv 1 \pmod{2}$, the check bit is 1.

5.5.3. a. The sum of the known digits is even, so the keep the sum of all digits even, we must have $? = 0$.

b. The sum of the known digits is odd, so $? = 1$.

c. The sum of the known digits is even, so $? = 0$.

5.5.5. a. We have $7 \cdot 1 + 3 \cdot 3 + 2 + 7 \cdot 9 + 3 \cdot 9 + 9 \equiv 7 \pmod{10}$, so the check digit is 7.

b. We have $7 \cdot 8 + 3 \cdot 0 + 5 + 7 \cdot 2 + 3 \cdot 3 + 7 \equiv 1 \pmod{10}$, so the check digit is 1.

c. We have $7 \cdot 6 + 3 \cdot 4 + 5 + 7 \cdot 1 + 3 \cdot 5 + 3 \equiv 4 \pmod{10}$, so the check digit is 4.

5.5.7. Here, transposition means that adjacent digits are in the wrong order. Suppose, first, that the first two digits, x_1 and x_2, or equivalently, the fourth and fifth digits are exchanged, and the error is not detected. Then $x_7 \equiv 7x_1 + 3x_2 + x_3 + 7x_4 + 3x_5 + x_6 \equiv 7x_2 + 3x_1 + x_3 + 7x_4 + 3x_5 + x_6 \pmod{10}$. It follows that $7x_1 + 3x_2 \equiv 7x_2 + 3x_1 \pmod{10}$ or $4x_1 \equiv 4x_2 \pmod{10}$. By Corollary 3.4.1, we see that $x_1 \equiv x_2 \pmod{5}$. This is equivalent to $\mid x_1 - x_2 \mid = 5$, as x_1 and x_2 are single digits. Similarly, if the second and third (or fifth and sixth) digits are transposed, we find that $2x_2 \equiv 2x_3 \pmod{10}$, which again reduces to $x_2 \equiv x_3 \pmod{5}$ by Corollary 3.4.1. Also, if the third and fourth digits are transposed, we find that $6x_3 \equiv 6x_4 \pmod{10}$ and $x_3 \equiv x_4 \pmod{5}$, similarly as before. The reverse argument will complete the proof.

5.5.9. a. We have $x_{10} \equiv 2 \cdot 1 + 1 \cdot 2 + 1 \cdot 3 + 3 \cdot 4 + 5 \cdot 5 + 4 \cdot 6 + 0 \cdot 7 + 0 \cdot 8 + 1 \cdot 9 \equiv 0 \pmod{11}$.

b. We have $x_{10} \equiv 0 \cdot 1 + 1 \cdot 2 + 9 \cdot 3 + 0 \cdot 4 + 8 \cdot 5 + 1 \cdot 6 + 0 \cdot 7 + 8 \cdot 8 + 2 \cdot 9 \equiv 3 \pmod{11}$.

c. We have $x_{10} \equiv 1 \cdot 1 + 2 \cdot 2 + 1 \cdot 3 + 2 \cdot 4 + 3 \cdot 5 + 9 \cdot 6 + 9 \cdot 7 + 4 \cdot 8 + 0 \cdot 9 \equiv 4 \pmod{11}$.

d. We have $x_{10} \equiv 0 \cdot 1 + 0 \cdot 2 + 7 \cdot 3 + 0 \cdot 4 + 3 \cdot 5 + 8 \cdot 6 + 1 \cdot 7 + 3 \cdot 8 + 3 \cdot 9 \equiv 10 (= X) \pmod{11}$.

5.5.11. a. We have $x_{10} \equiv 0 \cdot 1 + 3 \cdot 2 + 9 \cdot 3 + 4 \cdot 4 + 3 \cdot 5 + 8 \cdot 6 + 0 \cdot 7 + 4 \cdot 8 + 9 \cdot 9 \equiv 5 \pmod{11}$, which matches the check digit, so the ISBN is valid.

b. We have $x_{10} \equiv 1\cdot 1 + 0\cdot 2 + 9\cdot 3 + 2\cdot 4 + 3\cdot 5 + 1\cdot 6 + 2\cdot 7 + 2\cdot 8 + 1\cdot 9 \equiv 8 \pmod{11}$, so the ISBN is not valid.

c. We have $x_{10} \equiv 0\cdot 1 + 8\cdot 2 + 2\cdot 3 + 1\cdot 4 + 8\cdot 5 + 0\cdot 6 + 1\cdot 7 + 2\cdot 8 + 3\cdot 9 \equiv 6 \pmod{11}$, so the ISBN is valid.

d. We have $x_{10} \equiv 0\cdot 1 + 4\cdot 2 + 0\cdot 3 + 4\cdot 4 + 5\cdot 5 + 0\cdot 6 + 8\cdot 7 + 7\cdot 8 + 6\cdot 9 \equiv 10 \pmod{11}$, so the ISBN is valid.

e. We have $x_{10} \equiv 9\cdot 1 + 0\cdot 2 + 6\cdot 3 + 1\cdot 4 + 9\cdot 5 + 1\cdot 6 + 7\cdot 7 + 0\cdot 8 + 5\cdot 9 \equiv 0 \pmod{11}$, so the ISBN is not valid.

5.5.13. Computing the check digit for the incorrect ISBN yields $y_{10} \equiv 0\cdot 1 + 0\cdot 2 + 7\cdot 3 + 2\cdot 4 + 8\cdot 5 + 9\cdot 6 + 0\cdot 7 + 9\cdot 8 + 5\cdot 9 \equiv 9 \pmod{11}$. Then using the notation in the text, we have $(j-k)(x_k - x_j) \equiv 9 \pmod{11}$. Without loss of generality, assume that $j > k$. There are a number of possibilities to check. Let's suppose first that $j - k = 3$ and $x_k - x_j = 3$. We search for two digits which are three places apart and such that the second digit is 3 more than the first. Finding none, we suppose $j - k = 1$ and $x_k - x_j = 9$. We search for two consecutive digits such that the second is 9 more than the first. We find that the 7th and 8th digits satisfy these conditions and conclude that the correct ISBN is 0-07-289905-0. Had we been unsuccessful, we might have tried $j - k = 9$ or we might have replaced 9 by -2 or 20 or some other integer congruent to 9 modulo 11.

5.5.15. a. Using the congruence from Exercise 14, we compute $x_{12} \equiv -3(0 + 7 + 0 + 0 + 1 + 3) - (4 + 0 + 0 + 0 + 8) \equiv 5 \pmod{10}$. Since 5 is not the check digit for this UPC, the code is invalid.

b. Using the congruence from Exercise 14, we compute $x_{12} \equiv -3(3 + 1 + 0 + 0 + 0 + 8) - (1 + 0 + 0 + 1 + 3) \equiv 9 \pmod{10}$. Since 9 is the check digit for this UPC, the code is valid.

c. Using the congruence from Exercise 14, we compute $x_{12} \equiv -3(0 + 8 + 0 + 0 + 1 + 7) - (5 + 0 + 0 + 0 + 2) \equiv 5 \pmod{10}$. Since 5 is the check digit for this UPC, the code is valid.

d. Using the congruence from Exercise 14, we compute $x_{12} \equiv -3(2 + 6 + 0 + 0 + 1 + 9) - (2 + 5 + 0 + 1 + 7) \equiv 1 \pmod{10}$. Since 1 is not the check digit for this UPC, the code is invalid.

5.5.17. Let $x_1x_2x_3x_4x_5x_6x_7x_8x_9x_{10}x_{11}x_{12}$ be a correct UPC. Suppose that when the product is scanned, the numbers are read as $y_1y_2y_3y_4y_5y_6y_7y_8y_9y_{10}y_{11}y_{12}$, where $x_i = y_i$ if $i \neq k$, but $x_k \neq y_k$, for some k. Then, from the congruence in Exercise 14, $0 \equiv x_{12} - y_{12} \equiv -3(x_1 + x_3 + x_5 + x_7 + x_9 + x_{11}) - (x_2 + x_4 + x_6 + x_8 + x_{10}) + 3(y_1 + y_3 + y_5 + y_7 + y_9 + y_{11}) + (y_2 + y_4 + y_6 + y_8 + y_{10}) \equiv a\cdot(y_k - x_k) \pmod{10}$, where $a = 3$ or 1 according as k is odd or even. In either case, $(a, 10) = 1$, so we can divide the congruence by a to obtain $(y_k - x_k) \equiv 0 \pmod{10}$, which contradicts the assumption that $x_k \neq y_k$. We conclude that this code will always detect a single error.

5.5.19. a. Yes. If x_i is entered as y_i, then for both codewords to be valid, $x_i \equiv y_i \pmod{11}$. As x_i and y_i are single digits (less than 11), $x_i = y_i$.

b. No. We cannot detect any transpositions, as addition is commutative.

5.5.21. a. $x_{10} \equiv 8\cdot 1 + 4\cdot 1 + 5\cdot 0 + 10\cdot 4 + 3\cdot 9 + 2\cdot 1 + 7\cdot 2 + 6\cdot 3 + 9\cdot 8 \equiv 9 \pmod{11}$. $x_{11} \equiv 6\cdot 1 + 7\cdot 1 + 8\cdot 0 + 9\cdot 4 + 4\cdot 9 + 5\cdot 1 + 6\cdot 2 + 7\cdot 3 + 8\cdot 8 + 9\cdot 9 \equiv 4 \pmod{11}$.

b. If x_i is misentered as y_i, then if the congruence defining x_{10} holds, we see that $ax_i \equiv ay_i \pmod{11}$ by setting the two definitions of x_{10} congruent. From this, it follows by Corollary 3.4.1 that $x_i \equiv y_i \pmod{11}$ and so $x_i = y_i$. If the last digit, x_{11} is misentered as y_{11}, then the congruence defining x_{11} will hold if and only if $x_{11} = y_{11}$.

c. Suppose that x_i is misentered as y_i and x_j is misentered as y_j, with $i < j < 10$. Suppose both of the congruences defining x_{10} and x_{11} hold. Then by setting the two versions of each congruence congruent to each other we obtain $ax_i + bx_j \equiv ay_i + by_j \pmod{11}$ and $cx_i + dx_j \equiv cy_i + dy_j \pmod{11}$ where $a \neq b$ and $c \neq d$. If it is the case that $ad - bc \not\equiv 0 \pmod{11}$, then the coefficient matrix is invertible and we can multiply both sides of this system of congruences by the inverse to obtain $x_i =$

5.5.23. a. When we divide 00032781811224 by 7 we get a remainder of 1, so the check digit is $a_{15} = 1$.

b. When we divide 10238544122339 by 7 we get a remainder of 1, so the check digit is $a_{15} = 1$.

c. When we divide 00611133123278 by 7 we get a remainder of 6, so the check digit is $a_{15} = 6$.

5.5.25. Suppose an undetectable error is made in the ith digit, so that the incorrect digit b is written in place of the correct digit a_i. Then we must have $a_1 a_2 \cdots a_{14} \equiv a_1 a_2 \cdots b \cdots a_{14} \pmod{7}$ which reduces to $a_i 10^i \equiv b 10^i \pmod 7$. Since $(7, 10) = 1$, we can divide out the power of 10 and we have $a_i \equiv b \pmod 7$. Since $0 \leq b \leq 9$, the only undetectable errors are when we have one of the following substitutions: 0 for 7, 1 for 8, 2 for 9 or vice versa.

5.5.27. a. Since $3 \cdot 0 + 4 \cdot 3 + 5 \cdot 1 + 6 \cdot 7 + 7 \cdot 8 + 8 \cdot 4 + 9 \cdot 7 = 210 \equiv 1 \pmod{11}$, the check digit is 1.

b. Since $3 \cdot 0 + 4 \cdot 4 + 5 \cdot 2 + 6 \cdot 3 + 7 \cdot 5 + 8 \cdot 5 + 9 \cdot 5 = 164 \equiv 10 \pmod{11}$, the check digit is X.

c. Since $3 \cdot 1 + 4 \cdot 0 + 5 \cdot 6 + 6 \cdot 3 + 7 \cdot 6 + 8 \cdot 6 + 9 \cdot 9 = 222 \equiv 2 \pmod{11}$, the check digit is 2.

d. Since $3 \cdot 1 + 4 \cdot 3 + 5 \cdot 6 + 6 \cdot 3 + 7 \cdot 8 + 8 \cdot 3 + 9 \cdot 7 = 206 \equiv 8 \pmod{11}$, the check digit is 8.

5.5.29. Suppose two consecutive digits are transposed and the error is undetected. Then $3d_1 + 4d_2 + \cdots + 9d_7 \equiv 3d_1 + \cdots + (i+2)d_{i+1} + (i+3)d_i + \cdots + 9d_7 \pmod{11}$, which reduces to $(i+2)d_i + (i+3)d_{i+1} \equiv (i+2)d_{i+1} + (i+3)d_i \pmod{11}$, which in turn simplifies to $d_{i+1} \equiv d_i \pmod{11}$. So no error in fact existed. We conclude that all single transpositions of consecutive digits are detectable. (In fact all single transpositions are detectable.)

CHAPTER 6

Some Special Congruences

6.1. Wilson's Theorem and Fermat's Little Theorem

6.1.1. Note that $10! + 1 = 1(2 \cdot 6)(3 \cdot 4)(5 \cdot 9)(7 \cdot 8)10 + 1 = 1 \cdot 12 \cdot 12 \cdot 45 \cdot 56 \cdot 10 + 1 \equiv 1 \cdot 1 \cdot 1 \cdot 1 \cdot 1 \cdot (-1) + 1 \equiv 0 \pmod{11}$. Therefore 11 divides $10! + 1$.

6.1.3. By Wilson's theorem, we have $18 \equiv 18! \equiv 16!(17)(18) \equiv 16!(-2)(-1) \equiv 16!2 \pmod{19}$. Since $(2, 19) = 1$, we can divide both sides by 2 and get $9 \equiv 16! \pmod{19}$.

6.1.5. We see that $8 \cdot 9 \cdot 10 \cdot 11 \cdot 12 \cdot 13 \equiv 1 \cdot 2 \cdot 3 \cdot 4 \cdot 5 \cdot 6 = 6! \equiv -1 \pmod 7$, using Wilson's theorem for the last congruence.

6.1.7. Note that $437 = 19 \cdot 23$. From Wilson's theorem we have $18! \equiv -1 \pmod{19}$ and $22 \equiv 22! \pmod{23}$. Then $22 \equiv 22! \equiv 18!(-4)(-3)(-2)(-1) \equiv 18!(1) \pmod{23}$ Hence, $18! \equiv 22 \pmod{23}$. Now applying the Chinese remainder theorem to the system $x \equiv -1 \pmod{19}, x \equiv 22 \pmod{23}$ yields $x \equiv 436 \equiv -1 \pmod{437}$.

6.1.9. By the Division algorithm, we have $100 = 6 \cdot 16 + 4$. Then by Fermat's little theorem, $5^{100} \equiv 5^{6 \cdot 16 + 4} \equiv (5^6)^{16} \cdot 5^4 \equiv 1^{16} \cdot 5^4 \equiv 25^2 \equiv 4^2 \equiv 16 \equiv 2 \pmod 7$.

6.1.11. Since 999999999 is an odd multiple of 3, we know it is congruent to 3 modulo 6. So by Fermat's Little Theorem, we have $3^{999999999} \equiv 3^3 \equiv 27 \equiv -1 \pmod 7$.

6.1.13. We have $(3^5)^2 \equiv 243^2 \equiv 1^2 \equiv 1 \pmod{11^2}$.

6.1.15. a. Multiply both sides of $7x \equiv 12 \pmod{17}$ by 7^{15} to obtain $7^{16}x \equiv 7^{15} \cdot 12 \pmod{17}$. Since $7^{16} \equiv 1 \pmod{17}$ this gives $x \equiv 7^{15} \cdot 12 \equiv (7^3)^5 \cdot 12 \equiv 343^5 \cdot 12 \equiv 3^5 \cdot 12 \equiv 243 \cdot 12 \equiv 5 \cdot 12 \equiv 60 \equiv 9 \pmod{17}$.

b. Multiply both sides of $4x \equiv 11 \pmod{19}$ by 4^{17} to obtain $4^{18}x \equiv 4^{17} \cdot 11 \pmod{19}$. Since $4^{18} \equiv 1 \pmod{19}$, this gives $x \equiv 4^{17} \cdot 11 \equiv (4^2)^8 \cdot 4 \cdot 11 \equiv (-3)^8 \cdot 4 \cdot 11 \equiv ((-3)^4)^2 \cdot 4 \cdot 11 \equiv 81^2 \cdot 44 \equiv 5^2 \cdot 44 \equiv 6 \cdot 6 \equiv 17 \pmod{19}$.

6.1.17. Suppose that p is an odd prime. Then Wilson's theorem tells us that $(p-1)! \equiv -1 \pmod p$. Since $(p-1)! = (p-3)!(p-1)(p-2) \equiv (p-3)!(-1)(-2) \equiv 2 \cdot (p-3)! \pmod p$ this implies that $2 \cdot (p-3)! \equiv -1 \pmod p$.

6.1.19. Since $(a, 35) = 1$, we have $(a, 7) = (a, 5) = 1$, so we may apply Fermat's little theorem to get $a^{12} - 1 \equiv (a^6)^2 - 1 \equiv 1^2 - 1 \equiv 0 \pmod 7$, and $a^{12} - 1 \equiv (a^4)^3 - 1 \equiv 1^3 - 1 \equiv 0 \pmod 5$. Since both 5 and 7 divide $a^{12} - 1$, then 35 must also divide it.

6.1.21. When n is even, so is n^7, and when n is odd, so is n^7. It follows that $n^7 \equiv n \pmod 2$. Furthermore, since $n^3 \equiv n \pmod 3$, it follows that $n^7 = (n^3)^2 \cdot n \equiv n^2 \cdot n \equiv n^3 \equiv n \pmod 3$. We also know by Fermat's little theorem that $n^7 \equiv n \pmod 7$. since $42 = 2 \cdot 3 \cdot 7$, it follows that $n^7 \equiv n \pmod{42}$.

6.1.23. By Fermat's little theorem, $\sum_{k=1}^{p-1} k^{p-1} \equiv \sum_{k=1}^{p-1} 1 \equiv p - 1 \pmod p$.

6.1.25. By Fermat's little theorem we have $a \equiv a^p \equiv b^p \equiv b \pmod p$, hence $b = a + kp$ for some integer k. Then by the binomial theorem, $b^p = (a + kp)^p = a^p + \binom{p}{1}a^{p-1}kp + p^2N$ where N is some integer. Then

$b^p \equiv a^p + p^2ak + p^2N \equiv a^2 \pmod{p^2}$, as desired.

6.1.27. Using computational software, we find $r_2 = 4, r_3 = 64, r_4 \equiv 2114982 \pmod{7331117}, r_5 \equiv 2937380 \pmod{7331117}, r_6 \equiv 6924877 \pmod{7331117}, r_7 \equiv 3828539 \pmod{7331117}$, and $r_8 \equiv 4446618 \pmod{7331117}$. We have $(r_i - 1, 7331117) = 1$, for $i = 1, 2, \ldots, 7$, but $(r_8 - 1, 7331117) = 641$, so this is a factor of 7331117.

6.1.29. Suppose that p is prime. Then by Fermat's little theorem for every integer $a, a^p \equiv a \pmod{p}$ and by Wilson's theorem $(p-1)! \equiv -1 \pmod{p}$ so that $a(p-1)! \equiv -a \pmod{p}$. It follows that $a^p + (p-1)!a \equiv a + (-a) \equiv 0 \pmod{p}$. Consequently $p \mid [a^p + (p-1)!a]$.

6.1.31. Since $p - 1 \equiv -1, p - 2 \equiv -2, \ldots, (p-1)/2 \equiv (p-1)/2 \pmod{p}$, we have $((p-1)/2)!^2 \equiv -(p-1)! \equiv 1 \pmod{p}$, (since $p \equiv 3 \pmod 4$ the minus signs work out.) If $x^2 \equiv 1 \pmod{p}$, then $p \mid x^2 - 1 = (x-1)(x+1)$, so $x \equiv \pm 1 \pmod{p}$.

6.1.33. Suppose that $p \equiv 1 \pmod 4$. Let $y = \pm[(p-1)/2]!$. Then $y^2 \equiv [(p-1)/2]!^2 \equiv [(p-1)/2]!^2(-1)^{(p-1)/2} \equiv (1 \cdot 2 \cdot 3 \cdots (p-1)/2)(-1 \cdot (-2) \cdot (-3) \cdots (-(p-1)/2)) \equiv 1 \cdot 2 \cdot 3 \cdots (p-1)/2 \cdot (p+1)/2 \cdots (p-3)(p-2)(p-1) = p! \equiv -1 \pmod{p}$, where we have used Wilson's theorem. Now suppose that $x^2 \equiv -1 \pmod{p}$. Then $x^2 \equiv y^2 \pmod{p}$ where $y = [(p-1)/2]!$. Hence $(x^2 - y^2) = (x-y)(x+y) \pmod{p}$. It follows that $p \mid (x-y)$ or $p \mid (x+y)$ so that $x \equiv \pm \pmod{p}$.

6.1.35. If n is composite and $n \neq 4$, then Exercise 16 shows that $(n-1)!/n$ is an integer, so $[((n-1)! + 1)/n - [(n-1)!/n]] = [(n-1)!/n + 1/n - (n-1)!/n] = [1/n] = 0$ and if $n = 4$, then the same expression is also equal to 0. But if n is prime, then by Wilson's Theorem $(n-1)! = Kn - 1$ for some integer K. So $[((n-1)! + 1)/n - [(n-1)!/n]] = [(Kn - 1 + 1)/n - [(Kn - 1)/n]] = [K - (K - 1)] = 1$. Therefore, the sum increases by 1 exactly when n is prime, so it must be equivalent to $\pi(x)$.

6.1.37. Suppose that n and $n + 2$ are twin primes. Then since n is prime by Wilson's theorem we know that $(n-1)! \equiv -1 \pmod{n}$. Hence $4[(n-1)! + 1] + n \equiv 4 \cdot 0 + n \equiv 0 \pmod{n}$. Also, since $n + 2$ is prime by Wilson's theorem it follows that $(n+1)! \equiv -1 \pmod{n+2}$, so that $(n+1)n \cdot (n-1)! \equiv (-1)(-2)(n-1)! \equiv 2(n-1)! \equiv -1 \pmod{n+2}$. Hence $4[(n-1)! + 1] + n \equiv 2(2 \cdot (n-1)!) + 4 + n \equiv 2 \cdot (-1) + 4 + n = n + 2 \equiv 0 \pmod{n+2}$. Since $(n, n+2) = 1$ it follows that $4[(n-1)! + 1] + n \equiv 0 \pmod{n(n+2)}$.

6.1.39. We have $1 \cdot 2 \cdots (p-1) \equiv (p+1)(p+2) \cdots (2p-1) \pmod{p}$. Each factor is prime to p, so $1 \equiv ((p+1)(p+2) \cdots (2p-1))/(1 \cdot 2 \cdots (p-1)) \pmod{p}$. Thus $2 \equiv ((p+1)(p+2) \cdots (2p-1)2p)/(1 \cdot 2 \cdots (p-1)p) \equiv \binom{2p}{p} \pmod{p}$.

6.1.41. We first note that $1^p \equiv 1 \pmod{p}$. Now suppose that $a^p \equiv a \pmod{p}$. then by Exercise 40 we see that $(a+1)^p \equiv a^p + 1 \pmod{p}$. But by the inductive hypothesis $a^p \equiv a \pmod{p}$ we see that $a^p + 1 \equiv a + 1 \pmod{p}$. Hence $(a+1)^p \equiv a + 1 \pmod{p}$. This completes the inductive step of the proof.

6.1.43. a. If $c < 26$ then c cards are put into the deck above the card, so it ends up in the $2c$th position and $2c < 52$, so $b = 2c$. If $c \geq 26$ then $c - 1$ cards are put into the deck above the card, but 26 cards are taken away above it, so it ends up in the $b = (c - 26 + c - 1)$th place. Then $b = 2c - 25 \equiv 2c \pmod{53}$.

b. 52.

6.1.45. Assume without loss of generality that $a_p \equiv b_p \equiv 0 \pmod{p}$. Then, by Wilson's theorem, $a_1 a_2 \cdots a_{p-1} \equiv b_1 b_2 \cdots b_{p-1} \equiv -1 \pmod{p}$. Then $a_1 b_1 \cdots a_{p-1} b_{p-1} \equiv (-1)^2 \equiv 1 \pmod{p}$. If the set were a complete system, the last product would be $\equiv -1 \pmod{p}$.

6.1.47. The basis step is omitted. Assume $(p-1)^{p^{k-1}} \equiv -1 \pmod{p^k}$. Then, $(p-1)^{p^k} \equiv ((p-1)^{p^{k-1}})^p \equiv (-1 + mp^k)^p \equiv -1 + \binom{p}{1}mp^k + \cdots + (mp^k)^p \equiv -1 \pmod{p^{k+1}}$, where we have used the fact that $p \mid \binom{p}{j}$ for $j \neq 0$ or p.

6.2. Pseudoprimes

6.2.1. We find that $3^{90} = (3^4)^{22} \cdot 3^2 = 81^4 \cdot 9 \equiv (-10) \cdot 9 = -90 \equiv 1 \pmod{91}$. Hence 91 is a pseudoprime modulo 3.

6.2.3. Note that $2^{262} \equiv 2 \pmod{161038}$. Then $2^{161038} \equiv 2^{262 \cdot 614 + 170} \equiv 2^{614 + 170} \equiv 2 \pmod{161038}$.

6.2.5. From the Binomial Theorem, $(n-a)^n \equiv (-a)^n \equiv -(a^n) \equiv -a \equiv (n-a) \pmod{n}$, where we used $a^n \equiv a \pmod{n}$.

6.2.7. Raise the congruence $2^{2^n} \equiv -1 \pmod{F_n}$ to the $2^{2^n - n}$th power.

6.2.9. Suppose that n is a pseudoprime to the bases a and b. Then $b^n \equiv b \pmod{n}$ and $a_n \equiv a \pmod{n}$. It follows that $(ab)^n \equiv a^n b^n \equiv ab \pmod{n}$. Hence n is a pseudoprime to the base ab.

6.2.11. a. If $(ab)^{n-1} \equiv 1 \pmod{n}$, then, $1 \equiv a^{n-1} b^{n-1} \equiv 1 \cdot b^n \pmod{n}$ which implies that n is a pseudoprime to the base b.

  b. Let $a_1, a_2, \ldots, a_r$ be the bases to which n is a pseudoprime and for which $(a_i, n) = 1$ for each i. Then, by part (a), we know that, for each i, n is not a pseudoprime to the base ba_i. Thus, we have $2r$ different elements relatively prime to n. Then by the definition of $\phi(n)$, we have $r \leq \phi(n)/2$.

6.2.13. From $2^{18} \equiv 1 \pmod{1387}$ we get $2^{1387} \equiv 2 \pmod{1387}$ so 1387 is a pseudoprime. But $1387 - 1 = 2 \cdot 693$ and $2^{693} \equiv 512 \pmod{1387}$, which is all that must be checked, since $s = 1$. Thus 1387 fails Miller's test and hence is not a strong pseudoprime.

6.2.15. $25326001 = 2^4 1582875 = 2^s t$ and with this value of t, $2^t \equiv -1 \pmod{25326001}$, $3^t \equiv -1 \pmod{25326001}$, and $5^t \equiv 1 \pmod{25326001}$.

6.2.17. Suppose $c = 7 \cdot 23 \cdot q$, with q and odd prime, is a Carmichael number. Then by Theorem 6.7 we must have $(7-1)|(c-1)$, so $c = \equiv 7 \cdot 23 \cdot q \equiv 1 \pmod{6}$. Solving this yields $q \equiv 5 \pmod{6}$. Also, we must have $(23-1)|(c-1)$, so $c = \equiv 7 \cdot 23 \cdot q \equiv 1 \pmod{22}$. Solving this yields $q \equiv 19 \pmod{22}$ If we apply the Chinese remainder theorem to these two congruences we obtain $q \equiv 41 \pmod{66}$, that is $q = 41 + 66k$. Then we must have $(q-1)|(c-1)$, which is $(40 + 66k)|(7 \cdot 23 \cdot (41 + 66k) - 1)$. So there is an integer m such that $m(40 + 66k) = 6600 + 10626k = 160 + 6440 + 10626k = 160 + 161(40 + 66k)$. Therefore 160 must be a multiple of $40 + 66k$, which happens only when $k =$. Therefore $q = 41$ is the only such prime.

6.2.19. We have $321197185 - 1 = 321197184 = 4 \cdot 80299296 = 18 \cdot 17844288 = 22 \cdot 14599872 = 28 \cdot 11471328 = 36 \cdot 8922144 = 136 \cdot 2361744$, so $p-1|321197185 - 1$ for every prime p which divides 321197185. Therefore, by Theorem 6.7, 321197185 is a Carmichael number.

6.2.21. We can assume that $b < n$. Then b has fewer than $\log_2 n$ bits. Also, $t < n$ so it has fewer than $\log_2 n$ bits. It takes at most $\log_2 n$ multiplications to calculate b^{2^s} so it takes $O(\log_2 n)$ multiplications to calculate $b^{2^{\log_2 t}} = b^t$. Each multiplication is of two $\log_2 n$ bit numbers, and so takes $O((\log_2 n)^2)$ operations. So all together we have $O((\log_2 n)^3)$ operations.

6.3. Euler's Theorem

6.3.1. a. The set $1, 5$ is a reduced residue set modulo 6.

  b. The set $1, 2, 4, 5, 7, 8$ is a reduced residue set modulo 9.

  c. The set $1, 3, 7, 9$ is a reduced residue set modulo 10.

  d. The set $1, 3, 5, 9, 11, 13$ is a reduced residue set modulo 14.

e. The set $1, 3, 5, 7, 9, 11, 13, 15$ is a reduced residue set modulo 16.

f. The set $1, 2, 3, 4, 5, 6, 7, 8, 9, 10, 11, 12, 13, 14, 15, 16$ is a reduced residue set modulo 17.

6.3.3. If $(a, m) = 1$, then $(-a, m) = 1$, so $-c_i$ must appear among the c_j. Also $c_i \not\equiv -c_i \pmod{m}$, else $2c_i \equiv 0 \pmod{m}$ and so $(c_i, m) \neq 1$.

6.3.5. Since $\phi(10) = 4$, we have, by Euler's theorem, $3^{1000} \equiv (3^4)^{250} \equiv 1^{250} \equiv 1 \pmod{10}$. Therefore the last decimal digit of 3^{1000} is 1.

6.3.7. By Euler's theorem $3^{\phi(35)} = 3^{28} \equiv 1 \pmod{35}$. Since $100000 = 2857 \cdot 35 + 5$, it follows that $3^{100000} \equiv (3^{28})^{2857} \cdot 3^5 \equiv 1 \cdot 3^5 = 81 \equiv 11 \pmod{35}$.

6.3.9. Since $a^2 \equiv 1 \pmod{8}$ whenever a is odd, it follows that $a^{12} \equiv 1 \pmod{8}$ whenever $(a, 32760) = 1$. Euler's theorem tells us that $a^{\phi(9)} = a^6 \equiv 1 \pmod{9}$ whenever $(a, 9) = 1$, so that $a^{12} = (a^6)^2 \equiv 1 \pmod{9}$ whenever $(a, 32760) = 1$. Furthermore, Fermat's little theorem tells us that $a^4 \equiv 1 \pmod{5}$ whenever $(a, 5) = 1$, $a^6 \equiv 1 \pmod{7}$ whenever $(a, 7) = 1$, and $a^{12} \equiv 1 \pmod{13}$ whenever $(a, 13) = 1$. It follows that $a^{12} = (a^4)^3 \equiv 1 \pmod{5}$, $a^{12} = (a^6)^2 \equiv 1 \pmod{7}$, and $a^{12} \equiv 1 \pmod{13}$ whenever $(a, 32760) = 1$. Since $32760 = 2^3 3^2 \cdot 5 \cdot 7 \cdot 13$ and the moduli $8, 9, 5, 7,$ and 13 are pairwise relatively prime, we see that $a^{12} \equiv 1 \pmod{32760}$.

6.3.11. a. We multiply both sides of the congruence $5x \equiv 3 \pmod{14}$ by $5^{\phi(14)-1} = 5^5$ to obtain $5^6 x \equiv 5^5 \cdot 3 \pmod{14}$. Since $5^6 \equiv 1 \pmod{14}$ by Euler's theorem, it follows that $x \equiv 5^5 \cdot 3 \equiv (5^2)^2 \cdot 5 \cdot 3 \equiv 11^2 \cdot 15 \equiv 9 \cdot 1 \equiv 9 \pmod{14}$.

b. We multiply both sides of the congruence $4x \equiv 7 \pmod{15}$ by $4^{\phi(15)-1} = 4^7$ to obtain $4^8 x \equiv 4^8 \cdot 7 \pmod{15}$. since $4^8 \equiv 1 \pmod{15}$ by Euler's theorem, it follows that $x \equiv 4^7 \cdot 7 \equiv (4^2)^3 \cdot 4 \cdot 7 \equiv 1 \cdot 28 \equiv 13 \pmod{15}$.

c. We multiply both sides of the congruence $3x \equiv 5 \pmod{16}$ by $3^{\phi(16)-1} = 3^7$ to obtain $3^8 x \equiv 3^7 \cdot 5 \pmod{16}$. since $3^8 \equiv 1 \pmod{16}$ by Euler's theorem, it follows that $x \equiv 3^7 \cdot 5 \equiv 3^4 \cdot 3^3 \cdot 5 \equiv 1 \cdot 27 \cdot 5 \equiv 11 \cdot 5 \equiv 7 \pmod{16}$.

6.3.13. a. We have $x \equiv 4 \cdot 17^{10} + 3 \cdot 11^{16} \equiv 27 \pmod{187}$.

b. We have $x \equiv 1 \cdot 15^1 + 2 \cdot 10^2 + 3 \cdot 6^4 \equiv 23 \pmod{30}$.

c. We have $x \equiv 0 \cdot 105^1 + 0 \cdot 70^2 + 1 \cdot 42^4 + 6 \cdot 30^6 \equiv 6 \pmod{210}$.

d. We have $x \equiv 2 \cdot 50388^{10} + 3 \cdot 46189^4 + 4 \cdot 42636^{12} + 5 \cdot 32604^{16} + 6 \cdot 29172^{18} \equiv 150999 \pmod{554268}$.

6.3.15. We have $\phi(10) = 4$, so $7^4 \equiv 1 \pmod{10}$ and $7^{1000} \equiv (7^4)^{250} \equiv 1^{250} \equiv 1 \pmod{10}$.

6.3.17. We note that $\phi(p) = p - 1$ if p is prime, so $\phi(13) = 12, \phi(17) = 16,$ and $\phi(19) = 18$. Since the integers relatively prime to 16 are the odd integers, we see that $\phi(16) = 8$. The integers relatively prime to 14 are the odd integers not divisible by 7. We see that $\phi(14) = 6$. The integers relatively prime to 15 are those not divisible by either 3 or 5. we see that $\phi(15) = 8$. The integers relatively prime to 18 are the odd integers not divisible by 3. It follows that $\phi(18) = 6$. the integers relatively prime to 20 are the odd integers not divisible by 5. It follows that $\phi(20) = 8$.

6.3.19. If $(a, b) = 1$ and $(a, b - 1) = 1$ then $a \mid (b^{k\phi(a)} - 1)/(b - 1)$ which is a base b repunit. If $(a, b - 1) = d > 1$, then d divides any repunit of length $k(b - 1)$, and $(a/d) \mid (b^{k\phi(a/d)} - 1)/(b - 1)$ and these sets intersect infinitely often.

CHAPTER 7
Multiplicative Functions

7.1. The Euler Phi-Function

7.1.1. a. Since for all positive integers m and n, $f(mn) = 0 = 0 \cdot 0 = f(m) \cdot f(n)$, f is completely multiplicative.

  b. Since $f(6) = 2$, but $f(2) \cdot f(3) = 2 \cdot 2 = 4$, f is not completely multiplicative.

  c. Since $f(6) = 3$, but $f(2) \cdot f(3) = \frac{2}{2} \cdot \frac{3}{2} = \frac{3}{2}$, f is not completely multiplicative.

  d. Since $f(4) = \log(4) > 1$, but $f(2) \cdot f(2) = \log(2) \cdot \log(2) < 1$, f is not completely multiplicative.

  e. Since for any positive integers m and n, $f(mn) = (mn)^2 = m^2 n^2 = f(m) \cdot f(n)$, f is completely multiplicative.

  f. Since $f(4) = 4! = 24$, but $f(2) \cdot f(2) = 2!2! = 4$, f is not completely multiplicative.

  g. Since $f(6) = 7$, but $f(2) \cdot f(3) = 4 \cdot 3 = 12$, f is not completely multiplicative

  h. Since $f(4) = 4^4 = 256$, but $f(2) \cdot f(2) = 2^2 2^2 = 16$, f is not completely multiplicative.

  i. Since for any positive integers m and n, $f(mn) = \sqrt{mn} = \sqrt{m}\sqrt{n} = f(m) \cdot f(n)$, f is completely multiplicative.

7.1.3. We have the following prime factorizations of 5186, 5187, and 5188: $5186 = 2 \cdot 2593$, $5187 = 3 \cdot 7 \cdot 13 \cdot 19$, and $5188 = 2^2 1297$. Hence $\phi(5186) = \phi(2)\phi(2593) = 1 \cdot 2592 = 2592$, $\phi(5187) = \phi(3)\phi(7)\phi(13)\phi(19) = 2 \cdot 6 \cdot 12 \cdot 18 = 2592$, and $\phi(5188) = \phi(2^2)\phi(1297) = 2 \cdot 1296 = 2592$. It follows that $\phi(5186) = \phi(5187) = \phi(5188)$.

7.1.5. If $\phi(n) = 6$, and suppose k distinct primes divide n. Then either $k = 2$ and $p_1^{a_1} - p_1^{a_1-1} = 3$, which is impossible, or $k = 1$ and $p_1^{a_1} - p_1^{a_1-1} = 6$, so $p_1^{a_1} = 9$ and $n = 18$ or $p_1^{a_1} = 7$ and $n = 14$, or $k = 0$ and $p_1^{a_1} - p_1^{a_1-1} = 6$ and $p_1^{a_1} = 9 = n$ or $p_1^{a_1} = 7$ and $n = 7$. So the only solutions are $n = 7, 9, 14,$ or 18.

7.1.7. If $\phi(n) = 24$, we have 5 cases as $k = 0, 1, 2, 3$ or 4. Note that if $p^a - p^{a-1} = 2^m$, then $a = 1$ and $p = 2^m = 1$. Also note that $p^a - p^{a-1}$ is always even. In every case, $3 \mid p_1^{a_1} - p_1^{a_1-1}$. Every other factor in the formula for $\phi(n)$ is of the form 2^{k-1} or $p - 1$ where p is a Mersenne prime. If $k = 4$, then $p_1^{a_1} - p_1^{a_1-1} = 3$ which is impossible. If $k = 3$, then $\phi(n) = \phi(8 \cdot m) = \phi(8)\phi(m) = 4\phi(m)$, so $\phi(m) = 6$. By part (d) $m = 7$ or 9 so $n = 56$ or 72. If $k = 2$, then $\phi(n) = \phi(4 \cdot m) = \phi(4)\phi(m) = 2\phi(m)$, so $\phi(m) = 12$. Since $p_1^{a_1} - p_1^{a_1-1} \neq 3$, $p_1^{a_1} - p_1^{a_1-1} = 6$ or 12 so $p_1^{a_1} = 13$ and $n = 52$ or $p_1^{a_1} = 9$ and $p_2^{a_2} - p_2^{a_2-1} = 2$ which is impossible. If $k = 1$, then $\phi(n) = \phi(2m) = \phi(m)$ so the case $k = 0$ is covered here also. We have $p_1^{a_1} - p_1^{a_1-1} = 24$ or $(p_1^{a_1} - p_1^{a_1-1})(p_2^{a_2} - p_2^{a_2-1}) = 24$. In the first case we have $p_1 = 3$ since $3 \mid 24$ and this leads to $p_2 = 13$ so $n = 39$ or 78. In the second case, either $p_1 = 3$ and we have $(3-1)(p_2^{a_2} - p_2^{a_2-1}) = 24$ so $p_2 = 13$ as in the last case, or $p_1 = 5$ and $(5-1)(p_2^{a_2} - p_2^{a_2-1}) = 24$, so $p_2^{a_2} = 9$ or 7 which leads to $n = 45$ and 35 respectively if $k = 0$ and 90 and 70 if $k = 1$. Then the totality of all solutions is 35, 39, 45, 52, 56, 70, 72, 78, 84, and 90.

7.1.9. Studying Table E.2 on page 609 and 610, we discover that the nth term of this sequence is given by $\phi(2n)$.

7.1.11. Let $n = 3^k m$, where $(3, m) = 1$. If $k = 0$, then $\phi(3n) = 2\phi(n) \neq 3\phi(n)$. On the other hand, if $k \geq 1$, then $\phi(3n) = \phi(3^{k+1}m) = (3^{k+1} - 3^k)\phi(m) = 3(3^k - 3^{k-1})\phi(m) = 3\phi(3^k m) = 3\phi(n)$. Therefore, $\phi(3n) = 3\phi(n)$ if and only if $3 \mid n$.

7.1.13. If $n = 2^k p_1^{a_1} \cdots p_r^{a_r}$ then $\phi(n) = n(p_1 - 1)/p_1 \cdots (p_r - 1)/p_r$. If $\phi(n) = n/2$, we have $(p_1 - 1)/p_1 \cdots (p_r - 1)/p_r = 1/2$. Let p_r be the largest prime dividing n, then p_r divides none of $p_1 - 1, p_2 - 1, \cdots, p_r - 1$, so it must appear in the denominator of $(p_1 - 1)/p_1 \cdots (p_r - 1)/p_r$ in lowest terms. But $1/2$ is in lowest terms, therefore n has no odd prime divisors. Since $\phi(2^k) = 2^{k-1} = 2^k/2$, for $k = 1, 2, \cdots$ we have $n = 2, 2^2, \cdots$ as the only solutions.

7.1.15. If n is odd, then $(2, n) = 1$ and $\phi(2n) = \phi(2)\phi(n) = 1 \cdot \phi(n) = \phi(n)$. If n is even, say $n = 2^s t$ with t odd. Then $\phi(2n) = \phi(2^{s+1}t) = \phi(2^{s+1})\phi(t) = 2^s \phi(t) = 2(2^{s-1}\phi(t)) = 2(\phi(2^s)\phi(t)) = 2(\phi(2^s t)) = 2\phi(n)$.

7.1.17. If $\phi(n)$ is a power of 2 then every factor $p_i^{a_i} - p_i^{a_i - 1} = p_i^{a_i - 1}(p_i - 1)$ must be a power of 2. Then either $p_i = 2$ or $a_i = 1$ and $p_i - 1 = 2^{k_i}$ and so p_i is a Mersenne prime. Therefore $\phi(n)$ is a power of 2 if and only if $n = 2^k p_1 p_2 \cdots p_r$ where each p_i is a Mersenne prime.

7.1.19. Let $n = p_1^{a_1} \cdots p_r^{a_r}$ be the factorization for n. If $n = 2\phi(n)$ then $p_1^{a_1} \cdots p_r^{a_r} = 2\prod_{j=1}^{r} p_j^{a_j - 1}(p_j - 1)$. Cancelling the powers of all p_j's yields $p_1 \cdots p_r = 2\prod_{j=1}^{r}(p_j - 1)$. If any p_j is an odd prime, then the factor $(p_j - 1)$ is even and must divide the product on the left-hand side. But there can be at most one factor of 2 on the left-hand side and it is accounted for by the factor of 2 in front of the product on the right hand side. Therefore, no odd primes appear in the product. That is, $n = 2^j$ for some j.

7.1.21. Since $(m, n) = p$, p divides one of the terms, say n exactly once, so $n = kp$ with $(m, k) = 1 = (n, k)$. Then $\phi(n) = \phi(kp) = \phi(k)\phi(p) = \phi(k)(p - 1)$, and $\phi(mp) = p\phi(m)$ by the formula in Example 7.7. Then, $\phi(mn) = \phi(mkp) = \phi(mp)\phi(k) = (p\phi(m))(\phi(n)/(p - 1))$.

7.1.23. Let $p_1, \cdots, p_r$ be those primes dividing a but not b. Let $q_1, \cdots, q_s$ be those primes dividing b but not a. Let $r_1, \cdots r_t$ be those primes dividing a and b. Let $P = \prod(1 - \frac{1}{p_i})$, $Q = \prod(1 - \frac{1}{q_i})$ and $R = \prod(1 - \frac{1}{r_i})$. Then we have $\phi(ab) = abPQR = \frac{aPRbQR}{R} = \frac{\phi(a)\phi(b)}{R}$. But $\phi((a,b)) = (a,b)R$ so $R = \frac{\phi((a,b))}{(a,b)}$ and we have $\phi(ab) = \frac{\phi(a)\phi(b)}{R} = \frac{(a,b)\phi(a)\phi(b)}{\phi((a,b))}$ as desired.

7.1.25. From the formula for the ϕ function, we see that if $p \mid n$, then $p - 1 \mid k$. Since k has only finitely many divisors, there are only finitely many possibilities for prime divisors of n. Further, if p is prime and $p^a \mid n$, then $p^{a-1} \mid k$. Hence, $a \leq \log_p(k) + 1$. Therefore, each of the finitely many primes which might divide n may appear to only finitely many exponents. Therefore, there are only finitely many possibilities for n.

7.1.27. From the formula for the ϕ function, we see that if $p \mid n$, then $p - 1 \mid k$. Since k has only finitely many divisors, there are only finitely many possibilities for prime divisors of n. Further, if p is prime and $p^a \mid n$, then $p^{a-1} \mid k$. Hence, $a \leq \log_p(k) + 1$. Therefore, each of the finitely many primes which might divide n may appear to only finitely many exponents. Therefore, there are only finitely many possibilities for n.

7.1.29. As suggested, we take $k = 2 \cdot 3^{6j+1}$ with $j \geq 1$, and suppose that $\phi(n) = k$. From the formula for $\phi(n)$ we see that $\phi(n)$ has a factor of $(p - 1)$, which is even for every odd prime that divides n. Since there is only one factor of 2 in k, there is at most one odd prime divisor of n. Further, since $2 \parallel k$, we know that $4 \nmid n$. Since k is not a power of 2, we know that an odd prime p must divide n. So n is of the form p^a or $2p^a$. Recall that $\phi(p^a) = \phi(2p^a)$. It remains to discover the value of p. If $a = 1$, then $\phi(p^a) = p - 1 = 2 \cdot 3^{6j+1}$. But then, $p = 2 \cdot 3^{6j+1} + 1 \equiv 6 \cdot (3^6)^j + 1 \equiv (-1)(1)^j + 1 \equiv 0 \pmod 7$. Hence $p = 7$. But $\phi(7) = 6 = 2 \cdot 3^{6j+1}$ implies that $j = 0$, contrary to hypothesis, so this is not a solution. Therefore $a > 1$ and we have $\phi(p^a) = (p - 1)p^{a-1} = 2 \cdot 3^{6j+1}$, from which we conclude that $p = 3$ and $a = 6j + 2$. Therefore the only solutions are $n = p^{6j+2}$ and $n = 2p^{6j+2}$.

7.1. THE EULER PHI-FUNCTION

7.1.31. If $n = p^r m$, then $\phi(p^r m) = (p^r - p^{r-1})\phi(m) \mid (p^r m - 1)$, hence $p \mid 1$ or $r = 1$. So n is square-free. If $n = pq$, then $\phi(pq) = (p-1)(q-1) \mid (pq-1)$. Then $(p-1) \mid (pq-1) - (p-1)q = q - 1$. Similarly $(q-1) \mid (p-1)$, a contradiction.

7.1.33. Let $n = p_1^{a_1} p_2^{a_2} \cdots p_k^{a_k}$. Let P_i be the property that an integer is divisible by p_i. Let S be the set $\{1, 2, \ldots, n-1\}$. To compute $\phi(n)$ we need to correct the elements of S with more of the properties $P_1, P_2, \cdots, P_k$. Let $n(P_{i_1}, P_{i_2}, \cdots, P_{i_m})$ be the number of elements of S with all of properties $P_{i_1}, P_{i_2}, \cdots, P_{i_m}$. Then $n(p_{i_1}, \cdots P_{i_m}) = \frac{n}{p_{i_1} p_{i_2} \cdots p_{i_m}}$. By Exercise 18 of Section 1.4, we have $\phi(n) = n - (\frac{n}{p_1} + \frac{n}{p_2} + \cdots + \frac{n}{p_k}) + (\frac{n}{p_1 p_2} + \cdots + \frac{n}{p_{k-1} p_k}) + \cdots + (-1)^k (\frac{n}{p_1 \cdots p_k}) = n(1 - \sum_{p_i \mid n} \frac{1}{p_i} + \sum_{p_{i_1} p_{i_2} \mid n} \frac{1}{p_{i_1} p_{i_2}} - \sum_{p_{i_1} p_{i_2} p_{i_3}} \frac{1}{p_{i_1} p_{i_2} p_{i_3}} + \cdots + (-1)^k \frac{n}{p_1} \cdots p_k)$. On the other hand, notice that each term in the expansion of $(1 - \frac{1}{p_1})(1 - \frac{1}{p_2}) \cdots (1 - \frac{1}{p_k})$ is obtained by choosing either 1 or $-\frac{1}{p_i}$ from each factor and multiply the choice together. This gives each term the form $\frac{(-1)^m}{p_{i_1} p_{i_2} \cdots p_{i_m}}$. Note that each term can occur in only one way. Thus $n(1 - \frac{1}{p_1})(1 - \frac{1}{p_2}) \cdots (1 - \frac{1}{p_k}) = n(1 - \sum_{p_i \mid n} \frac{1}{p_i} + \sum_{p_{i_1} p_{i_2}} \frac{1}{p_{i_1} p_{i_2}} - \cdots (-1)^k \frac{n}{p_1 \cdots p_k}) = \phi(n)$.

7.1.35. Note that $1 \leq \phi(m) \leq m - 1$ for $m > 1$. Hence if $n \geq 2$, $n > n_1 > n_2 > \cdots \geq 1$ where $n_i = \phi(n)$ and $n_i = \phi(n_{i-1})$ for $i > 1$. Since $n_i, i = 1, 2, 3, \ldots$ is a decreasing sequence of positive integers, there must be a positive integer r such that $n_r = 1$.

7.1.37. Note that the definition of $f * g$ can also be expressed as $(f * g)(n) = \sum_{a \cdot b = n} f(a)g(b)$. Then the fact that $f * g = g * f$ is evident.

7.1.39. a. If either $m > 1$ or $n > 1$ then $mn > 1$ and one of $\iota(m)$ or $\iota(n)$ is equal to zero. Then $\iota(mn) = 0 = \iota(m)\iota(n)$. Otherwise, $m = n = 1$ and we have $\iota(mn) = 1 = 1 \cdot 1 = \iota(m)\iota(n)$. Therefore $\iota(n)$ is multiplicative.

b. $(\iota * f)(n) = \sum_{d \mid n} \iota(d) f(\frac{n}{d}) = \iota(1) f(\frac{n}{1}) = f(n)$ since $\iota(d) = 0$ except when $d = 1$. $(f * \iota)(n) = (\iota * f)(n) = f(n)$ by Exercise 37.

7.1.41. Let $h = f * g$ and let $(m, n) = 1$. Then $h(mn) = \sum_{d \mid mn} f(d) g(\frac{mn}{d})$. Since $(m, n) = 1$, each divisor d of mn can be expressed in exactly one way as $d = ab$ where $a \mid m$ and $b \mid n$. Then $(a, b) = 1$ and $(\frac{m}{a}, \frac{n}{b}) = 1$. Then there is a one-to-one correspondence between the divisors d of mn and the pairs of products ab where $a \mid m$ and $b \mid n$. Then

$$h(mn) = \sum_{\substack{a \mid m \\ b \mid n}} f(ab) g(\frac{mn}{ab}) = \sum_{\substack{a \mid m \\ b \mid n}} f(a) f(b) g(\frac{m}{a}) g(\frac{n}{b})$$

$$= \sum_{a \mid m} f(a) g(\frac{m}{a}) \sum_{b \mid n} f(b) g(\frac{n}{b}) = h(m) h(n)$$

as desired.

7.1.43. a. Since $12 = 2^2 3$, we have $\lambda(12) = (-1)^{2+1} = -1$.

b. Since $20 = 2^2 5$, we have $\lambda(20) = (-1)^{2+1} = -1$.

c. Since $210 = 2 \cdot 3 \cdot 5 \cdot 7$, we have $\lambda(210) = (-1)^{1+1+1+1} = 1$.

d. Since $1000 = 2^3 5^3$, we have $\lambda(1000) = (-1)^{3+3} = 1$.

e. Since $1001 = 7 \cdot 11 \cdot 13$, we have $\lambda(1001) = (-1)^{1+1+1} = -1$.

f. Since $10! = 2^8 3^4 5^2 7$, we have $\lambda(10!) = (-1)^{8+4+2+1} = -1$.

g. Since $20! = 2^{18} 3^8 5^4 7^2 11 \cdot 13 \cdot 17 \cdot 10$, we have $\lambda(20!) = (-1)^{18+8+4+2+1+1+1+1} = 1$.

7.1.45. Let $f(n) = \sum_{d|n} \lambda(d)$. Then f is the Dirichlet product of λ and the constant function $h(n) = 1$. Since h and λ are both multiplicative, so is f, by Exercise 41. Now, $f(p^t) = \lambda(1) + \lambda(p) + \lambda(p^2) + \cdots + \lambda(p^t) = 1 - 1 + 1 - \cdots + (-1)^t = 0$ if t is odd and $= 1$ if t is even. Then $f(p_1^{a_1} p_2^{a_2} \cdots p_r^{a_r}) = \prod f(p_i^{a_i}) = 0$ if any a_i is odd and $= 1$ if all a_i are even and hence n is a square.

7.1.47. If f and g are completely multiplicative and m and n are positive integers we have $(fg)(mn) = f(mn)g(mn) = f(m)f(n)g(m)g(n) = f(m)g(m)f(n)g(n) = (fg)(m)(fg)(n)$, so fg is also completely multiplicative.

7.1.49. We have $f(mn) = \log mn = \log m + \log n = f(m) + f(n)$. Hence $f(n) = \log n$ is completely additive.

7.1.51. a. Since $12 = 2^2 3$, we have $\omega(12) = 2$.

b. Since $30 = 2 \cdot 2 \cdot 5$, we have $\omega(30) = 3$.

c. Since $32 = 2^5$, we have $\omega(32) = 1$.

d. Since the primes that divide 10! are exactly those primes less than or equal to 10, namely 2, 3, 5 and 7, we have $\omega(10!) = 4$.

e. Since the primes that divide 20! are exactly those primes less than or equal to 20, namely 2, 3, 5, 7, 11, 13, 17 and 19, we have $\omega(20!) = 8$.

f. Since the primes that divide 50! are exactly those primes less than or equal to 50, and $\pi(50) = 15$, we have $\omega(50!) = 15$.

7.1.53. Let $(m,n) = 1$, then by the additivity of f we have $f(mn) = f(m) + f(n)$. Then $g(mn) = 2^{f(mn)} = 2^{f(m)+f(n)} = 2^{f(m)} 2^{f(n)} = g(m)g(n)$, so g is multiplicative.

7.2. The Sum and Number of Divisors

7.2.1. a. Since $35 = 5 \cdot 7$ and σ is multiplicative, we see that $\sigma(35) = (1+5)(1+7) = 6 \cdot 8 = 48$.

b. Since $196 = 2^2 7^2$ and σ is multiplicative, we see that $\sigma(196) = (1+2+2^2)(1+7+7^2) = 7 \cdot 57 = 399$.

c. Since $1000 = 2^3 5^3$ and σ is multiplicative, we see that $\sigma(1000) = (1 + 2 + 2^2 + 2^3) \cdot (1 + 5 + 5^2 + 5^3) = 15 \cdot 156 = 2340$.

d. By Lemma 7.1 we have $\sigma(2^{100}) = \frac{2^{101}-1}{2-1} = 2^{101} - 1$.

e. Since $\sigma(n)$ is a multiplicative function, we have $\sigma(2 \cdot 3 \cdot 5 \cdot 7 \cdot 11) = (1+2)(1+3)(1+5) \cdot (1+7)(1+11) = 6912$.

f. Since σ is multiplicative, it follows that $\sigma(2^5 \cdot 3^4 \cdot 5^3 \cdot 7^2 \cdot 11) = (1 + 2 + 2^2 + 2^3 + 2^4 + 2^5) \cdot (1 + 3 + 3^2 + 3^3 + 3^4) \cdot (1 + 5 + 5^2 + 5^3) \cdot (1 + 7 + 7^2) \cdot (1 + 11) = 63 \cdot 121 \cdot 156 \cdot 57 \cdot 12 = 813404592$.

g. The prime factorization of 10! is $10! = 2^8 3^4 5^2 7$. By Theorem 6.8, we conclude that $\sigma(10) = \frac{2^9-1}{2-1} \cdot \frac{3^5-1}{3-1} \cdot \frac{5^2-1}{5-1} \cdot \frac{7^2-1}{7-1} = 511 \cdot 242 \cdot 6 \cdot 8 = 5935776$.

h. The prime factorization of 20! is $20! = 2^{18} \cdot 3^8 \cdot 5^4 \cdot 7^2 \cdot 11 \cdot 13 \cdot 17 \cdot 19$. By Theorem 6.8 it follows that $\sigma(20) = \frac{2^{19}-1}{2-1} \cdot \frac{3^9-1}{3-1} \cdot \frac{5^5-1}{5-1} \cdot \frac{7^3-1}{7-1} \cdot \frac{11^2-1}{11-1} \cdot \frac{13^2-1}{13-1} \cdot \frac{17^2-1}{17-1} \cdot \frac{19^2-1}{19-1} = 9841 \cdot 781 \cdot 57 \cdot 12 \cdot 14 \cdot 18 \cdot 20 = 26495791882560$.

7.2.3. Let $n = p_1^{a_1} p_2^{a_2} \cdots p_s^{a_s}$. We need to find when $\tau(n)$ is odd. By Theorem 7.9 $\tau(n) = (a_1+1)(a_2+1)\cdots(a_s+1)$, so each factor a_i+1 must be odd, hence each a_i must be even. Therefore n is a perfect square.

7.2.5. a. For each part of this exercise let the prime factorization of n be $p_1^{a_1} p_2^{a_2} \cdots p_r^{a_r}$. Then since σ is multiplicative, we have $\sigma(n) = \prod_{i=1}^{r}(1 + p_i + \cdots + p_i^{a_i})$.

Suppose that $\sigma(n) = 12$. Each factor in the formula for $\sigma(n)$ must divide 12. The only ways to get factors, other than 1, of 12 for sums of this type are $(1+2) = 3, (1+3) = 4, (1+5) = 6, (1+11) = 12$. Hence the only values of n for which $\sigma(n) = 12$ are $n = 2 \cdot 3 = 6$ and $n = 11$.

b. Suppose that $\sigma(n) = 18$. Each factor in the formula for $\sigma(n)$ must divide 18 and the product of these factors must be 18. The only ways to get factors, other that 1, of 18 for sums of this type are $(1+2) = 3, (1+5) = 6,$ and $(1+17) = 18$. It follows that the only solutions of $\sigma(n) = 18$ are $n = 2 \cdot 5 = 10$ and $n = 17$.

c. Suppose that $\sigma(n) = 24$. Each factor in the formula for $\sigma(n)$ must divide 24 and the product of these factors must be 24. The only ways to get factors, other than 1, of 24 for sums of this type are $(1+2) = 3, (1+3) = 4, (1+5) = 6, (1+7) = 8, (1+11) = 12,$ and $(1+23) = 24$. It follows that the only solutions of $\sigma(n) = 24$ are $n = 2 \cdot 7 = 14, n = 3 \cdot 5 = 15,$ and $n = 23$.

d. Suppose that $\sigma(n) = 48$. Each factor in the formula for $\sigma(n)$ must divide 48 and the product of these factors must be 48. The only ways to get factors, other than 1, of 48 for sums of this type are $(1+2) = 3, (1+3) = 4, (1+5) = 6, (1+7) = 8, (1+11) = 12, (1+23) = 24,$ and $(1+47) = 48$. If follows that the only solutions of $\sigma(n) + 48$ are $n = 3 \cdot 11 = 33, n = 5 \cdot 7 = 35,$ and $n = 47$.

e. Suppose that $\sigma(n) = 52$. Each factor in the formula for $\sigma(n)$ must divide 52 and the product of these factors must be 52. The only ways to get factors, other than 1, of 52 for sums of this type are $(1+3) = 4$ and $(1+3+9) = 13$. Since only one factor for each prime can be included, there are no solutions of $\sigma(n) = 52$.

f. Suppose that $\sigma(n) = 84$. Each factor in the formula for $\sigma(n)$ must divide 84 and the product of these factors must be 84. The only ways to get factors, other than 1, of 84 for sums of this type are $(1+2) = 3, (1+3) = 4, (1+5) = 6, (1+2+4) = 7, (1+11) = 12, (1+13) = 14,$ and $(1+83) = 84$. It follows that the only solutions of $\sigma(n) = 84$ are $n = 5 \cdot 13 = 65, n = 4 \cdot 11 = 44,$ and $n = 83$.

7.2.7. Note that $\tau(p^{k-1}) = k$ whenever p is prime and k is a positive integer $k > 1$. Hence the equation $\tau(n) = k$ has infinitely many solutions.

7.2.9. The only positive integers with exactly three prime divisors are those integers of the form p^2 where p is prime. We see this using the formula given in Theorem 7.9. We have $\tau(p_1^{a_1} \cdots p_t^{a_t}) = (a_1 + 1)(a_2 + 1) \cdots (a_t + 1)$. Since the terms on the right-hand side are all at least 2, this product can equal 3 if and only if there is precisely one term on the right-hand side that is equal to 3.

7.2.11. We first suppose that n is not a perfect square. Then the divisors of n come in pairs with product n, that is, when d is a divisor, so is n/d and conversely. Since there are $\tau(n)/2$ such pairs, the product of all divisors is $n^{\tau(n)/2}$. Now suppose that n is a perfect square. Then there are $\frac{\tau(n)-1}{2}$ pairs with product n and the extra divisor $\sqrt{n}$. Hence the product of all the divisors of n is $n^{(\tau(n)-1)/2} \cdot n^{1/2} = n^{\tau(n)/2}$.

7.2.13. a. The nth term is given by $\sigma(2n)$.

b. The nth term is given by $\sigma(n) - \tau(n)$.

c. The nth term of this sequence is the least positive integer m with $\tau(m) = n$.

d. The nth term is the number of solutions k to the equation $\sigma(k) = n$.

7.2.15. If we list the values of $\tau(n)$ for $n = 2, 3, 4, \ldots$, in order, we can identify highly composite integers by noting the first occurrence of a value which is larger than all previous values. From Table 2 in Appendix E we see that $\tau(2) = 2$ is the first occurrence of 2, and is larger than all previous values. This is first exceeded when we find $\tau(4) = 3$. This is exceeded when we find $\tau(6) = 4$. This is exceeded when we find $\tau(12) = 6$. This is exceeded when we find $\tau(24) = 8$. This is exceeded when we find $\tau(36) = 9$. So the first six highly composite numbers are $2, 4, 6, 12, 24$ and 36.

7.2.17. Let a be the largest highly composite integer less than or equal to n. Note that $2a$ is less than or equal to $2n$ and has more divisors than a and hence $\tau(2a) > \tau(a)$. By Exercise 16, there must be a highly composite integer b with $a < b \leq 2a$. If $b \leq n$, this contradicts the choice of a. Therefore $n < b \leq 2n$. It follows that there must be a highly composite integer k with $2^m < k \leq 2^{m+1}$ for every nonnegative integer m. Therefore, there are at least m highly composite integers less than or equal to 2^m. Thus the mth highly composite integer is less than or equal to 2^m.

7.2.19. If $n = 2^a 3^b$ is highly composite, then by Exercise 18 we have $a \geq b$. Since $2^a 3^b > 2^{a-1} b 3^{b-1} 5$, we must have $\tau(2^a 3^b) > \tau(2^{a-1} b 3^{b-1} 5)$, that is $(a+1)(b+1) > 2ab$. Rearranging the inequality yields $(a-1)(b-1) < 2$. Hence, either $a = b = 2$ or $b < 2$. In the first case we have $n = 36$, which is highly composite. If $b = 1$, assume $a > 3$. We have $n = 2^a 3 > 2^{a-1} 3 \cdot 5$. Then $\tau(2^a 3) = 2(a+1) > 4(a-2) = \tau(2^{a-1} 3 \cdot 5)$. This reduces to $a < 5$, so we need only check $a = 1, 2, 3$ and 4, which correspond to the numbers $6, 12, 24$ and 48, all of which are highly composite. Finally, if $b = 0$ assume $a > 2$. Then $n = 2^a > 2^{a-2} 3$, and so $tau(2^a) = a + 1 > 2(a-1) = \tau(2^{a-2} 3)$, which reduces to $a < 5$. So we need only check the cases $a = 0, 1, 2, 3$ and 4, which correspond to the numbers $1, 2, 4, 8$ and 16. Of these, only $1, 2,$ and 4 are highly composite. The complete list of the highly composite numbers of the form $2^a 3^b$ is $1, 2, 4, 6, 12, 24, 36$ and 48.

7.2.21. We find that $\sigma_k(p) = \sum_{d|p} d^k = 1^k + p^k = 1 + p^k$.

7.2.23. Suppose that a and b are positive integers with $(a, b) = 1$. Then $\sum_{d|ab} d^k = \sum_{d_1|a, d_2|b} (d_1 d_2)^k = \sum_{d_2|a} d_1^k \sum_{d_2|a} d_2^k = \sigma_k(a)\sigma_k(b)$.

7.2.25. Let $n = p_1^{a_1} \cdots p_r^{a_r}$. Then $\phi(n) = (p_1^{a_1} - p_1^{a_1-1}) \cdots (p_r^{a_r} - p_r^{a_r-1}) = \sum T_j$. Where $\sum T_j$ is this product is expanded. Each term T_j is of the form $T_j = (-1)^k p_1^{b_1} \cdots p_r^{b_r}$ where $b_i = a_i$ or $a_i - 1$. Note that each one of these terms is a divisor of n, and note that one of the terms is $p_1^{a_1} \cdots p_r^{a_r} = n$. Now since $\sigma(n)$ is the sum of the divisors of n, each of the terms T_j above appears in the sum $\sigma(n) = \sum_{d|n} d$, without the $(-1)^k$. Note that n also appears in this sum. Then we have $\sigma(n) + \phi(n) = \sum_{d|n} d + \sum T_j = 2n + \sum_{\substack{d|n \\ d<n}} d + \sum_{T_j \neq n} T_j$. Now if T_j is negative, then the $|T_j|$ appearing in the first sum will cancel it. But if T_j is positive we get two terms T_j in the last sum. Then we have $\sigma(n) + \phi(n) = 2n + \sum_{d|n, d<n, d\neq T_j} d + 2\sum_{T_j > 0, t_j \neq n} T_j$. Since both of these last sums are nonnegative, we need them both to be zero in order to have a solution. In particular the expansion of $\phi(n) = (p_1^{a_1} - p_1^{a_1-1}) \cdots (p_r^{a_r} - p_r^{a_r-1})$ can have no positive terms other than n, and therefore we must have $\phi(n) = (p_1^{a_1} - p_1^{a_1-1})$. Now the term $d = 1$ appears in the first sum unless $T_j = -1$ for some j. Therefore $\phi(n) = p_1 - 1$ and so n is prime.

7.2.27. Let $n = p_1^{a_1} p_2^{a_2} \cdots p_r^{a_r}$ and let x and y be integers such that $[x, y] = n$. then $x \mid n$ and $y \mid n$ so we have $x = p_1^{b_1} p_2^{b_2} \cdots p_r^{a_r}$ and $y = p_1^{c_1} p_2^{c_2} \cdots p_r^{c_r}$, where b_i and $c_i = 0, 1, 2, \ldots, a_i$. Since $[x, y] = n$, we must have $\max\{b_i, c_i\} = a_i$ for each i. Then one of b_i and c_i must be equal to a_i and the other can range over $0, 1, \ldots, a_i$. Therefore we have $2a_i + 1$ ways to choose the pair (b_i, c_i) for each i. Then in total, we can choose the exponents $b_1, b_2, \ldots b_r, c_1, \ldots, c_r$ in $(2a_1 + 1)(2a_2 + 1) \cdots (2a_r + 1) = \tau(n^2)$ ways.

7.2.29. Suppose that n is composite. Then $n = ab$ where a and b are integers with $1 < a \leq b < n$. It follows that either $a \geq \sqrt{n}$ or $b \geq \sqrt{n}$. Consequently $\sigma(n) \geq 1 + a + b + n > 1 + \sqrt{n} + n > n + \sqrt{n}$. Conversely, suppose that n is prime. Then $\sigma(n) = n + 1$ so that $\sigma(n) \leq n + \sqrt{n}$. Hence $\sigma(n) > n + \sqrt{n}$ implies that n is composite.

7.2.31. For $n = 1$, the statement is true. Suppose that $\sum_{j=1}^{n-1} \tau(j) = 2\sum_{j=1}^{[\sqrt{n-1}]} \left[\frac{n-1}{j}\right] - [\sqrt{n-1}]^2$. For the induction step, it suffices to show that $\tau(n) = 2\sum_{j=1}^{[\sqrt{n-1}]} \left(\left[\frac{n}{j}\right] - \left[\frac{n-1}{j}\right]\right) = 2\sum_{\substack{j \leq [\sqrt{n-1}] \\ j | n}} 1$, which is true by the definition of $\tau(n)$, since there is one factor less than $\sqrt{n}$ for every factor greater than $\sqrt{n}$. Note that if n is a perfect square, we must add the term $2\sqrt{n} - (2\sqrt{n} - 1) = 1$ to the last two sums. For $n = 100$, we have $\sum_{j=1}^{100} \tau(j) = 2\sum_{j=1}^{10} \left[\frac{n}{j}\right] - 100 = 482$.

7.2.33. We use the identity $\sum_{j=0}^{\min\{a,b\}} (p^{a+b-j} + p^{a+b-j-1} + \cdots + p^j) = (p^a + p^{a-1} + \cdots + 1)(p^b + p^{b-1} + \cdots + 1)$. If $a = \prod p^{a_i}$ and $b = \prod p^{b_i}$, then $\sigma(a)\sigma(b) = \prod (p^{a_i} + p^{a_i-1} + \cdots + 1)(p^{b_i} + p^{b_i-1} + \cdots + 1) = \prod \sum_{j=0}^{\min\{a_i,b_i\}} (p_i^{a_i+b_i-j} + p_i^{a_i+b_i-j-1} + \cdots + p_i^j)$

7.2.35. From Exercises 52 and 53 in Section 7.1 we know that the arithmetic function $f(n) = 2^{\omega(n)}$ is multiplicative. Further, since the Dirichlet product $h(n) = \sum_{d|n} 2^{\omega(d)} = f * g(n)$, where $g(n) = 1$ is also multiplicative, we know that $h(n)$ is also multiplicative. See Exercise 41 in Section 7.1. Since $\tau(n)$ and n^2 are multiplicative, so is $\tau(n^2)$. Therefore, it sufficient to prove the identity for n equal to a prime power, p^a. We have $\tau(p^{2a}) = (2a+1)$. On the other hand we have $\sum_{d|p^a} 2^{\omega(d)} = \sum_{i=0}^{a} 2^{\omega(p^i)} = 1 + \sum_{i=1}^{a} 2^1 = 2a + 1$, which completes the proof.

7.2.37. Let **M** be the matrix. Let **D** be the matrix with entries $\phi(1), \phi(2), \ldots, \phi(n)$ on the diagonal and zeros elsewhere. Let **A** be the matrix of 0's and 1's defined by the rule: If i divides j then the (i,j) entry is 1, otherwise, it is 0. Then **A** has all 0's below the main diagonal, and 1's on the main diagonal, therefore $\det(\mathbf{A}) = 1$. Check that $\mathbf{M} = \mathbf{A}\mathbf{D}\mathbf{A}^T$. Then $\det(\mathbf{M}) = 1 \cdot \det(\mathbf{D}) \cdot 1 = \phi(1)\phi(2)\cdots\phi(n)$.

7.2.39. Suppose there are infinitely many pairs of twin primes. Let p and $p+2$ be a pair of twin primes. Then $\sigma(p) = p + 1$ and $\phi(p+2) = p + 2 - 1 = p + 1$. So each pair of twin primes is a solution to the equation. Next Suppose there are infinitely many primes of the form $2^q - 1$ with q prime. Then $\phi(2^{q+1}) = 2^q$ and $\sigma(2^q - 1) = 2^q - 1 + 1 = 2^q$. So once again we have infinitely many solutions.

7.3. Perfect Numbers and Mersenne Primes

7.3.1. From the table on page 264, the first six Mersenne primes are given by $2^p - 1$ with $p = 2, 3, 5, 7, 13$, and 17. Then Theorem 6.9 gives the first six even perfect numbers as $2^1(2^2 - 1) = 6$; $2^2(2^3 - 1) = 28$; $2^4(2^5 - 1) = 496$; $2^6(2^7 - 1) = 8,128$; $2^{12}(2^{13} - 1) = 33,550,336$; and $2^{16}(2^{17} - 1) = 8589869056$.

7.3.3. a. By the difference of cubes factorization we have $2^{15} - 1 = (2^5)^3 - 1 = (2^5 - 1)(2^{10} + 2^5 + 1)$, so $2^5 - 1 = 31$ is a factor.

b. Since $7 | 91$, $127 = 2^7 - 1 | 2^{91} - 1$.

c. Since $7 | 1001$, $127 = 2^7 - 1 | 2^{1001} - 1$.

7.3.5. We have $\sigma(12) = 28, \sigma(18) = 39, \sigma(20) = 42, \sigma(24) = 60, \sigma(30) = 72$ and $\sigma(36) = 91$.

7.3.7. Suppose that $n = p^k$ where p is prime and k is a positive integer. Then $\sigma(p^k) = \frac{p^{k+1}-1}{p-1}$. Note that $2p^k - 1 < p^{k+1}$ since $p \geq 2$. It follows that $p^{k+1} - 1 < 2(p^{k+1} - p^k) = 2p^k(p-1)$, so that $\frac{(p^{k+1}-1)}{p-1} < 2p^k = 2n$. It follows that $n = p^k$ is deficient.

7.3.9. Suppose that n is abundant or perfect. Then $\sigma(n) \geq 2n$. Suppose that $n | m$. Then $m = nk$ for some integer k. The divisors of m include the integers kd and $d | n$. Hence $\sigma(m) \geq \sum_{d|n} (k+1)d = (k+1)\sum_{d|n} d = (k+1)\sigma(n) \geq (k+1)2n > 2kn = 2m$. Hence m is abundant.

7.3.11. If p is any prime, then $\sigma(p) = p + 1 < 2p$, so p is deficient. Since there are infinitely many primes, we must have infinitely many deficient numbers.

7.3.13. See Exercises 6 and 9. For a positive integer a let $n = 3^a 5 \cdot 7$ and compute $\sigma(n) = \sigma(3^a 5 \cdot 7) = (3^{a+1} - 1)/(3-1)(5+1)(7+1) = (3^{a+1} - 1)24 = 3^{a+1}24 - 24 = 2 \cdot 3^a(36) - 24 = 2 \cdot 3^a(35) + 2 \cdot 3^a - 24 = 2n + 2 \cdot 3^a - 24$, which will be greater than $2n$ whenever $a \geq 3$. This demonstrates infinitely many odd abundant integers.

7.3.15. a. The prime factorizations of 220 and 284 are $220 = 2^2 \cdot 5 \cdot 11$ and $284 = 2^2 \cdot 71$. Hence $\sigma(220) = \sigma(2^2)\sigma(5)\sigma(11) = 7 \cdot 6 \cdot 12 = 504$ and $\sigma(284) = \sigma(2^2)\sigma(71) = 7 \cdot 72 = 504$. Since $\sigma(220) = \sigma(284) = 220 + 284 = 504$, it follows that 220 and 284 form an amicable pair.

b. The prime factorizations of 1184 and 1210 are $1184 = 2^5 \cdot 37$ and $1210 = 2 \cdot 5 \cdot 11^2$. Hence $\sigma(1184) = \sigma(2^5)\sigma(37) = 63 \cdot 38 = 2394$ and $\sigma(1210) = \sigma(2)\sigma(5)\sigma(11^2) = 3 \cdot 6 \cdot 133 = 2394$. Since $\sigma(1184) = \sigma(1210) = 1184 + 1210 = 2394$, 1184 and 1210 form an amicable pair.

c. The prime factorizations of 79750 and 88730 are $79750 = 2 \cdot 5^3 \cdot 11 \cdot 29$ and $88730 = 2 \cdot 5 \cdot 19 \cdot 467$. Hence $\sigma(79750) + \sigma(2)\sigma(5^3)\sigma(11)\sigma(29) = 3 \cdot 156 \cdot 12 \cdot 30 = 168480$ and similarly $\sigma(88730) = \sigma(2)\sigma(5)\sigma(19)\sigma(467) = 3 \cdot \cdot 6 \cdot 20 \cdot 468 = 168480$. Since $\sigma(79750) = \sigma(88730) = 79750 + 88730 = 168480$ it follows that 79750 and 88730 form an amicable pair.

7.3.17. Since $120 = 2^3 \cdot 3 \cdot 5$ and σ is multiplicative, we have $\sigma(120) = \sigma(2^3 \cdot 3 \cdot 5) = \sigma(2^3)\sigma(3)\sigma(5) = 15 \cdot 4 \cdot 6 = 360$. Since $\sigma(120) = 360 = 3 \cdot 120$, it follows that 120 is 3-perfect.

7.3.19. $\sigma(2^7 3^4 5 \cdot 7 \cdot 11^2 \cdot 17 \cdot 19) = \frac{2^8-1}{2-1} \cdot \frac{3^5-1}{3-1}(5+1)(7+1)\frac{11^3-1}{11-1}(17+1)(19+1) = 255 \cdot 121 \cdot 6 \cdot 8 \cdot 133 \cdot 18 \cdot 20 = 5 \cdot 14182439040$.

7.3.21. Suppose that n is 3-perfect and 3 does not divide n. Then $\sigma(3n) = \sigma(3)\sigma(n) = 4 \cdot 3n = 12n = 4 \cdot 3n$. Hence $3n$ is 4-perfect.

7.3.23. For example, $\sigma(2^6 3^4 5^2 7^2 11 \cdot 13) = \sigma(908107200) = 4561786152 > 5 \cdot 908107200$.

7.3.25. We have $\sigma(\sigma(16)) = \sigma(31) = 32 = 2 \cdot 16$. Hence 16 is superperfect.

7.3.27. Certainly if r and s are integers, then $\sigma(rs) \geq rs + r + s + 1$. Suppose $n = 2^q t$ is superperfect with t odd and $t > 1$. Then $2n = 2^{q+1}t = \sigma(\sigma(2^q t)) = \sigma\left((2^{q+1} - 1)\sigma(t)\right) \geq (2^{q+1} - 1)\sigma(t) + (2^{q+1} - 1) + \sigma(t) + 1 > 2^{q+1}\sigma(t) \geq 2^{q+t}(t+1)$. Then $t > t+1$, a contradiction. Therefore we must have $n = 2^q$, in which case we have $2n = 2^{q+1} = \sigma(\sigma(2^q)) = \sigma\left(2^{q+1} - 1\right) = \sigma(2n - 1)$. Therefore $2n - 1 = 2^{q+1} - 1$ is prime.

7.3.29. a. By Theorem 7.12 any divisor of $M_7 = 127$ must be of the form $14k + 1$ where k is a positive integer. There are no primes of this form less than $\sqrt{127}$ so $M_7 = 127$ is prime.

b. By Theorem 7.12 any divisor of $M_{11} = 2047$ must be of the form $22k + 1$ where k is a positive integer. The only prime of this form less than $\sqrt{2047}$ is 23. Since 23 does not divide 2047, it follows that $M_{11} = 2047$ is prime.

c. By Theorem 7.12 any divisor of $M_{17} = 131071$ must be of the form $34k + 1$ where k is a positive integer. The primes of this form less than $\sqrt{131071} < 363$ are 103, 137, 239, and 307, but none of these divide 131071. Hence $M_{17} = 131071$ is prime.

d. By Theorem 7.12 any divisor of $M_{29} = 536870911$ must be of the form $58k + 1$ where k is a positive integer. We first note neither that 59 nor 107 divides 536870911. However, $233 = 58 \cdot 4 = 1$ does divide 536870911 since $536870911 = 233 \cdot 2304167$. Hence M_{29} is not prime.

7.3.31. $M_n(M_n + 2) = (2^n - 1)(2^n + 1) = 2^{2n} - 1$. If $2n + 1$ is prime then $\phi(2n+1) = 2n$ and $2^{2n} \equiv 1 \pmod{2n+1}$. Then $(2n + 1) \mid 2^{2n} - 1 = M_n(M_n + 2)$. Therefore $(2n+1) \mid M_n$ or $(2n+1) \mid (M_n + 2)$.

7.3.33. Since m is odd, $m^2 \equiv 1 \pmod{8}$, so $n = p^a m^2 \equiv p^a \pmod 8$. By Exercise 32 (a), $a \equiv 1 \pmod 4$, so $p^a \equiv p^{4k}p \equiv p \pmod 8$, since p^{4k} is an odd square. Therefore $n \equiv p \pmod 8$.

7.3.35. First suppose that $n = p^a$ where p is prime and a is a positive integer. Then $\sigma(n) = \frac{p^{a+1}-1}{p-1} < \frac{p^{a+1}}{p-1} = \frac{np}{p-1} = \frac{n}{1-\frac{1}{p}} \leq \frac{n}{\frac{2}{3}} < \frac{3n}{2}$ so that $\sigma(n) \neq 2n$ and n is not perfect. Next suppose that $n = p^a q^b$ where a and b are primes and a and b are positive integers. Then $\sigma(n) = \frac{p^{a+1}-1}{p-1} \cdot \frac{q^{b+1}-1}{q-1} < \frac{p^{a+1}q^{b+1}}{(p-1)(q-1)} = \frac{npq}{(p-1)(q-1)} = \frac{n}{(1-\frac{1}{p})(1-\frac{1}{q})} \leq \frac{n}{(\frac{2}{3})(\frac{4}{5})} = \frac{15n}{8} < 2n$. Hence $\sigma(n) \neq 2n$ and n is not perfect.

7.3.37. By Exercise 11 of Section 7.2 it follows that the product of all positive divisors of an integer n is $n^{\frac{\tau(n)}{2}}$. If the product of all divisors of n other than n is n^2 then $n^{\frac{\tau(n)}{2}-1} = n^2$ so that $\frac{\tau(n)}{2} = 3$. This implies that $\tau(n) = 6$. The integers with $\tau(n) = 6$ are those of the form p^5 and $p^2 q$ where p and q are primes.

7.3.39. Suppose $M_n = 2^n - 1 = a^k$, with n and k integers greater than 1. Then a must be odd. If $k = 2j$, then $2^n - 1 = (a^j)^2$. Since $n > 1$ and the square of an odd integer is congruent to 1 modulo 4, reduction of the last equation modulo 4 yields the contradiction $-1 \equiv 1 \pmod 4$, therefore k must be odd. Then $2^n = a^k + 1 = (a+1)(a^{k-1} - a^{k-2} + \cdots + 1)$. So $a + 1 = 2^m$ for some integer m. Then $2^n - 1 = (2^m - 1)^k$. Now $n > mk$ so reduction modulo 2^{2m} gives $-1 \equiv -k 2^m - 1 \pmod{2^{2m}}$ or, since k is odd, $2^m \equiv 0 \pmod{2^{2m}}$, a contradiction.

7.4. Möbius Inversion

7.4.1. a. Since $12 = 2^2 3$ is not squarefree, $\mu(12) = 0$.

b. Since $15 = 3 \cdot 5$ is the product of two primes, $\mu(15) = (-1)^2 = 1$.

c. Since $30 = 2 \cdot 3 \cdot 5$ is the product of three primes, $\mu(30) = (-1)^3 = -1$.

d. Since $50 = 2 \cdot 5^2$ is not squarefree, $\mu(50) = 0$.

e. Since $1001 = 7 \cdot 11 \cdot 13$ is the product of three primes, $\mu(1001) = (-1)^3 = -1$.

f. Since $2 \cdot 3 \cdot 5 \cdot 7 \cdot 11 \cdot 13$ is the product of six primes, $\mu(2 \cdot 3 \cdot 5 \cdot 7 \cdot 11 \cdot 13) = (-1)^6 = 1$.

g. Since $4|10!$, we know $10!$ is not squarefree, so we have $\mu(10!) = 0$.

7.4.3. Since 4 divides $100, 104$, and 108, the value of μ for each of these is 0. Since $101, 103, 107$, and 109 are prime, μ for each of these values is -1. Then $\mu(102) = \mu(2 \cdot 3 \cdot 17) = (-1)^3 = -1, \mu(105) = \mu(3 \cdot 5 \cdot 7) = (-1)^3 = -1, \mu(106) = \mu(2 \cdot 53) = (-1)^2 = 1$, and $\mu(110) = \mu(2 \cdot 5 \cdot 11) = (-1)^3 = -1$.

7.4.5. Such n must be the product of an even number of distinct primes. The only product of zero primes is 1. The products of 2 primes which are less than or equal to 100, are $2 \cdot 3 = 6, 2 \cdot 5 = 10, 2 \cdot 7 = 14, 2 \cdot 11 = 22, 2 \cdot 13 = 26, 2 \cdot 17 = 34, 2 \cdot 19 = 38, 2 \cdot 23 = 46, 2 \cdot 29 = 58, 2 \cdot 31 = 62, 2 \cdot 37 = 74, 2 \cdot 41 = 82, 2 \cdot 43 = 86, 2 \cdot 47 = 94, 3 \cdot 5 = 15, 3 \cdot 7 = 21, 3 \cdot 11 = 33, 3 \cdot 13 = 39, 3 \cdot 17 = 51, 3 \cdot 19 = 57, 3 \cdot 23 = 69, 3 \cdot 29 = 87, 3 \cdot 31 = 93, 5 \cdot 7 = 35, 5 \cdot 11 = 55, 5 \cdot 13 = 65, 5 \cdot 17 = 85, 5 \cdot 19 = 95, 7 \cdot 11 = 77$, and $7 \cdot 13 = 91$. Since the product of the four smallest primes is $210 > 100$, the above list is exhaustive.

7.4.7. Starting with the values $\mu(1) = \mu(6) = \mu(10) = 1, \mu(2) = \mu(3) = \mu(5) = \mu(7) = -1$, and $\mu(4) = \mu(8) = \mu(9) = 0$, we compute $M(1) = 1, M(2) = 1 + (-1) = 0, M(3) = 0 + (-1) = -1, M(4) = -1 + 0 = -1, M(5) = -1 + (-1) = -2, M(6) = -2 + 1 = -1, M(7) = -1 + -1 = -2, M(8) = -2 + 0 = -2, M(9) = -2 + 0 = -2$, and $M(10) = -2 + 1 = -1$.

7.4.9. Since $\mu(n)$ is 0 for nonsquarefree n, 1 for n a product of an even number of distinct primes and -1 for n a product of a odd number of distinct primes, the sum $M(n) = \sum_{i=1}^n \mu(i)$ is unaffected by the non-squarefree numbers, but counts 1 for every even product and -1 for every odd product. Thus $M(n)$

7.4.11. For any nonnegative integer k, the numbers $n = 36k + 8$ and $n + 1 = 36k + 9$ are consecutive and divisible by $4 = 2^2$ and $9 = 3^2$ respectively. Therefore $\mu(36k + 8) + \mu(36k + 9) = 0 + 0 = 0$.

7.4.13. Since every multiple of 4 is nonsquarefree, we can have at most 3 consecutive integers for which μ takes on nonzero values.

7.4.15. Let $h(n) = n$ be the identity function. Then from Theorem 7.7 we have $h(n) = n = \sum_{d|n} \phi(n)$. Then by the Möbius inversion formula, we have $\phi(n) = \sum_{d|n} \mu(d)h(n/d) = \sum_{d|n} \mu(d)(n/d) = n\sum_{d|n} \mu(d)/d$, as desired.

7.4.17. Since μ and f are multiplicative, then so is their product μf, by Exercise 46 of Section 7.1. Further, the summatory function $\sum_{d|n} \mu(d)f(d)$ is also multiplicative by Theorem 7.17. Therefore it suffices to prove the proposition for n a prime power. We compute $\sum_{d|p^a} \mu(d)f(d) = \mu(p^a)f(p^a) + \mu(p^{a-1})f(p^{a-1}) + \cdots + \mu(p)f(p) + \mu(1)f(1)$. But for exponents greater than 1, $\mu(p^j) = 0$, so the above sum equals $\mu(p)f(p) + \mu(1)f(1) = -f(p) + 1$, as desired.

7.4.19. Here we let $1/n$ play the role of $f(n)$ in the identity in Exercise 17. This gives $\sum_{d|n} \mu(d)/d = \prod_{j=1}^{k}(1 - 1/p_j)$. We might note that this resembles the formula for $\phi(n)$, indeed, it equals $\phi(n)/n$. Compare Exercise 15.

7.4.21. Here we let σ play the role of f in the identity. Then the sum equals $\prod_{j=1}^{k}(1 - \sigma(p_j)) \prod_{j=1}^{k}(1 - (p_j + 1)) = \prod_{j=1}^{k} p_i$.

7.4.23. Since both sides of the equation are known to be multiplicative, (see Exercise 35 in Section 7.2) it suffices to prove the identity for $n = p^a$, a prime power. On one hand we have $\sum_{d|p^a} \mu^2(d) = \mu^2(p) + \mu^2(1) = 1 + 1 = 2$. On the other hand, we have $\omega(p^a) = 1$, so the right side is $2^1 = 2$, which equals the left side.

7.4.25. Let λ play the role of f in the identity of Exercise 17. Then the left side equals $\prod_{j=1}^{k}(1 - \lambda(p_j)) = \prod_{j=1}^{k}(1 - (-1)) = 2^k$. But $\omega(n) = k$ by definition, so we're done.

7.4.27. We compute $\mu * \nu(n) = \sum_{d|n} \mu(d)\nu(n/d) = \sum_{d|n} \mu(d) = \iota(n)$, by Theorem 7.15.

7.4.29. Since $\nu(n)$ is identically 1, we have $F(n) = \sum_{d|n} f(d) = \sum_{d|n} f(d)\nu(n/d) = f * \nu(n)$. If we Dirichlet multiply both sides by μ, we have $F * \mu = f * \nu * \mu = f * \iota = f$. as desired.

7.4.31. From the Möbius inversion formula, we have $\Lambda(n) = \sum_{d|n} \mu(d)\log(n/d) = \sum_{d|n} \mu(d)(\log n - \log d) = \sum_{d|n} \mu(d)\log(n) - \sum_{d|n} \mu(d)\log(d) = \log n \sum_{d|n} \mu(d) - \sum_{d|n} \mu(d)\log(d) = \log n\nu(n) - \sum_{d|n} \mu(d)\log(d) = -\sum_{d|n} \mu(d)\log(d)$, since $\nu(n) = 0$ if n is not 1, and $\log n = 0$ if $n = 1$.

CHAPTER 8

Cryptology

8.1. Character Ciphers

8.1.1. We translate ATTACK AT DAWN into the corresponding numbers. We obtain 0 19 19 0 2 10 0 19 3 0 22 13. When encrypting this message using the Caesar cipher we obtain the numbers 3 22 22 3 5 13 3 22 6 3 25 16. Translating this into letters give DWWDF NDWGD ZQ.

8.1.3. We first translate the message SURRENDER IMMEDIATELY into the corresponding numbers. We obtain 18 20 17 17 4 13 3 4 17 8 12 12 4 3 8 0 19 4 11 24. We encipher each of these numbers using the transformation $C \equiv 11P + 18 \pmod{26}$. This gives 8 4 23 23 10 5 25 10 23 2 20 20 10 25 2 18 19 10 9 22. Translating back to letters gives IEXXK FZKXC UUKZC STKJW.

8.1.5. Since 5 is an inverse for 21 modulo 26, we have $P \equiv 5(C - 5) \equiv 5C + 1 \pmod{26}$ as the deciphering transformation. Converting the ciphertext to numbers gives us $24, 11, 5, 16, 23, 15, 2, 17, 8, 19$. We apply the deciphering transform to each number, for instance, $5 \cdot 24 + 1 \equiv 17 \pmod{26}$. Continuing in this fashion gives us 17, 4, 0, 3, 12, 24, 11, 8, 15, 18 which are the numerical equivalents of READM YLIPS.

8.1.7. Since E is the most common letter suppose that E is sent to Q. Since E corresponds to 4 and Q corresponds to 16, we have $4 + k \equiv \pmod{16}$. Hence $k = 12$.

8.1.9. By counting letter frequencies, we find that M is the most common letter, occuring 8 times. We guess that M stands for E, which would be a shift of 8. We subtract 8 from each letter and get ANIDE AISLI KEACH ILDNO NEISB ETTER THANY OUROW NFROM CHINE SEFOR TUNEC OOKIE

8.1.11. Since E=4 and T=19 are the most common letters in plaintext and W=22 and B=1, we have $22 \equiv a4 + b$ and $1 \equiv a19 + b \pmod{26}$. By Theorem 3.14, the solution to the system is $a \equiv 9, b \equiv 12 \pmod{26}$.

8.1.13. We count the frequencies of letters in the ciphertext and discover that A, B, T, and N appear most often, namely 6 times each. Let $P = D(C) \equiv cC + d \pmod{26}$. Then $D(A) = d$ is one of A, E, N, or S. From which we deduce that $d = 0, 4, 13$ or 18. Also $D(B) = c + d$, must be another one of these numbers. Since A, B, T, N is not a simple shift of A, E, N, S, we see that c is not 0. Assuming that d is also not zero, the possible pairs for (c, d) are $(9, 4), (14, 4), (5, 13), (17, 13), (12, 18)$, and $(21, 18)$. We try various of these and discover that $P \equiv 5C + 13$ is the deciphering transformation. Applying this to the ciphertext gives us THISM ESSAG EWASE NCIPH EREDU SINGA NAFFI NETRA NSFOR MATIO N

8.1.15. We have $C \equiv 17(5P + 13) + 3 \equiv 85P + 224 \equiv 7P + 16 \pmod{26}$.

8.2. Block and Stream Ciphers

8.2.1. We first translate the letters of the message DO NOT OPEN THIS ENVELOPE into the corresponding numbers, grouping letters into blocks of six. This gives 3 14 13 14 19 14 15 4 13 19 7 8 18 4 13 21 4 11 14 15 4. The letters of the word SECRET, which is the key, translate to 184217419. For each block $p_1 p_2 p_3 p_4 p_5 p_6$ we find $c_i \equiv p_i + k_i \pmod{26}$ where $0 \leq c_i \leq 25$ where $k_1 = 18, k_2 = 4, k_3 = 2, k_4 = 17, k_5 = 4$, and $k_6 = 19$. This gives 21 18 15 5 23 7 7 8 15 10 11 1 10 8 15 12 8 4 6 19 6. Translating this back to letters gives VSPFXH HIPLKB KIPMIE GTG.

8.2.3. The numerical equivalents for the key TWAIN are 19 22 0 8 13. The numerical equivalents for ANENGLISHMAN are 0 13 4 13 6 11 8 18 7 12 0 13. Adding the key numbers to the corresponding first five plaintext numbers yields 19 9 4 21 19, (mod 26) which stand for TJEVT. Adding the key numbers to the corresponding next five plaintext numbers yields 4 4 18 15 25, (mod 26) which stand for EESPZ. Adding the numbers for TA to the last two letters yields TJ. Continuing in this fashion we find the cipher text to be TJEVT EESPZ TJIAN IARAB GSHWQ HASBU BJGAO XYACF XPHML AWVMO XANLB GABMS HNEIA TIEZV VWNQF TLEZF HJWPB WKEAG AENOF UACIH LATPR RDADR GKTJR XJDWA XXENB KA

8.2.5. Let n be the key length, and suppose $k_1, k_2, \ldots, k_n$ are the numerical equivalents of the letters of the keyword. If $p_i = p_j$ are two plaintext characters separated by a multiple of the key length, when we separate the plaintext into blocks of length n, p_i and p_j will be in the same position in their respective blocks, say the mth position. So when we encrypt them, we get $c_i \equiv p_i + k_m \equiv p_j + k_m \equiv c_j \pmod{26}$.

8.2.7. Searching the ciphertext, we find two occurrences of KMK which are 42 positions apart, and two occurrences of PWQW which are 39 positions apart, so we guess that the period is $(42, 39) = 3$. The index of coincidence for the letters in positions $1, 4, 7, \ldots$ is 0.064, the index of coincidence for the letters in positions $2, 5, 8, \ldots$ is 0.072, and the index of coincidence for the letters in positions $3, 6, 9, \ldots$ is 0.068. Since these indexes are all about 0.065, we are sure that the period is 3. Counting frequencies of the letters in positions $1, 4, 7, \ldots$, we find 9 R's and 7 C's, and since $R - C = 15$, we suspect that $R \to T$ and $C \to E$, which implies that $l_1 = Y$. Counting frequencies for the letters in positions $2, 5, 8, \ldots$, we find 11 E's, 6 M's, 5 W's and 5 I's. We seek a difference among these letters which is the same as a difference among the commonly occurring letters E, T, N, R, I, and O. We find that $M - I = 4$ and $E - I = 4$, so we suspect that $M \to I$ and $I \to E$, which implies that $l_2 = E$. Counting frequencies for the letters in positions $3, 6, 9, \ldots$ we find 8 S's, 7 W's and 6 K's. We try successively assuming that each of these has plaintext E, and discover that $W \to E$ yields a sensible message with key YES. The plaintext is MISTA KESAR EAPAR TOFBE INGHU MANAP PRECI ATEYO URMIS TAKES FORWH ATTHE YAREP RECIO USLIF ELESS ONSTH ATCAN ONLYB ELEAR NEDTH EHARD WAYUN LESSI TISAF ATALM ISTAK EWHIC HATLE ASTOT HERSC ANLEA RNFRO M.

8.2.9. Searching the ciphertext, we find two occurrences of UPRW, which are 16 positions apart. We also find two occurrences of UQ, which are 12 positions apart, so we guess that the period is $(16, 12) = 4$. We compute the indexes of coincidence for each of the four letter groups and get 0.059, 0.055, 0.058, and 0.043, which are all significantly greater than 0.038, so we believe the keyword has 4 letters. Counting frequencies of the letters in positions $1, 5, 9, \ldots$, we find 6 F's and 5 U's, and since $U - F = 15$ and $T - E = 15$, we guess that $U \to T$ and so $l_1 = B$. Counting frequencies of the letters in positions $2, 6, 10, \ldots$, we find 6 Q's, 5 P's and 4 W's. Since $W - Q = O - I = 6$, we guess that $W \to O$, so that $l_2 = I$. Counting frequencies of letters in positions $3, 7, 11, \ldots$, we find 6 E's and 5 V's and R's. Since $V - E = 17 = E - N$, we guess that $V \to E$, and so $l_3 = R$. Counting frequencies of letters in positions $4, 8, 12, \ldots$, we find 5 D's, 4 K's and 3 each of B, H, J, O, U, and W. (Since the index of coincidence for this group was relatively small, we are not surprised at the more random-seeming distribution of letters.) After several attempts we guess that $W \to T$ and so $l_4 = D$. Using the keyword BIRD, we find the plaintext to be IONCE HADAS PARRO WALIG HTUPO NMYSH OULDE RFORA MOMEN TWHIL EIWAS HOEIN GINAV ILLAG EGARD ENAND IFELT THATI WASMO REDIS TINGU ISHED BYTHA TCIRC UMSTA NCETH ATISH OULDH AVEBE ENBYA NYEPA ULETI COULD HAVEW ORN.

8.2.11. Searching the ciphertext, we find two occurrences of ZEELN which are 40 positions apart. We also find two occurrences of SUMHR which are 45 positions apart, so we guess that the period is $(40, 45) = 5$. The indexes of coincidence for the five letter groups are 0.073, 0.062, 0.062, 0.059, and 0.089, which are all significantly greater than 0.038, so we are confirmed in our guess that the keyword has length 5. Counting frequencies of the letters in positions $1, 6, 11, \ldots$, we find 5 W's, S's and M's, and 4 G's and L's. Since $W - L = E - T \equiv 15 \pmod{26}$, we guess that $W \to E$, and so $l_1 = S$. Counting frequencies of the letters in positions $2, 7, 12, \ldots$, we find 6 A's, 5 T's and 4 E's. Since these letters should be frequent in the plaintext, we suspect that $A \to A$ and that $l_2 = A$. Counting frequencies of the letters in positions $3, 8, 13, \ldots$, we find 5 K's and 5 Z's, and since $Z - K = 15 = T - E$, we guess that $Z \to T$ and so $l_3 = G$. Counting frequencies of the letters in positions $4, 9, 14, \ldots$, we find 5 T's, 4 S's and H's, and

3 E's and N's. Since these letters should be frequent in the plaintext, we suspect that $A \to A$ and that $l_4 = A$. Counting frequencies of the letters in positions 5, 10, 15,..., we find 7 R's, 6 U's and 5 G's. Since $R - G = E - T = 15$, we guess that $R \to E$ and so $l_5 = N$. Using the keyword SAGAN we discover the plaintext to be BUTTH EFACT THATS OMEGE NIUSE SWERE LAUGH EDATD OESNO TIMPL YTHAT ALLWH OAREL AUGHE DATAR EGENI USEST HEYLA UGHED ATCOL UMBUS THEYL AUGHE DATFU LTONT HEYLA UGHED ATTHE WRIGH TBROT HERSB UTTHE YALSO LAUGH EDATB OZOTH ECLOW N.

8.2.13. We first translate BEWARE OF THE MESSENGER into numerical equivalents. This gives 01 04 22 00 17 04 14 05 19 07 04 12 04 18 18 04 13 06 04 17. We now encipher each block. We have $3 \cdot 1 + 10 \cdot 4 \equiv 17 \pmod{26}$, $9 \cdot 1 + 7 \cdot 4 \equiv 11 \pmod{26}$; $3 \cdot 22 + 10 \cdot 0 \equiv 14 \pmod{26}$, $9 \cdot 22 + 7 \cdot 0 \equiv 16 \pmod{26}$; $3 \cdot 17 + 10 \cdot 4 \equiv 13 \pmod{26}$, $9 \cdot 17 + 7 \cdot 4 \equiv 25 \pmod{26}$; $3 \cdot 14 + 10 \cdot 5 \equiv 14 \pmod{26}$, $9 \cdot 14 + 7 \cdot 5 \equiv 5 \pmod{26}$; $3 \cdot 19 + 10 \cdot 7 \equiv 23 \pmod{26}$, $9 \cdot 19 + 7 \cdot 7 \equiv 12 \pmod{26}$; $3 \cdot 4 + 10 \cdot 12 \equiv 2 \pmod{26}$, $9 \cdot 4 + 7 \cdot 12 \equiv 16 \pmod{26}$; $3 \cdot 4 + 10 \cdot 18 \equiv 10 \pmod{26}$, $9 \cdot 4 + 7 \cdot 18 \equiv 6 \pmod{26}$; $3 \cdot 18 + 10 \cdot 4 \equiv 16 \pmod{26}$, $9 \cdot 18 + 7 \cdot 4 \equiv 8 \pmod{26}$; $3 \cdot 13 + 10 \cdot 6 \equiv 21 \pmod{26}$, $9 \cdot 13 + 7 \cdot 6 \equiv 3 \pmod{26}$; $3 \cdot 4 + 10 \cdot 17 \equiv 0 \pmod{26}$, $9 \cdot 4 + 7 \cdot 17 \equiv 25 \pmod{26}$. This gives the enciphered values 17 11 14 16 13 25 14 5 23 12 2 16 10 6 16 8 21 3 0 25. Translating back to letters gives R L O Q N Z O F X M C Q K G Q I V D A Z.

8.2.15. The matrix $\begin{pmatrix} 13 & 4 \\ 9 & 1 \end{pmatrix}$ has inverse $\begin{pmatrix} 9 & 16 \\ 23 & 13 \end{pmatrix}$. The numerical values of R D are 17 and 3. Then $\begin{pmatrix} 9 & 16 \\ 23 & 13 \end{pmatrix} \begin{pmatrix} 17 \\ 3 \end{pmatrix} \equiv \begin{pmatrix} 19 \\ 14 \end{pmatrix} \pmod{26}$, and 19 14 are the numerical values for TO. Continuing in this fashion we have TO SL EE PP UR CH AN CE TO DR EA MX

8.2.17. RH NI TH and HE correspond to 17 7 13 8 19 7 and 7 4 respectively, so we have $\begin{pmatrix} 17 & 13 \\ 7 & 8 \end{pmatrix} \equiv \begin{pmatrix} a & b \\ c & d \end{pmatrix} \begin{pmatrix} 19 & 7 \\ 7 & 4 \end{pmatrix} \pmod{26}$. Since $\begin{pmatrix} 4 & 19 \\ 19 & 19 \end{pmatrix}$ is an inverse for $\begin{pmatrix} 19 & 7 \\ 7 & 4 \end{pmatrix}$ we have $\begin{pmatrix} a & b \\ c & d \end{pmatrix} \equiv \begin{pmatrix} 17 & 13 \\ 7 & 8 \end{pmatrix} \begin{pmatrix} 4 & 19 \\ 19 & 19 \end{pmatrix} = \begin{pmatrix} 3 & 24 \\ 24 & 25 \end{pmatrix} \pmod{26}$.

8.2.19. We have $\mathbf{C} \equiv \mathbf{AP} \pmod{26}$. Multiplying both sides on the left by $\mathbf{A}$ gives $\mathbf{AC} \equiv \mathbf{A}^2\mathbf{P} \equiv \mathbf{IP} \equiv \mathbf{P} \pmod{26}$. The congruence $\mathbf{A}^2 \equiv \mathbf{I} \pmod{26}$ follows since $\mathbf{A}$ is involutory. It follows that $\mathbf{A}$ is also a deciphering matrix.

8.2.21. We have $C \equiv C_1 C_2 P \equiv \begin{pmatrix} 5 & 1 \\ 25 & 4 \end{pmatrix} \begin{pmatrix} 2 & 1 \\ 1 & 4 \end{pmatrix} P \equiv \begin{pmatrix} 11 & 32 \\ 54 & 143 \end{pmatrix} P \equiv \begin{pmatrix} 11 & 6 \\ 2 & 13 \end{pmatrix} P$. Hence the product cipher is given by $C \equiv AP \pmod{26}$ where $A = \begin{pmatrix} 11 & 6 \\ 2 & 13 \end{pmatrix}$.

8.2.23. If the plaintext is grouped into blocks of size m, we may take $\frac{[m,n]}{m}$ of these blocks to form a super-block of size $[m,n]$. If $\mathbf{A}$ is the $m \times m$ enciphering matrix, form the $[m,n] \times [m,n]$ matrix $\mathbf{B}$ with $\frac{[m,n]}{m}$ copies of $\mathbf{A}$ on the diagonal and zeros elsewhere: $\mathbf{B} = \begin{pmatrix} \mathbf{A} & 0 & \cdots & 0 \\ 0 & \mathbf{A} & \cdots & 0 \\ \vdots & & \ddots & \vdots \\ 0 & \cdots & & \mathbf{A} \end{pmatrix}$ Then $\mathbf{B}$ will encipher $\frac{[m,n]}{m}$ blocks of size m at once. Similarly, if $\mathbf{C}$ is the $n \times n$ enciphering matrix, form the corresponding $[m,n] \times [m,n]$ matrix $\mathbf{D}$. Then by Exercise 8, $\mathbf{BD}$ is an $[m,n] \times [m,n]$ enciphering matrix which does everything at once.

8.2.25. Multiplication of $(0\cdots 010\cdots 0)\begin{pmatrix} p_1 \\ p_2 \\ \vdots \\ p_n \end{pmatrix}$ with the 1 in the ith place yields the 1×1 matrix (P_i). So if the jth row of a matrix $\mathbf{A}$ is $(0\cdots 010\cdots 0)$ then $\mathbf{A}\begin{pmatrix} p_1 \\ \vdots \\ p_n \end{pmatrix} = \begin{pmatrix} C_1 \\ \vdots \\ C_n \end{pmatrix}$ gives $C_j = P_i$. So if every row of $\mathbf{A}$ has its 1 in a different column, then each C_j is equal to a different P_i. Hence $\mathbf{A}$ is a "permutation" matrix.

8.2.27. The matrix $\begin{pmatrix} 3 & 2 \\ 7 & 11 \end{pmatrix}$ has inverse $\begin{pmatrix} 17 & 4 \\ 1 & 7 \end{pmatrix}$ modulo 26. We compute $\mathbf{P} \equiv \begin{pmatrix} 17 & 4 \\ 1 & 7 \end{pmatrix}\left(C - \begin{pmatrix} 8 \\ 19 \end{pmatrix}\right) \equiv \begin{pmatrix} 17 & 4 \\ 1 & 7 \end{pmatrix}\mathbf{C} + \begin{pmatrix} 22 \\ 15 \end{pmatrix}$ (mod 26).

8.2.29. The matrix $\begin{pmatrix} 5 & 2 \\ 11 & 15 \end{pmatrix}$ has inverse $\begin{pmatrix} 15 & 24 \\ 15 & 5 \end{pmatrix}$ modulo 26. We compute $\mathbf{P} \equiv \begin{pmatrix} 15 & 24 \\ 15 & 5 \end{pmatrix}\left(C - \begin{pmatrix} 14 \\ 3 \end{pmatrix}\right) \equiv \begin{pmatrix} 17 & 4 \\ 1 & 7 \end{pmatrix}\mathbf{C} + \begin{pmatrix} 4 \\ 9 \end{pmatrix}$ (mod 26). Applying this deciphering transformation to the numeric equivalents of the ciphertext and converting back to letters gives TOXIC WASTE as the plaintext.

8.2.31. Make a frequency count of the trigraphs and use a published English language count of frequencies of trigraphs. Then proceed as in problem 18. There are 12 variables to determine, so 4 guesses are needed.

8.2.33. Let $\mathbf{A}$ be an $m \times m$ matrix, $\mathbf{B}$ be an $m \times 1$ matrix, $\mathbf{D}$ be an $n \times n$ matrix, and $\mathbf{E}$ be an $n \times 1$ matrix. Form $mn \times mn$ matrices $\mathbf{X}$ and $\mathbf{Y}$ by placing n copies of $\mathbf{A}$ along the diagonal of $\mathbf{X}$ and m copies of $\mathbf{D}$ along the diagonal of $\mathbf{Y}$. Form $mn \times 1$ matrices $\mathbf{Z}$ and $\mathbf{W}$ by stringing n copies of $\mathbf{B}$ together and m copies of $\mathbf{E}$, respectively. Then the product transformation is given by $\mathbf{C} = \mathbf{YXP} + \mathbf{YZ} + \mathbf{W}$ which is an affine transformation based on a block size of mn.

8.2.35. To decrypt the string, we need to add corresponding bits of the string to the keystream to produce the string 21 1121 1012 and then reduce modulo 2 to get 01 1101 1010 as the plaintext.

8.2.37. We first convert the ciphertext to its numerical equivalents: 25 21 17 16 7 3 20 9 8 12. The seed is I which has numerical value 8. We subtract 8 from 25 to get 17 which stands for R. Then we subtract 17 from 21 to get 4 which stands for E. Then we subtract 4 from 17 to get 13 which stands for N. Then we subtract 13 from 16 to get 3 which stands for D. Then we subtract 3 from 7 to get 4 which stands for E. Then we subtract 4 from 3 to get $-1 \equiv 25$ (mod 26) which stands for Z. Then we subtract 25 from 20 to get $-5 \equiv 21$ (mod 26) which stands for V. Then we subtract 21 from 9 to get $-12 \equiv 14$ (mod 26) which stands for O. Then we subtract 14 from 8 to get $-6 \equiv 20$ (mod 26) which stands for U. Then we subtract 20 from 12 to get $-8 \equiv 18$ (mod 26) which stands for S. So the plaintext is RENDE ZVOUS.

8.2.39. Let $p_1 p_2 \cdots p_m$ and $q_1 q_2 \cdots q_m$ be two different plaintext bit streams. Let $k_1, k_2, \ldots, k_m$ be the keystream by which the plaintexts are encrypted. Then note that for any $i = 1, 2, \ldots, m$, $E_{k_i}(p_i) + E_{k_i}(q_i) = k_i + p_i + k_i + q_i = 2k_i + p_i + q_i \equiv p_i + q_i$ (mod 2). Therefore, by adding corresponding bits of the ciphertext streams, we get the sums of the corresponding bits of the plaintext streams. This can lead to the discovery of portions of the keystream. For instance if $p_i + q_i$ is known to be 2, then it is known that both p_i and q_i are 1. Then if $E_{p_i} = 1$ we know that $k_i = 0$, but if $E_{p_i} = 0$ then $k_i = 1$. Likewise, if $p_i + q_i$ is known to be 0, then it is known that both p_i and q_i are 0. Then if $E_{p_i} = 0$ we know that $k_i = 0$, but if $E_{p_i} = 1$ then $k_i = 1$. If significant portions of the keystream are discovered in this way, then decoded parts of each message will aid in deducing further pieces of the keystream, perhaps resulting in complete cryptanalysis.

8.3. Exponentiation Ciphers

8.3.1. Since $25 < p < 2525, m = 1$ The numerical equivalents for GOOD MORNING, in blocks of $2 = 2m$ digits, are 06 14 14 03 12 14 17 13 08 12 06. Raising each of these 2-digit numbers to the 3rd power and reducing modulo 101 gives: 14 17 17 27 11 17 65 76 07 76 14.

8.3.3. We find that 17 is an inverse of 5 modulo $28 = \phi(29)$. We raise each block to the 17th power and reduce modulo 29 to get 01 04 00 12 12 04 20 15, which are the numerical equivalents of BEAM ME UP.

8.3.5. We encipher messages using the transformation $c \equiv P^{11}$ (mod 31). The deciphering exponent is the inverse of 11 modulo 30 since $\phi(31) = 30$. But 11 is its own inverse modulo 30 since $11 \cdot 11 \equiv 121 \equiv 1$ (mod 30). It follows that 11 is both the enciphering and deciphering exponent.

8.4. Public Key Cryptography

8.4.1. Suppose that $n = pq = 14647$ and $\phi(n) = 14400$. Since $\phi(n) = (p-1)(q-1) = pq - (p+q) + 1$, we have $14400 = 14647 - (p+q) + 1$ we have $p + q = 248$. Also, we have $p - q = \sqrt{(p+q)^2 - 4n} = \sqrt{248^2 - 4 \cdot 14647} = \sqrt{2916} = 54$. When we add $p + q = 248$ and $p - q = 54$ we see that $2_p = 302$. Hence $p = 151$ and $q = 97$.

8.4.3. Since a block of ciphertext p is less than n, we must have $(p, n) = p$ or q. Therefore the cryptanalyst has a factor of n.

8.4.5. We first translate the letters of BEST WISHES into their numerical equivalents. We group together numbers into blocks of four digits since $n = 2669$. This gives 01041819220818070418. We use the transformation $C \equiv P^3$ (mod 2669) to encipher the message. We have $104^3 \equiv 1215$ (mod 2669), $1819^3 \equiv 1224$ (mod 2669), $2208^3 \equiv 1471$ (mod 2669), $1807^3 \equiv 23$ (mod 2669), $418^3 \equiv 116$ (mod 2669). Hence the ciphertext is 1215 1224 1471 0023 0116.

8.4.7. Since $2747 = 41 \cdot 67$, we have $\phi(2747) = 40 \cdot 66 = 2640$. An inverse for 13 modulo 2640 is 2437, so we raise each ciphertext block to the 2437 power modulo 2747. For instance, $2206^{2437} \equiv 0617$ (mod 2747). The entire plain text is 0617 0404 1908 1306 1823, which corresponds to the message GR EE TI NG SX.

8.4.9. We convert the plaintext into numerical equivalents and group into blocks of 4 (appending an X) to get 1804 1111 1314 2223. Applying the enciphering algorithm to the first block yields $C \equiv 1804 \cdot 1809 \equiv 2145$ (mod 2573). We encrypt the other blocks the same way. The ciphertext is 0872 2263 1537 2392.

8.4.11. No. It is as if the encryption key were $(e_1 e_2, n)$, and it is no more difficult (or easy) to discover the inverse of $e = e_1 e_2$ than it would be to discover the inverse of either of the factors modulo $\phi(n)$.

8.4.13. Suppose P is a plaintext message and the two encrypting exponents are e_1 and e_2. Let $a = (e_1, e_2)$. Then there exist integers x and y such that $e_1 x + e_2 y = 1$. Let $C_1 \equiv P^{e_1}$ (mod n) and $C_2 \equiv P^{e_2}$ (mod n) be the two cipher texts. Since $C_1, C_2, e_1,$ and e_2 are known to the decipherer, and since x and y are relatively easy to compute, then it is also easy to compute $C_1^x C_2^y \equiv P^{e_1 x} P^{e_2 y} \equiv P^{e_1 x + e_2 y} \equiv P^a$ (mod n). If $a = 1$, then P has been recovered. If a is fairly small, then it may not be too difficult to compute ath roots of P^a and thereby recover P.

8.4.15. Encryption works the same as for the two prime case. For decryption, we must compute an inverse d for e modulo $\phi(n) = (p-1)(q-1)(r-1)$ where $n = pqr$ the product of three primes. Then we proceed as in the two prime case.

8.5. Knapsack Ciphers

8.5.1. a. We have $3 < 5, 3 + 5 = 8 < 9, 3 + 5 + 9 = 17 < 19$, and $3 + 5 + 9 + 19 = 36 < 40$. Hence the sequence is super-increasing.

b. We have $2 < 6, 2+6 = 8 < 10$, but $2+6+10 = 18 > 15$. Hence the sequence is not super-increasing.

c. We have $3 < 7, 3+7 = 10 < 17, 3+7+17 = 27 < 30$, and $3+7+17+30 = 57 < 59$. Hence the sequence is super-increasing.

d. We have $11 < 21, 11+21 = 32 < 41, 11+21+41 = 73 < 81$, but $11+21+41+81 = 154 > 151$. Hence the sequence is not super-increasing.

8.5.3. Proceed by induction. Certainly $a_1 < 2a_1 < a_2$. Suppose $\sum_{j=1}^{n-1} a_j < a_n$. Then $\sum_{j=1}^{n} a_j = \sum_{j=1}^{n-1} a_j + a_n < a_n + a_n = 2a_n < a_{n+1}$. This completes the induction step.

8.5.5. We multiply each element by 17 and reduce modulo 163 to get: $(17, 51, 85, 7, 14, 45, 73)$.

8.5.7. 273 is the inverse of 17 modulo 464. We multiply each ciphertext element by 273 and reduce modulo 464 to get 242 59 280 101. Then $242 = 22 + 41 + 179$ and so corresponds to 01101 which is the binary equivalent for N. $59 = 18 + 41$ and 10100 stands for U. $280 = 18 + 83 + 179$ and 10011 stands for T. $101 = 18 + 83$ and 10010 stands for S. So the plaintext message is NUTS.

8.5.9. If the multipliers and moduli are $(w_1, m_1), (w_2, m_2), \ldots, (w_r, m_r)$, the inverse $\overline{w_1}, \overline{w_2}, \ldots, \overline{w_r}$ can be computed with respect to their corresponding moduli. Then we multiply and reduce succesively by $(\overline{w_r}, m_r), (\overline{w_{r-1}}, m_{r-1}), \ldots, (\overline{w_1}, m_1)$. The result will be the plaintext sequence of easy knapsack problems.

8.5.11. Since $5 \mid 15960$ the product must contain 95, the only element divisible by 5, so $15960 = 95 \cdot 168$. 8 is the only even element so we must have $95 \cdot 8 \cdot 21$ as the only possibility.

8.5.13. For $i = 1, 2, \ldots, n$, we have $b^{\alpha_i} \equiv a_i \pmod{m}$. Then $b^S \equiv P \equiv (b^{\alpha_1})^{x_1} (b^{\alpha_2})^{x_2} \cdots (b^{\alpha_n})^{x_n} \equiv b^{\alpha_1 x_1 + \cdots + \alpha_n x_n} \pmod{n}$. Then $S \equiv \alpha_1 x_1 + \cdots + \alpha_n x_n \pmod{\phi(m)}$. Since $S + k\phi(m)$ is also a logarithm of P to the base b we may take the congruence to be an equation. Since the $x_i = 0$ or 1, this becomes an additive knapsack problem on the sequence $(\alpha_1, \alpha_2, \ldots, \alpha_n)$.

8.6. Cryptographic Protocols and Applications

8.6.1. The first party, having chosen $k_1 = 27$, computes $y_1 \equiv 5^{27} \equiv 94 \pmod{103}$ and sends it to the second party. The second party, having chosen $k_2 = 31$, computes $K \equiv 94^{31} \equiv 90 \pmod{103}$.

8.6.3. We compute $K = ((7^3)^{10})^5 \equiv 7^{150} \pmod{601}$. Using a calculator or computational software we find $K \equiv 7^{150} \equiv 476 \pmod{601}$.

8.6.5. Let $k_1, k_2, \ldots, k_n$ be the private keys for parties 1 through n respectively. There are n steps in this protocol. The first step is for each of the parties 1 through n to compute the least positive residue of $r^{k_i} \pmod{p}$ and send this value y_i to the $i + 1$st party. (The nth party sends his value to the 1st party.) Now the ith party has the value y_{i-1} (where we take y_0 to be y_n.) The second step is for each party to compute the least positive residue of $y_{i-1}^{k_i} \pmod{p}$ and send this value to the $i + 1$st party. Now the ith party has the least positive residue of $r^{k_{i-1} + k_{i-2}} \pmod{p}$. This process is continued for a total of n steps. However, at the nth step, the computed value is not sent on to the next party. Then the ith party will have the least positive residue of $r^{k_{i-1} + k_{i-2} + \cdots + k_1 + k_n + k_{n-1} + \cdots k_{i+1} + k_i} \pmod{p}$, which is exactly the value of K desired.

8.6.7. a. We have $\phi(23 \cdot 47) = 1012$ and 675 an inverse for 3 (mod 1012). The numerical equivalents for CHEERS HAROLD are 02 07 04 04 17 18 07 00 17 14 11 03, using blocks of 2 since $25 < 1081 < 2525$. The first step to perform the transformation D$\equiv$P^{675} (mod 1081) on each block, which gives us 0867 1003 0394 0394 0521 0625 1003 0000 0521 0477 1022 0357. Next we perform the transformation C$\equiv$d^7 (mod 1829) which gives 0371 0354 0858 0858 0087 1369 0354 0000 0087 1543 1797 0535.

b. We have $\phi(31 \cdot 59) = 1740$ and 1243 an inverse for 7 modulo 1740. The numerical equivalents for SINCERELY AUDREY are 18 08 13 02 04 17 04 11 24 0 20 03 17 04 24. We take each and perform the transformation $D \equiv P^{1243}$ (mod 1829) and perform the transformation $C \equiv D^3$ (mod 1081). This gives 0833 0475 0074 0323 0621 0105 0621 0865 0421 0000 0746 0803 0105 0621 0421.

8.6.9. a. If $n_i < n_j$, the block sizes are chosen small enough so that each block is unique modulo n_i. Since $n_i < n_j$, each block will be unique modulo n_j after applying the transformation D_{k_i}. Therefore we can apply E_{k_j} to $D_{k_i}(P)$ and retain uniqueness of blocks. If $n_i > n_j$ the argument is similar.

b. If $n_i < n_j$, individual j receives $E_{k_j}(D_{k_i}(P))$ and know an inverse for e_j modulo $\phi(n_i)$. So he can apply $D_{k_j}(E_{k_j}(D_{k_i}(P))) = D_{k_i}(P)$. Since he also knows e_i, he can apply $E_{k_i}(D_{k_i}(P)) = P$ and discover the plaintext P. If $n_i > n_j$, individual j receives $D_{k_i}(E_{k_j}(P))$. Since he knows e_i he can apply $E_{k_i}(D_{k_i}(E_{k_j}(P))) = E_{k_j}(P)$. Since he also knows $\overline{e_j}$ he can apply $D_{k_j}(E_{k_j}(P)) = P$ and discover the plaintext P.

c. Since only individual i knows $\overline{e_i}$, only he can apply the transformation D_{k_i} and thereby make $E_{k_i}(D_{k_i}(P))$ intelligible.

d. $n_i = 2867 > n_j = 2537$, so we compute $D_{k_i}(E_{k_j}(P))$. Both n_i and $n_j > 2525$ so we use blocks of 4. REGARDS FRED becomes 1704 0600 1703 1805 1704 0323 (adding an X to fill out the last block.) $e_i = 11$ and $\phi(n_i) = 2760$, so $\overline{e_i} = 251$. We apply $E_{k_j} \equiv P^{e_j} \equiv P^{13}$ (mod 2537) to each block and get 1943 0279 0847 0171 1943 0088. Then we apply $D_{k_i}(E) \equiv E^{251}$ (mod 2867) and get 0479 2564 0518 1571 0479 1064. Now since $n_j < n_i$ individual j must send $E_{k_i}(D_{k_j}(P))$, $e_j = 13$, $\phi(2537) = 2436$ and $\overline{e_j} = 937$. Then $D_{k_j}(P) \equiv P^{937}$ (mod 2537) and $E_{k_i}(D) \equiv D^{11}$ (mod 2867). The cipher text is 1609 1802 0790 2508 1949 0267.

8.6.11. Suppose the master key $K = 3, p = 5, M_1 = 8, m_2 = 9, m_3 = 11$, and $t = 13$. Then $M = m_1 m_2 = 72 > p \cdot m_3 = 5 \cdot 11 = 55$ and $t = 13 < \frac{M}{p} = \frac{72}{5}$. We have $K_0 = K + tp; = 3 + 13 \cdot 5 = 68$. The shadows k_1, k_2, and k_3 are given by $k_1 \equiv 68 \equiv 4$ (mod 8), $k_2 \equiv 68 \equiv 5$ (mod 9), and $k_3 \equiv 68 \equiv 2$ (mod 11).

8.6.13. The 3 shadows from Exercise 11 are $k_1 = 4, k_2 = 5$ and $k_3 = 2$. If k_1 and k_2 are known, we solve the system of congruences $x \equiv 4$ (mod 8), $x \equiv 5$ (mod 9) to get $x = 68$. If k_1 and k_3 are known, we solve the system of congruences $x \equiv 4$ (mod 8), $x \equiv 2$ (mod 11) to get $x = 68$. If k_2 and k_3 are known, we solve the system of congruences $x \equiv 5$ (mod 9), $x \equiv 2$ (mod 11) to get $x = 68$. In all three cases we recover K_0. Then $K = K_0 - tp = 68 - 13 \cdot 5 = 3$.

CHAPTER 9

Primitive Roots

9.1. The Order of an Integer and Primitive Roots

9.1.1. a. Since the order of an integer modulo 5 divides $\phi(5) = 4$, the order of an integer modulo 5 must equal 1, 2, or 4. Since $2^2 \equiv 4 \not\equiv 1 \pmod 5$ the order of 2 modulo 5 is 4.

b. Since the order of an integer modulo 10 divides $\phi(10) = 4$, the order of an integer modulo 10 must equal 1, 2, or 4. Since $3^2 \equiv 9 \not\equiv 1 \pmod{10}$, the order of 3 modulo 10 is 4.

c. Since the order of an integer modulo 13 divides $\phi(13) = 12$, the order of an integer modulo 12 must equal 1, 2, 3, 4, 6, or 12. We have $10^2 = (-3)^2 \equiv 9 \pmod{13}$, $10^3 \equiv 9 \cdot (-3) \equiv -1 \pmod{13}$, $10^4 \equiv (-1) \cdot (-3) \equiv 3 \pmod{13}$, and $10^6 = 10^3 \cdot 10^3 \equiv (-1)^2 = 1 \pmod{13}$. If follows that the order of 10 modulo 13 is 6.

d. Since the order of an integer modulo 10 divides $\phi(10) = 4$, the order of an integer modulo 10 must equal 1, 2, or 4. We have $7^2 = 49 \equiv 9 \equiv 1 \pmod{10}$, hence the order of 7 modulo 10 is 4.

9.1.3. a. We have $\phi(6) = 2$, and $5^2 \equiv 1 \pmod 6$.

b. We have $\phi(11) = 10$, $2^2 \equiv 4$, $2^5 \equiv -1$, $2^{10} \equiv 1 \pmod{11}$.

9.1.5. Only 1, 5, 7, 11 are prime to 12. Each one squared is congruent to 1, but $\phi(12) = 4$.

9.1.7. Since $\phi(\phi(14)) = \phi(6) = 2$, there are 2: 3 and 5.

9.1.9. That $\text{ord}_n a = \text{ord}_n \bar{a}$ follows from the fact that $a^t \equiv 1 \pmod n$ if and only if $\bar{a}^t \equiv 1 \pmod n$. To see this, suppose that $a^t \equiv 1 \pmod n$. Then $\bar{a}^t \equiv (\bar{a}^t a^t)(a^t) \equiv (a\bar{a})^t a^t \equiv 1^t \cdot 1 \equiv 1 \pmod n$. The converse is shown in a similar manner.

9.1.11. We have $[r,s]/(r,s) \le \text{ord}_n ab \le [r,s]$

9.1.13. Let $r = \text{ord}_m a^t$, then $a^{tr} \equiv 1 \pmod m$, hence $tr \ge ts$ and $r \ge s$. Since $1 \equiv a^{st} \equiv (a^t)^s \pmod n$, we have $s \ge r$.

9.1.15. Suppose that r is a primitive root modulo the odd prime p. Then $r^{(p-1)/q} \not\equiv 1 \pmod p$ for all prime divisors q of $p-1$ since no smaller power than the $(p-1)$st of r is congruent to 1 modulo p. Conversely, suppose that $r^{(p-1)/q} \not\equiv 1 \pmod p$ for all prime divisors of $p-1$. Suppose that r is not a primitive root of p. Then there is an integer t such that $r^t \equiv 1 \pmod p$ with $t < p-1$. Since t must divide $p-1$, we have $p-1 = st$ for some positive integer s greater than 1. Then $(p-1)/s = t$. Let q be a prime divisor of s. Then $(p-1)/q = t(s/q)$, so that $r^{(p-1)/q} = r^{t(s/q)} = (r^t)^{s/q} \equiv 1 \pmod p$. This contradicts the original assumption, so r is a primitive root modulo p.

9.1.17. Since $2^{2^n} + 1 \equiv 0 \pmod{F_n}$, then $2^{2^n} \equiv -1 \pmod{F_n}$. Squaring gives $(2^{2^n})^2 \equiv 1 \pmod{F_n}$. Thus, $\text{ord}_{F_n} 2 \le 2^n 2 = 2^{n+1}$.

9.1.19. Note that $a^t < m = a^n - 1$ whenever $1 \le t < n$. Hence a^t cannot be congruent to 1 modulo m when t is a positive integer less than n. However, $a_n \equiv 1 \pmod m$ since $m = (a^n - 1) \mid (a^n - 1)$. It follows that ord$_m a = n$. Since ord$_m a \mid m$, we see that $n \mid \phi(m)$.

9.1.21. First suppose that pq is a pseudoprime to the base 2. By Fermat's Little Theorem, $2^p \equiv 2 \pmod{p}$, so there exists an integer k such that $2^p - 2 = kp$. Then $2^{M_p-1} - 1 = 2^{2^p-1} - 1 = 2^{kp} - 1$. This last expression is divisible by $2^p - 1 = M_p$ by Lemma 6.1. Hence, $2^{M_p-1} \equiv 1 \pmod{M_p}$, or $2^{M_p} \equiv 2 \pmod{M_p}$. Since pq is a pseudoprime to the base 2, we have $2^{pq} \equiv 2 \pmod{pq}$, so $2^{pq} \equiv 2 \pmod{p}$. But $2^{pq} \equiv (2^p)^q \equiv 2^q \pmod{p}$. Therefore $2^q \equiv 2 \pmod{p}$. Then there exists an integer l such that $M_q - 1 = 2^q - 2 = lp$. Then $2^{M_q-1} - 1 = 2^{2^q-2} = 2^{lp} - 1$, so $2^p - 1 = M_p$ divides $2^{M_q-1} - 1$. Therefore $2^{M_q} \equiv 2 \pmod{M_p}$. Then we have $2^{M_p M_q} \equiv (2^{M_p})^{M_q} \equiv 2^{M_q} \equiv 2 \pmod{M_p}$. Similarly, $2^{M_p M_q} \equiv 2 \pmod{M_q}$. By the Chinese remainder theorem, noting that M_p and M_q are relatively prime, we have $2^{M_p M_q} \equiv 2 \pmod{M_p M_q}$. Therefore $M_p M_q$ is a pseudoprime to the base 2. Conversely, suppose $M_p M_q$ is a pseudoprime to the base 2. From the reasoning in the proof of Theorem 6.6, we have that $2^{M_p} \equiv 2 \pmod{p}$. Therefore $2^{M_p M_q} \equiv 2^{(M_p-1)M_q + M_q} \equiv 2^{M_q} \equiv 2 \pmod{p}$. But since $M_p = 2^p - 1 \equiv 0 \pmod{M_p}$, we have that the order of 2 modulo M_p is p. Therefore $p | M_q - 1$. In other words, $2^q \equiv 2 \pmod{p}$. Then $2^{pq} \equiv 2^q \equiv 2 \pmod{p}$. Similarly, $2^{pq} \equiv 2 \pmod{q}$. Therefore, by the Chinese remainder theorem, $2^{pq} \equiv 2 \pmod{pq}$. Therefore, since pq is composite, it is a pseudoprime to the base 2.

9.1.23. Let $j = $ ord$_{\phi(n)} e$. Then $e^j \equiv 1 \pmod{\phi(n)}$. Since ord$_n P \mid \phi(n)$, we have $e^j \equiv 1 \pmod{\text{ord}_n P}$. Then by Theorem 8.2, $P^{e^j} \equiv P \pmod{n}$, so $C^{e^{j-1}} \equiv (P^e)^{e^{j-1}} \equiv P^{e^j} \equiv P \pmod{n}$ and $C^{e^j} \equiv P^e \equiv C \pmod{n}$.

9.2. Primitive Roots for Primes

9.2.1. a. By Lagrange's Theorem, there are at most 2 roots. Since $(\pm 3)^2 + 2 \equiv 0 \pmod{11}$, we have found all the roots.

b. By Lagrange's Theorem, there are at most 2 roots. Since $(\pm 1)^2 + 10 \equiv 0 \pmod{11}$, we have found all the roots.

c. By Lagrange's Theorem, there are at most 3 roots. Note that the polynomial factors thus $x^3 + x^2 + 2x + 2 = x^2(x+1) + 2(x+1) = (x^2+2)(x+1)$. So $x = -1$ is a solution, and the two solutions from part (a) are solutions, and this must be all.

d. By Lagrange's Theorem, there are at most 4 roots. Since the polynomial is an even function it suffices to check only the numbers $0, 1, 2, 3, 4$, and 5 as roots. We find that none of these work, so there are no roots of the polynomial modulo 11.

9.2.3. a. There are $\phi(7-1) = \phi(6) = 2$ primitive roots modulo 7.

b. There are $\phi(13-1) = \phi(12) = 4$ primitive roots of 13.

c. There are $\phi(17-1) = \phi(16) = 8$ primitive roots of 17.

d. There are $\phi(19-1) = \phi(18) = 6$ primitive roots of 19.

e. There are $\phi(29-1) = \phi(28) = 12$ primitive roots of 29.

f. There are $\phi(47-1) = \phi(46) = 22$ primitive roots of 47.

9.2.5. There must be $\phi(\phi(13)) = 4$ primitive roots modulo 13. Since 2 is one, the others must be 2 raised to a power relatively prime to $\phi(13) = 12$. So we take $2^5 \equiv 6 \pmod{13}$, $2^7 \equiv 6 \cdot 4 \equiv 11 \pmod{13}$, and $2^{11} \equiv 6 \cdot 6 \cdot 2 \equiv 7 \pmod{13}$. So a complete set of primitive roots is $2, 6, 7, 11$

9.2.7. Since $\phi(19) = 18$ and $\phi(18) = 6$, we seek 6 primitive roots for 19. Since 2 is one, we raise 2 to the powers which are relatively prime to 18, namely, $2^5, 2^7, 2^{11}, 2^{13}$, and 2^{17}. Reducing modulo 19 gives us 2, 3,

10, 13, 14, 15, as a complete set of primitive roots.

9.2.9. By Lagrange's Theorem there are at most two solutions to $x^2 \equiv 1 \pmod{p}$, and we know $x \equiv \pm 1$ are the two solutions. Since $p \equiv 1 \pmod 4$, $4 \mid (p-1) = \phi(p)$ so there is an element x of order 4 modulo p. Then $x^4 = (x^2)^2 \equiv 1 \pmod{p}$, so $x^2 \equiv \pm 1 \pmod p$. If $x^2 \equiv 1 \pmod p$ then x does not have order 4. Therefore $x^2 \equiv -1 \pmod p$.

9.2.11. a. Let $f(x) = a_n x^n + a_{n-1} x^{n-1} + \cdots a_0$ and let k be the largest integer such p does not divide a_k. Let $g(x) = a_k x^k + a_{k-1} x^{k-1} + \cdots a_0$. Then $f(x) \equiv g(x) \pmod p$ for every value of x. In particular $g(x)$ has the same set of roots as $f(x)$. Since the number of roots is greater than $n > k$, this contradicts Lagrange's theorem. Therefore, no such k exists and p must divide every coefficient of $f(x)$.

b. Note that the degree of $f(x)$ is $p-2$. By Fermat's little theorem we have that $x^{p-1} - 1 \equiv 0 \pmod p$, for $x = 1, 2, \ldots, p-1$. Further, each x in the same range is a zero for $(x-1)(x-2)\cdots(c-p+1)$. Therefore, each $x = 1, 2, \ldots, p-1$ is a root of $f(x)$. Since $f(x)$ has degree $p-2$ and $p-1$ roots, part (a) tells us that all the coefficients of $f(x)$ are divisible by p.

c. From part (b) we know that the constant term of $f(x)$ is divisible by p. The constant term is given by $f(0) = (-1)(-2)\cdots(-p+1) + 1 \equiv (-1)^{p-1}(p-1)! + 1 \equiv (p-1)! + 1 \equiv 0 \pmod p$, which is Wilson's theorem.

9.2.13. a. Since $q_i^{t_i} \mid \phi(p) = p-1$, by Theorem 9.8 there exists $\phi(q_i^{t_i})$ elements of order $q_i^{t_i}$ for each $i = 1, 2, \ldots, r$. Let a_i be a fixed element of this order.

b. Using induction and Exercise 10 of Section 9.1, we have $\operatorname{ord}_p(a) = \operatorname{ord}_p(a_1 a_2 \cdots a_r) = \operatorname{ord}_p(a_1 \cdots a_{r-1})$ $\operatorname{ord}_p(a_r) = \cdots = \operatorname{ord}_p(a_1) \cdots \operatorname{ord}_p(a_r)$ since $\{\operatorname{ord}_p(a_1), \operatorname{ord}_p(a_2), \ldots, \operatorname{ord}_p(a_r)\} = \{q_1^{t_1}, \ldots, q_r^{t_r}\}$ are pairwise relatively prime.

c. $\phi(29) = 28 = 2^2 7$, and $12^4 \equiv 1 \pmod{29}$, so $\operatorname{ord}_{29}(12) = 4$. Also, $16^7 \equiv 1 \pmod{29}$ so $\operatorname{ord}_{29}(16) = 7$. Then by part (b), $\operatorname{ord}_{29}(12 \cdot 16) = 4 \cdot 7 = 28$. Therefore $12 \cdot 16 = 192 \equiv 18 \pmod{29}$ is a primitive root modulo 29.

9.2.15. If n is odd, composite and not a power of 3, then the product in Exercise 14 is $\prod_{j=1}^r (n-1, p_j - 1) \geq (n-1, 3-1)(n-1, 5-1) \geq 2 \cdot 2 = 4$. So there must be two bases other than -1 and $+1$.

9.2.17. a. Suppose that $f(x)$ is a polynomial with integer coefficients of degree $n-1$. Suppose that $x_1, x_2, \cdots, x_n$ are incongruent modulo p where p is prime. Consider the polynomial $g(x) = f(x) - \sum_{j=1}^n \left(f(x_j) \prod_{i \neq j} (x - x_i) \overline{(x_j - x_i)} \right)$. Note that x_j, $j = 1, 2, \cdots, n$ is a root of this polynomial modulo p since its value at x_j is $f(x_j) - [0 + 0 + \cdots + f(x_j) \prod_{i \neq j} (x_j - x_i)\overline{(x_j - x_i)} + \cdots + 0] \equiv f(x_j) - f(x_j) \cdot 1 \equiv 0 \pmod p$. Since $g(x)$ has n incongruent roots modulo p and since it is of degree $n-1$ or less, we can easily us Lagrange's theorem (Theorem 8.6) to see that $g(x) \equiv 0 \pmod p$ for every integer x.

b. By part (a) we have $f(5) \equiv f(1)(5-2)\overline{(1-2)}(5-3)\overline{(1-3)} + f(2)(5-1)\overline{(2-1)}(5-3)\overline{(2-3)} + f(3)(5-1)\overline{(3-1)}(5-2)\overline{(3-2)} \equiv 8 \cdot 3 \overline{(-1)} 2 \cdot \overline{(-2)} + 2 \cdot 4 \cdot 1 \cdot 2 \cdot \overline{(-1)} + 4 \cdot 4 \cdot \overline{2} \cdot 3 \cdot 1 \equiv 8 \cdot 3 \cdot 10 \cdot 2 \cdot 5 + 2 \cdot 4 \cdot 1 \cdot 2 \cdot 10 + 4 \cdot 4 \cdot 6 \cdot 3 \cdot 1 \equiv 10 \pmod{11}$.

9.2.19. By Exercise 23 of Section 9.1, $j \mid \operatorname{ord}_{\phi(n)} e$. Here, $\phi(n) = \phi(pq) = 4p'q'$, so $j \mid \phi(4p'q') = 2(p'-1)(q'-1)$. Choose e to be a primitive root modulo p'. Then $p' - 1 = \phi(p') \mid \phi(\phi(n))$, so $p' - 1 \mid \operatorname{ord}_{\phi(n)} e$. The decrypter needs $e^j \equiv 1 \pmod n$, but this choice of e forces $j = p' - 1$, which will take quite some time to find.

9.3. The Existence of Primitive Roots

9.3.1. The positive integers that have a primitive root are 2,4 and integers of the form p^t, and $2p^t$ where p is prime and t is a positive integer. Hence the integers in the list that have a primitive root are 4, $10 = 2 \cdot 5$,

9.3.3. a. First note that 2 is a primitive root modulo 3. Since $2^{3-1} = 2^2 = 4 \not\equiv 1 \pmod{3^2}$, 2 is also a primitive root modulo 3^2.

b. First note that 2 is a primitive root modulo 5. Since $2^{5-1} = 2^4 = 16 \not\equiv 1 \pmod{5^2}$, 2 is also a primitive root modulo 5^2.

c. First note that 5 is a primitive root modulo 23. Since $5^{23-1} = 5^{22} \equiv 323 \not\equiv 1 \pmod{23^2}$, 5 is also a primitive root modulo 23^2.

d. First note that 2 is a primitive root modulo 29. Since $2^{29-1} = 2^{28} \equiv 30 \not\equiv 1 \pmod{29^2}$, 2 is also a primitive root modulo 29^2.

9.3.5. a. We know that 2 is a primitive root of 3 and also of 3^2 since $2^{(3-1)} = 4 \not\equiv 1 \pmod 9$. It follows that 2 is also a primitive root of 3^k for all positive integers k.

b. From Exercise 2(a) we know that 2 is a primitive root modulo 11^2. It follows that 2 is a primitive root modulo 11^k for all positive integers k.

c. From Exercise 2(b) we know that 2 is a primitive root modulo 13^2. It follows that 2 is a primitive root modulo 13^k for all positive integers k.

d. From Exercise 2(c) we know that 3 is a primitive root modulo 17^2. It follows that 3 is a primitive root modulo 17^k for all positive integers k.

9.3.7. a. Since 2 is even and primitive root for 5, we have by Theorem 9.14 that $5 + 2 = 7$ is a primitive root for 10.

b. Since 3 is odd and a primitive root for 17, we have by Theorem 9.14 that 3 is also a primitive root for 34.

c. Since 2 is even and a primitive root for 19, we have by Theorem 9.14 that $2 + 19 = 21$ is a primitive root for 38.

d. We have $50 = 2 \cdot 5^2$. By Exercise 3(b), 2 is a primitive root for 5^2. By Theorem 9.14, since 2 is even, $25 + 2 = 27$ is a primitive root for 50.

9.3.9. First note that 2 is primitive root of 11. Since 2 is even, Theorem 9.14 tells us that $2 + 11 = 13$ is a primitive root of 22. Hence the primitive roots of 22 are the least positive residues of 13^k where $1 \leq k < \phi(22) = 10$ and $(k, \phi(12)) = (k, 10) = 1$. These are the integers $13^1 = 13$, $13^3 \equiv 19 \pmod{22}$, $13^7 \equiv 7 \pmod{22}$, and $13^9 \equiv 17 \pmod{22}$. Hence the primitive roots of 22 are 7, 13, 17, and 19.

9.3.11. By Exercise 7 in Section 9.1, a complete set of primitive roots modulo 19 is $2, 3, 10, 13, 14, 15$. By Theorem 9.14, the odd numbers in this set are primitive roots of 38, and if we add 19 to each of the even numbers in this set, we also have primitive roots of 38. Thus we have $2 + 19, 3, 10 + 19, 13, 14 + 19, 15$ as all the primitive roots of 38. Reducing gives us $3, 13, 15, 21, 29, 33$.

9.3.13. Suppose that r is a primitive root of m and suppose further that $x^2 \equiv 1 \pmod m$. Let $x \equiv r^t \pmod m$ where $0 \leq t \leq p - 1$. Then $r^{2t} \equiv 1 \pmod m$. Since r is a primitive root, it follows that $\phi(m) \mid 2t$ so that $2t = k\phi(m)$ and $t = k\phi(m)/2$ for some integer k. We have $x \equiv r^t = r^{k\phi(m)/2} = r^{(\phi(m)/2)k} \equiv (-1)^k \equiv \pm 1 \pmod m$, since $r^{\phi(m)/2} \equiv -1 \pmod m$. Conversely, suppose that m has no primitive root. Then m is not of one of the forms $2, 4, p^a$, or $2p^a$ with with p and odd prime. So either 2 distinct primes divide m or $m = 2^b M$ with M and odd integer and $b > 1$ or $m = 2^b$ with $c > 2$. In each of these cases we have $\phi(m) = 2^c N$ with N odd and $c \geq 3$. From Theorem 9.12, we know there are at least 3 solutions y_1, y_2, y_2

to $y^2 \equiv 1 \pmod{2^c}$ and certainly $z \equiv 1 \pmod{N}$ is a solution of $x^2 \equiv 1 \pmod{N}$. By the Chinese remainder theorem, there is a unique solution modulo $2^c N$ of the system $x \equiv y_i \pmod{2^c}, z \equiv 1 \pmod{N}$ for $i = 1, 2, 3$. Since these solutions are distinct modulo m, at least one of them is not $\pm 1 \pmod{m}$.

9.3.15. By Theorem 9.12 we know that $\operatorname{ord}_{2^k} 5 = \phi(2^k)/2$. Hence the 2^{k-2} integers $5^j, j = 0, 1, \cdots, 2^{k-2} - 1$, are incongruent modulo $2k$. Similarly the 2^{k-2} integers $-5^j, j = 0, 1, \cdots, 2^{k-2} - 1$, are incongruent modulo 2^k. Note that 5^j cannot be congruent to -5^i modulo 2^k where i and j are integers since $5^j \equiv 1 \pmod 4$ but $-5^i \equiv 3 \pmod 4$. It follows that the integers $1, 5, \cdots, 5^{2^{k-2}-1}, -1, -5, \cdots, -5^{2^{k-2}-1}$ are 2^{k-1} incongruent integers modulo 2^k. Since $\phi(2^k) = 2^{k-1}$ and every integer of the form $(-1)^\alpha 5^\beta$ is relatively prime to 2^k, it follows that every odd integer is congruent to an integer of this form with $\alpha = 0$ or 1 and $0 \le \beta = 2^{k-2} - 1$.

9.4. Index Arithmetic

9.4.1. We first compute the least positive residues of the powers of 5 modulo 23. We have $5^1 \equiv 5 \pmod{23}$, $5^2 \equiv 2 \pmod{23}$, $5^3 \equiv 10 \pmod{23}$, $5^4 \equiv 4 \pmod{23}$, $5^5 \equiv 20 \pmod{23}$, $5^6 \equiv 8 \pmod{23}, 5^7 \equiv 17 \pmod{23}$, $5^8 \equiv 16 \pmod{23}$, $5^9 \equiv 11 \pmod{23}$, $5^{10} \equiv 9 \pmod{23}$, $5^{11} \equiv 22 \pmod{23}$, $5^{12} \equiv 18 \pmod{23}, 5^{13} \equiv 21 \pmod{23}$, $5^{14} \equiv 13 \pmod{23}$, $5^{15} \equiv 19 \pmod{23}$, $5^{16} \equiv 3 \pmod{23}, 5^{17} \equiv 15 \pmod{23}$, $5^{18} \equiv 6 \pmod{23}$, $5^{19} \equiv 7 \pmod{23}$, $5^{20} \equiv 12 \pmod{23}$, $5^{21} \equiv 14 \pmod{23}$, and $5^{22} \equiv 1 \pmod{23}$. Hence $\operatorname{ind}_5 1 = 22, \operatorname{ind}_5 2 = 2, \operatorname{ind}_5 3 = 16, \operatorname{ind}_5 4 = 4, \operatorname{ind}_5 5 = 1, \operatorname{ind}_5 6 = 18, \operatorname{ind}_5 7 = 19, \operatorname{ind}_5 8 = 6, \operatorname{ind}_5 9 = 10, \operatorname{ind}_5 10 = 3, \operatorname{ind}_5 11 = 9, \operatorname{ind}_5 12 = 20, \operatorname{ind}_5 13 = 14, \operatorname{ind}_5 14 = 21, \operatorname{ind}_5 15 = 17, \operatorname{ind}_5 16 = 8, \operatorname{ind}_5 17 = 7, \operatorname{ind}_5 18 = 12, \operatorname{ind}_5 19 = 15, \operatorname{ind}_5 20 = 5, \operatorname{ind}_5 21 = 13$, and $\operatorname{ind}_5 22 = 11$.

9.4.3. a. Suppose that $3^x \equiv 2 \pmod{23}$. We take indices with respect to the primitive root 5 of 23. This gives $\operatorname{ind}_5(3^x) = \operatorname{ind}_5 2$ which implies that $x \operatorname{ind}_5 3 \equiv \operatorname{ind}_5 2 \pmod{22}$. Since $\operatorname{ind}_5(3^x) = 16$ and $\operatorname{ind}_5 2 = 2$ it follows that $16 \equiv 2 \pmod{22}$. Hence $8x \equiv 1 \pmod{11}$. Since 7 is the inverse of 8 modulo 11, it follows that $x \equiv 7 \pmod{11}$, so that $x \equiv 7$ or $18 \pmod{22}$.

b. Taking indices gives us $x \operatorname{ind}_5 13 \equiv \operatorname{ind}_5 5 \pmod{22}$ or $14x \equiv 1 \pmod{22}$ which has no solutions since $(14, 22) = 2 \nmid 1$.

9.4.5. We use the table of indices on page 612 of the text. We see that 2 is a primitive root for 29. Taking indices base 2 of the congruence and expanding gives us $\operatorname{ind}_2 8 + 7 \operatorname{ind}_2 x \equiv \operatorname{ind}_2 b \pmod{28}$. From the table we have $3 + 7\operatorname{ind}_2 x \equiv \operatorname{ind}_2 b \pmod{28}$, which has a solution if and only if $(7, 28) = 7 \mid (\operatorname{ind}_2 b - 3)$. So $\operatorname{ind}_2 b - 3 = 0, 7, 14$, or 21, that is $\operatorname{ind}_2 b = 3, 10, 17$, or 24. This corresponds to $b = 8, 9, 21$, or 20, respectively.

9.4.7. Taking indices of the congruence gives us $x \operatorname{ind} x \equiv \operatorname{ind} x \pmod{22}$, so that $22 \mid (\operatorname{ind} x)(x - 1)$. If $(\operatorname{ind} x, 22) = 1$, then $22 \mid (x - 1)$, which is the case for $x = 5, 7, 10, 11, 14, 15, 17, 19, 20$, and 21, from the table on page 548. So any solution of the systems $x \equiv 1 \pmod{22}, x \equiv a \pmod{23}$, as a runs through the list above, is a solution to the congruence. If $(\operatorname{ind} x, 22) = 2$, then x is one of $2, 3, 4, 6, 8, 9, 12, 13, 16$, or 18, and $11 \mid (x - 1)$, so any solution to the systems $x \equiv 1 \pmod{11}, x \equiv b \pmod{23}$, as b runs through this list, is also a solution to the congruence. If $(\operatorname{ind} x, 22) = 11$, then $\operatorname{ind} x = 11$, so $x \equiv 22 \pmod{23}$, but this is not a solution. Finally, if $(\operatorname{ind} x, 22) = 22$, then $\operatorname{ind} x = 22$, so $x \equiv 1 \pmod{23}$. Since $23 \cdot 22 = 506$, we list the solutions modulo 506: 1, 12, 23, 24, 45, 46, 47, 67, 69, 70, 78, 89, 91, 92, 93, 100, 111, 115, 116, 133, 137, 138, 139, 144, 155, 161, 162, 177, 183, 184, 185, 188, 199, 207, 208, 210, 221, 229, 230, 231, 232, 243, 253, 254, 265, 275, 276, 277, 287, 299, 300, 309, 321, 322, 323, 331, 345, 346, 353, 367, 368, 369, 375, 386, 391, 392, 397, 413, 414, 415, 419, 430, 437, 438, 441, 459, 460, 461, 463, 483, 484, 485, 496, 505.

9.4.9. Suppose that $x^4 \equiv -1 \pmod{p}$ and let $y = \operatorname{ind}_r x$. Then, $-x$ is also a solution and by Exercise 8, $\operatorname{ind}_r(-x) \equiv \operatorname{ind}_r(-1) + \operatorname{ind}_r(x) \equiv (p-1)/2 + y \pmod{p-1}$. So without loss of generality we may take $0 < y < (p-1)/2$, or $0 < 4y < 2(p-1)$. Taking indices of both sides of the congruence yields $4y \equiv \operatorname{ind}_r(-1) \equiv (p-1)/2 \pmod{p-1}$, again using Exercise 8. So $4y = (p-1)/2 + m(p-1)$ for some m. But $4y < 2(p-1)$, so either $4y = (p-1)/2$ and so $p = 8y + 1$ or $4y = 3(p-1)/2$. In this last case, 3 must divide y, so we have $p = 8(y/3) + 1$. So in either case, p is of the desired form. Conversely, suppose $p = 8k + 1$ and let r be a primitive root of p. Take $x = r^k$. Then $x^4 \equiv r^{4k} \equiv r^{(p-1)/2} \equiv -1 \pmod{p}$ by Exercise 8. So this x is a solution.

9.4.11. We have $7 \equiv (-1)^1 5^2 \pmod{2^4}$ and $9 \equiv (-1)^0 5^2 \pmod{2^4}$. Hence the index systems of 7 and 9 modulo 16 are $(1,2)$ and $(0,2)$, respectively.

9.4.13. Since $7 \equiv (-1)5^2 \pmod{32}$ and $11 \equiv (-1)5^5 \pmod{32}$, we have that the index systems for 7 and 11 are $(1,2)$ and $(1,5)$ respectively. Let the index system for x be (α, β). Then by the rules in Exercise 12, the index system for $7x^9$ is $(1+9\alpha, 2+9\beta)$, which must equal the index system for 11. Therefore $1+9\alpha \equiv 1 \pmod 2$, so $\alpha = 0$. And $2+9\beta \equiv 5 \pmod 8$, so $\beta = 3$. Then $x \equiv (-1)^0 5^3 \equiv 29 \pmod{32}$. For the second congruence, we note that the index system for 3 and 17 are $(1,3)$ and $(0,4)$ respectively. Then the index system for 3^x is $(x, 3x)$ and we must have $x \equiv 0 \pmod 2$ while $3x \equiv 4 \pmod 8$. A solution to the second congruence is necessarily a solution to the first. So all solutions are given by $x \equiv 4 \pmod 8$.

9.4.15. We have $120 = 2^3 \cdot 3 \cdot 5$. The index system of 17 modulo 120 is $(\alpha, \beta, \gamma^1, \gamma^2)$ where $17 \equiv (-1)^\alpha 5^\beta \pmod{2^3}$, $17 \equiv 2^{\gamma_1} \pmod 3$, and $17 \equiv 2^{\gamma_2} \pmod 5$. We see that $17 \equiv 1 \equiv (-1)^0 5^0 \pmod{2^3}$, $17 \equiv 2^1 \pmod 3$, and $17 \equiv 2^1 \pmod 5$, so that $\alpha = 0$, $\beta = 0$, $\gamma_1 = 1$, and $\gamma_2 = 1$. Hence the index system of 17 modulo 120 is $(0,0,1,1)$. The index system of 41 modulo 120 is $(\alpha, \beta, \gamma_1, \gamma_2)$ where $41 \equiv (-1)^\alpha 5^\beta \pmod{2^3}$, $41 \equiv 2^{\gamma_1} \pmod 3$, and $41 \equiv 2^{\gamma_2} \pmod 5$. We see that $41 \equiv 1 \equiv (-1)^0 5^0 \pmod{2^3}$, $41 \equiv 2 \equiv 2^1 \pmod 3$, and $41 \equiv 1 \equiv 2^4 \pmod 5$. Hence $\alpha = 0$, $\beta = 0$, $\gamma_1 = 1$, and $\gamma_2 = 4$. Hence the index system of 41 modulo 120 is $(0,0,1,4)$.

9.4.17. We have $60 = 4 \cdot 3 \cdot 5$. We take $3, 2$, and 2 as primitive roots for $4, 3$, and 5 respectively. Then we find that the index system for 11 is $(1,1,0)$, while the index system for 43 is $(1,0,3)$. Let the index system for x be (α, β, γ). Applying the rules from Exercise 16, we have $(1+7\alpha, 1+7\beta, 0+7\gamma) = (1,0,3)$. Therefore $1+7\alpha \equiv 1 \pmod{\phi(4)}$, so $\alpha = 0$. Next, $1+7\beta \equiv 0 \pmod{\phi(3)}$, so $\beta = 1$. Next, $0+7\gamma \equiv 3 \pmod{\phi(3)}$, so $\gamma = 1$. Therefore the index system for x is $(0,1,1)$. Using the Chinese remainder theorem, we solve the system $x \equiv 3^0 \pmod 4$, $x \equiv 2^1 \pmod 3$, $x \equiv 2^1 \pmod 5$, to get that $x \equiv 17 \pmod{60}$.

9.4.19. We must have k odd for this exercise. We seek a solution to $x^k \equiv a \pmod{2^e}$. We take indices as described before Exercise 11. Suppose $a \equiv (-1)^\alpha 5^\beta$ and $x \equiv (-1)^\gamma 5^\delta$. Then we have $\text{ind} x^k = (k\gamma, k\delta)$ and $\text{ind} a = (\alpha, \beta)$, so $k\gamma \equiv \alpha \pmod 2$ and $k\delta \equiv \beta \pmod{2^{e-2}}$. Since k is odd, both congruences are solvable for γ and δ, which determine x.

9.4.21. First we show that $\text{ord}_{2^e} 5 = 2^{e-2}$. Indeed, $\phi(2^e) = 2^{e-1}$, so it suffices to show that the highest power of 2 dividing $5^{2^{e-2}} - 1$ is 2^e. We proceed by induction. The basis step is the case $e = 2$, which is true. Note that $5^{2^{e-2}} - 1 = (5^{2^{e-3}} - 1)(5^{2^{e-3}} + 1)$. The first factor is exactly divisible by 2^{e-1} by the induction hypothesis. The second factor differs from the first by 2, so it is exactly divisible by 2, therefore $5^{2^{e-2}} - 1$ is exactly divisible by 2^e, as desired. Hence, if k is odd, the numbers $\pm 5^k, \pm 5^{2k}, \ldots, \pm 5^{2^{e-2}k}$ are 2^{e-1} incongruent kth power residues, which is the number given by the formula. If 2^m exactly divides k, then $5^k \equiv -5^k \pmod{2^e}$, so the formula must be divided by 2, hence the factor $(k, 2)$ in the denominator. Further, 5^{2^m} has order $2^{e-2}/2^m$ if $m \le e-2$ and order 1 if $m > e-2$, so the list must repeat modulo 2^e every $\text{ord}_{2^e} 5^{2^m}$ terms, whence the other factor in the denominator.

9.4.23. a. From the first inequality in Case (*i*) of the proof of Theorem 6.10, if n is not square-free, the probability is strictly less than $2n/9$, which is substantially smaller than $(n-1)/4$ for large n. If n is square-free, the argument following inequality (9.6) shows that if n has 4 or more factors, then the probability is less than $n/8$. The next inequality shows that the worst case for $n = p_1 p_2$ is when $s_1 = s_2$ and s_1 is as small as possible, which is the case stated in this exercise.

b. We have $n - 1 = 2 \cdot 3^2 \cdot 7 \cdot 13^2 \cdot 29 \cdot 41 \cdot 197$, and $p_1 - 1 = 2 \cdot 3 \cdot 7 \cdot 29 \cdot 41$ and $p_2 - 1 = 2 \cdot 3 \cdot 7 \cdot 29 \cdot 41$. So that, using the notation in the proof of Case (*ii*) of Theorem 6.10, $t = 3^2 \cdot 7 \cdot 13^2 \cdot 29 \cdot 41 \cdot 197$, $t_1 = t_2 = \cdot 3 \cdot 7 \cdot 29 \cdot 41$, and $s_1 = 1$. Then the number of integers b with $1 \le b \le n-1$, for which n is a strong pseudoprime to the base b is $T_1 T_2 (1 + \sum_{j=0}^{0} 2^{j2}) = (3 \cdot 7 \cdot 29 \cdot 41)^2 (2)$. so the probability that n is a strong pseudoprime to the base b is $2(\cdot 3 \cdot 7 \cdot 29 \cdot 41)^2 / (n-1) = 2(\cdot 3 \cdot 7 \cdot 29 \cdot 41)^2 / (2 \cdot 3^2 \cdot 7 \cdot 13^2 \cdot$

$29 \cdot 41 \cdot 197) = 7 \cdot 29 \cdot 41/(13^2 \cdot 197) = 0.24999\ldots$.

9.5. Primality Tests Using Orders of Integers and Primitive Roots

9.5.1. We have $2^2 \equiv 4 \pmod{101}$, $2^5 \equiv 32 \pmod{101}$, $2^{10} \equiv (2^5)^2 \equiv 32^2 \equiv 14 \pmod{101}$, $2^{20} \equiv (2^{10})^2 \equiv 14^2 \equiv 95 \pmod{101}$, $2^{25} \equiv (2^5)^5 \equiv 32^5 \equiv (32^2)^2 32 \equiv 1024^2 32 \equiv 14^2 32 \equiv 196 \cdot 32 \equiv -6 \cdot 32 \equiv -192 \equiv 10 \pmod{101}$, $2^{50} \equiv (2^{25})^2 \equiv 10^2 \equiv 100 \equiv -1 \pmod{101}$, $2^{100} \equiv (2^{50})^2 \equiv (-1)^2 \equiv 1 \pmod{101}$. Since $2^{\frac{(101-1)}{q}} \not\equiv 1 \pmod{101}$ for every proper divisor q of 100, and $2^{\frac{(101-1)}{q}} \equiv 1 \pmod{101}$ it follows that 101 is prime.

9.5.3. Applying Corollary 9.18.1, we have $233 - 1 = 2^3 29$, $3^{116} \equiv -1 \pmod{233}$, and $3^8 \equiv 27 \not\equiv 1 \pmod{233}$. So 233 is prime.

9.5.5. The first condition implies $x^{F_n - 1} \equiv 1 \pmod{F_n}$. The only prime dividing $F_n - 1 = 2^{2^n}$ is 2, and $(F_n - 1)/2 = 2^{2^n - 1}$, so the second condition implies $2^{(F_n - 1)/2} \not\equiv 1 \pmod{F_n}$. Then by Theorem 9.18, F_n is prime.

9.5.7. Let p be a prime dividing n. By the hypotheses, $x_j^{n-1} \equiv 1 \pmod{n}$, but $(x_j^{(n-1)/q_j} - 1, n) = 1$, so we know that $\text{ord}_p x_j$ divides $n - 1$, but not $(n-1)/q_j$. Therefore $\text{ord}_p x_j$ is divisible by $p_j^{a_j}$ for some prime p_j dividing q_j. But since $\text{ord}_p x_j$ also divides $p - 1$ and since the q_j are pairwise relatively prime, it follows that $\prod_{j=1}^r p_j^{a_j}$ divides $p - 1$. Therefore, $p \geq 1 + \prod_{j=1}^r p_j^{a_j} \geq 1 + \prod_{j=1}^r b_j^{a_j} > \sqrt{n}$, by the last inequality of the hypotheses. Therefore, n can have only one such prime divisor, namely itself.

9.5.9. Since $n - 1 = 9928 = 2^3 17 \cdot 73$, we take $F = 2^3 17 = 136$ and $R = 73$, noting that $F > R$. We apply Pocklington's test with $a = 2$. We check (using a calculator or computational software) that $2^{9928} \equiv 1 \pmod{9929}$ and $(2^{9928/2} - 1, 9929) = 1$ and $(2^{9928/17} - 1, 9929) = 1$, since 2 and 17 are the only primes dividing F. Therefore n passes Pocklington's test and so is prime.

9.5.11. Note that $3329 = 2^8 13 + 1$ and $13 < 2^8$, so it is of the form which can be tested by Proth's test. We try $2^{(3329-1)/2} \equiv 2^{1664} \equiv 1 \pmod{3329}$ (using a calculator or computational software.) So Proth's test fails for $a = 2$. Next we try $a = 3$ and compute $3^{1664} \equiv -1 \pmod{3329}$, which shows that 3329 is prime.

9.5.13. We apply Pocklington's test to this situation. Note that $n - 1 = hq^k$, so we let $F = q^k$ and $R = h$ and observe that by hypothesis $F > R$. Since q is the only prime dividing F, we need only check that there is an integer a such that $a^{n-1} \equiv 1 \pmod{n}$ and $(a^{(n-1)/q} - 1, n) = 1$. But both of these conditions are hypotheses, therefore n is prime by Pocklington's test.

9.6. Universal Exponents

9.6.1. a. Since the prime factorization of 100 is $100 = 2^2 5^2$ we have $\lambda(100) = [\lambda(2^2), \phi(5^2)] = [2, 20] = 20$.

b. Since the prime factorization of 144 is $144 = 2^4 3^2$ we have $\lambda(144) = [\lambda(2^4), \phi(3^2)] = [4, 6] = 12$.

c. Since the prime factorization of 22 is $222 = 2 \cdot 3 \cdot 37$, we have $\lambda(222) = [\lambda(2), \phi(3), \phi(37)] = [1, 2, 36] = 36$.

d. Since the prime factorization of 884 is $884 = 2^2 \cdot 13 \cdot 17$, we have $\lambda(884) = [\lambda(2^2), \phi(13), \phi(17)] = [2, 12, 16] = 48$.

e. We have $\lambda(2^4 \cdot 3^3 \cdot 5^2 \cdot 7) = [\lambda(2^4), \phi(3^3), \phi(5^2), \phi(7)] = [4, 18, 20, 6] = 180$.

f. We have $\lambda(2^5 \cdot 3^2 \cdot 5^2 \cdot 7^3 \cdot 11^2 \cdot 13 \cdot 17 \cdot 19) = [\lambda(2^5), \phi(3^2), \phi(5^2), \phi(7^3), \phi(11^2), \phi(13), \phi(17), \phi(19)] = [8, 6, 20, 294, 110, 12, 16, 18] = [2^3, 2 \cdot 3, 2^2 \cdot 5, 2 \cdot 3 \cdot 7^2, 2 \cdot 5 \cdot 11, 2^2 \cdot 3, 2^4, 2 \cdot 3^2] = 2^4 \cdot 3^2 \cdot 5 \cdot 7^2 \cdot 11 = 388080$.

g. Since $10! = 2^8 \cdot 3^4 \cdot 5^2 \cdot$ we have $\lambda(10!) = [\lambda(2^8), \phi(3^4), \phi(5^2), \phi(7)] = [64, 54, 20, 6] = 8640$.

h. Since $20! = 2^{18} \cdot 3^8 \cdot 5^4 \cdot 7^2 \cdot 11 \cdot 13 \cdot 17 \cdot 19$, it follows that $\lambda(20!) = [\lambda(2^{18}), \phi(3^8), \phi(5^4), \phi(7^2), \phi(11), \phi(13), \phi(17), \phi(19)] = [65536, 4374, 500, 42, 10, 12, 16, 18] = [2^{16}, 2 \cdot 3^7, 2^2 5^3, 2 \cdot 3 \cdot 7, 2 \cdot 5, 2^2 \cdot 3, 2^4, 2 \cdot 3^2] = 2^{16} \cdot 3^7 \cdot 5^3 \cdot 7 = 125411328000$.

9.6.3. We seek $n = 2^{t_0} p_1^{t_1} \cdots p_m^{t_m}$ such that $\lambda(n) = [\lambda(2^{t_0}), \phi(p_1^{t_1}), \ldots, \phi(p_m^{t_m})] = 12$. So we must have $\lambda(2^{t_0}) \mid 12$. For $t_0 \geq 3$, we have $\lambda(2^{t_0}) = 2^{t_0-2} \mid 12$, so the largest t_0 can be is $t_0 = 4$. We also must have $\phi(p_i^{t_i}) = p_i^{t_i-1}(p_i - 1) \mid 12$, so $p_i - 1 = 1, 2, 3, 4, 6$, or 12, and $p_i = 2, 3, 4, 5, 7$, or 13. But p_i is an odd prime, so $p_i = 3, 5, 7$, or 13 are the only possibilities for odd prime divisors of n. Also, $p_i^{t_i-1} \mid 12$, so if $t_i > 1$, we have that $p_i = 3$ and $t_i = 2$. Therefore the largest such n is $2^4 3^2 \cdot 5 \cdot 7 \cdot 13 = 65520$.

9.6.5. Suppose that $m = 2^{t_0} p_1^{t_1} p_s^{t_s}$. Then $\lambda(m) = [\lambda(2^{t_0}), \phi(p_1^{t_1}), \ldots, \phi(p_{t_s})]$. Furthermore, $\phi(m) = \phi(2^{t_0})\phi(p_1^{t_1}) \cdots \phi(p_s^{t_s})$. Since $\lambda(2^{t_0}) = 1, 2$, or 2^{t_0-2} when $t_0 = 1, 2$, or $t_0 \geq 3$, respectively, it follows that $\lambda(2^{t_0}) \mid \phi(2^{t_0}) = 2^{t_0-1}$. Since the least common multiple of a set of numbers divides the product of these numbers, or their multiples, we see that $\lambda(m) \mid \phi(m)$.

9.6.7. For any integer x with $(x, n) = (x, m) = 1$ we have $x^a \equiv 1 \pmod{n}$ and $x^a \equiv 1 \pmod{m}$. Then the Chinese remainder theorem gives us $x^a \equiv 1 \pmod{[n, m]}$. But since n is the largest integer with this property, we must have $[n, m] = n$, so $m \mid n$.

9.6.9. Suppose that $ax \equiv b \pmod{m}$. Multiplying both sides of this congruence by $a^{\lambda(m)-1}$ gives $a^{\lambda(m)} x \equiv a^{\lambda(m)-1} b \pmod{m}$. Since $a^{\lambda(m)} \equiv 1 \pmod{m}$, it follows that $x \equiv a^{\lambda(m)-1} b \pmod{m}$. Conversely, let $x_0 \equiv a^{\lambda(m)-1} b \pmod{m}$. then $ax_0 \equiv aa^{\lambda(m)-1} b \equiv a^{\lambda(m)} b \equiv b \pmod{m}$, so x_0 is a solution

9.6.11. a. First suppose that $m = p^a$. Then we have $x(x^{c-1} - 1) \equiv 0 \pmod{p^a}$. Let s be a primitive root for p^a, then the solutions to $x^{c-1} \equiv 1$ are exactly the powers s^k with $(c-1)k \equiv 1 \pmod{\phi(p^a)}$, and there are $(c-1, \phi(p^a))$ of these. Also, 0 is a solution, so we have $1 + (c-1, \phi(p^a))$ solutions all together. Now if $m = p_1^{a_1} \cdots p_r^{a_r}$, we can count the number of solutions modulo $p_i^{a_i}$ for each i. There is a one-to-one correspondence between solutions modulo m and the set of r-tuples of solutions to the system of congruences modulo each of the prime powers. The correspondence is given by the Chinese Remainder Theorem.

b. Suppose $(c-1, \phi(m)) = 2$, then $c-1$ is even. Since $\phi(p^a)$ is even for all prime powers, except 2, we have $(c-1, \phi(p_i^{a_i})) = 2$ for each i. Then by part (a), we have the number of solutions $= 3^r$. If 2^1 is a prime factor, then $\phi(m) = \phi(m/2)$, and since x^c and x have the same parity, x is a solution modulo m if and only if it is a solution modulo $m/2$, so the proposition still holds.

9.6.13. Let $n = 3pq$, with $p < q$ odd primes, be a Carmichael number. Then by Theorem 9.27, $p-1 \mid 3pq - 1 = 3(p-1)q + 3q - 1$, so $p-1 \mid 3q - 1$, say $(p-1)a = 3q - 1$. Since $q > p$, we must have $a \geq 4$. Similarly, there is an integer b such that $(q-1)b = 3p - 1$. Solving these two equations for p and q yields $q = (2a + ab - 3)/(ab - 9)$, and $p = (2b + ab - 3)/(ab - 9) = 1 + (2b + 6)/(ab - 9)$. Then since p is an odd prime greater than 3, we must have $4(ab - 9) \leq 2b + 6$, which reduces to $b(2a - 1) \leq 21$. Since $a \geq 4$, this implies that $b \leq 3$. Then $4(ab - 9) \leq 2b + 6 \leq 12$, so $ab \leq 21/4$, so $a \leq 5$. Therefore $a = 4$ or 5. If $b = 3$, then the denominator in the expression for q is a multiple of 3, so the numerator must be a multiple of 3, but that is impossible since there is no choice for a which is divisible by 3. Thus $b = 1$ or 2. The denominator of q must be positive, so $ab > 9$, which eliminates all remaining possibilities except $a = 5$, $b = 2$, in which case $p = 11$ and $q = 17$. So the only Carmichael number of this form is $561 = 3 \cdot 11 \cdot 17$.

9.6.15. Assume $q < r$. By Theorem 9.23, $q - 1 \mid pqr - 1 = (q-1)pr + pr - 1$. Therefore $q - 1 \mid pr - 1$, say $a(q-1) = pr - 1$. Similarly $b(r-1) = pq - 1$. Since $q < r$, we must have $a > b$. Solving these two equations for q and r yields $r = (p(a-1) + a(b-1))/(ab - p^2)$ and $q = (p(b-1) + b(a-1))/(ab - p^2) = 1 + (p^2 + pb - p - b)/(ab - p^2)$. Since this last fraction must be an integer we have $ab - p^2 \leq p^2 + pb - p - b$ which reduces to $a(b-1) \leq 2p^2 + p(b-1)$ or $a - 1 \leq 2p^2/b + p(b-1)/b \leq 2p^2 + p$. So there are only finitely many values for a. Likewise, the same inequality gives us $b(a-1) \leq 2p^2 + pb - p$ or $b(a - 1 - p) \leq 2p^2 - p$. Since $a > b$ and the denominator of the expression for q must be positive, we have that $a \geq p + 1$.

If $a = p+1$, we have $(p+1)(q-1) = pq - p + q - 1 = pr - 1$, which implies that $p|q$, a contradiction. Therefore $a > p+1$, and so $a - 1 - p$ is a positive integer. The last inequality gives us $b \leq b(a-1-p) \leq 2p^2 - p$. Therefore there are only finitely many values for b. Since a and b determine q and r, we see that there can be only finitely many Carmichael numbers of this form.

9.6.17. We have $q_n(ab) \equiv ((ab)^{\lambda(n)} - 1)/n = (a^{\lambda(n)}b^{\lambda(n)} - a^{\lambda(n)} - b^{\lambda(n)} + 1 + a^{\lambda(n)} + b^{\lambda(n)} - 2)/n = (a^{\lambda(n)} - 1)(b^{\lambda(n)} - 1)/n + ((a^{\lambda(n)} - 1) + (b^{\lambda(n)} - 1))/n \equiv q_n(a) + q_n(b) \pmod{n}$. At the last step, we use the fact that n^2 must divide $(a^{\lambda(n)} - 1)(b^{\lambda(n)} - 1)$, since $\lambda(n)$ is the universal exponent.

CHAPTER 10

Applications of Primitive Roots and the Order of an Integer

10.1. Pseudorandom Numbers

10.1.1. First term: 69; second term: 76, since $69^2 = 4761$; third term: 77, since $76^2 = 5776$; fourth term: 92, since $77^2 = 5929$; fifth term: 46, since $92^2 = 8464$; sixth term: 11, since $46^2 = 2116$; seventh term: 12, since $11^2 = 0121$; eighth term: 14, since $12^2 = 0144$; ninth term: 19, since $14^2 = 0196$; tenth term: 36, since $19^2 = 0361$; eleventh term: 29, since $36^2 = 1296$; twelfth term: 84, since $29^2 = 0841$; thirteenth term: 05, since $84^2 = 7056$; fourteenth term: 02, since $5^2 = 0025$; fifteenth term: 00, since $02^2 = 0004$; sixteenth term and all remaining terms are 00, since $0^2 = 0000$.

10.1.3. We compute $x_0 = 2, x_1 = 15, x_2 = 17, x_3 = 0, x_4 = 7, x_5 = 10, x_6 = 22, x_7 = 20, x_8 = 12, x_9 = 15$, and $x_10 = 2 = x_0$. So the period length is 10.

10.1.5. a. From Theorem 10.2, we must have $a \equiv 1 \pmod{4}$ since $4 \mid 1000$, and since 5 is the only odd prime dividing 1000, we must also have $a \equiv 1 \pmod{5}$. By the Chinese remainder theorem, we have $a \equiv 1 \pmod{20}$.

b. We have $30030 = 2 \cdot 3 \cdot 5 \cdot 7 \cdot 11 \cdot 13$. So we solve the system $a \equiv 1 \pmod{m}$, for $m = 2, 3, 5, 7, 11, 13$, to get $a \equiv 1 \pmod{30030}$.

c. We have $10^6 - 1 = 3^2 7 \cdot 11 \cdot 13 \cdot 37$, so we must have $a \equiv 1 \pmod{m}$ for $m = 3, 7, 11, 13, 37$. This system has solutions $a \equiv 1 \pmod{111111}$, by the Chinese remainder theorem.

d. We have $2^{25} - 1 = 31 \cdot 601 \cdot 1801$, so we must have $a \equiv 1 \pmod{m}$ for $m = 31, 601, 1801$. By the Chinese remainder theorem, we have $a \equiv 1 \pmod{2^{25} - 1}$.

10.1.7. a. Since $2^{31} \equiv 1 \pmod{M_{31}}$, the order of 2 must be a divisor of 31. Since 31 is prime, the order must be 31, which is the period length.

b. Using computational software, we compute $3^{(M_{31}-1)/p} \pmod{M_{31}}$ for each prime power divisor p^k of $M_{31} - 1$. The residue is 1 only for $p = 3$, but not for $p^k = 3^2$. Therefore the period length is $(M_{31} - 1)/3 = 715827882$.

c. From part (a) we have $4^{31} = (2^{31})^2 \equiv 1 \pmod{M_{31}}$, so the order of 4, and hence the period length, must be 31.

d. Using computational software, we compute $5^{(M_{31}-1)/p} \pmod{M_{31}}$ for each prime power divisor p^k of $M_{31}-1$. The residue is 1 only for $p = 11$. Therefore the period length is $(M_{31}-1)/11 = 195225786$.

e. Using computational software, we compute $13^{(M_{31}-1)/p} \pmod{M_{31}}$ for each prime power divisor p^k of $M_{31} - 1$. The residue is 1 only for $p = 2$. Therefore the period length is $(M_{31} - 1)/2 = 1073741823$.

f. Using computational software, we compute $17^{(M_{31}-1)/p} \pmod{M_{31}}$ for each prime power divisor p^k of $M_{31} - 1$. The residue is 1 only for $p = 2$. Therefore the period length is $(M_{31} - 1)/2 =$

1073741823.

10.1.9. We compute $x_1 \equiv 8^2 \equiv 64 \pmod{77}$, $x_2 \equiv 64^2 \equiv 15 \pmod{77}$, $x_3 \equiv 15^2 \equiv 71 \pmod{77}$, $x_4 \equiv 71^2 \equiv 36 \pmod{77}$, and $x_5 \equiv 36^2 \equiv 64 \equiv x_1 \pmod{77}$. So the sequence of numbers is $8, 64, 15, 71, 36, 64, \ldots$.

10.1.11. First we compute $\text{ord}_{77} 8$. Since $8 \equiv 1 \pmod{7}$ and $8^{10} \equiv 1 \pmod{11}$ by Fermat's Little Theorem, the Chinese remainder theorem shows that $8^{10} \equiv 1 \pmod{77}$. Since $8^5 \equiv 43 \pmod{77}$, we know that $\text{ord}_{77} 8 = 10$. Therefore $t = 1$ and $s = 5$. Since 2 is a primitive root modulo 5, we know that $\text{ord}_5 2 = 4$. So by Theorem 10.4, the period length is 4.

10.1.13. Using the notation of Theorem 10.4, we have $\phi(77) = 60$, so $\text{ord}_{77} x_0$ is a divisor of $60 = 2^2 3 \cdot 5$. Then the only possible values for s are the odd divisors of 60, which are 3, 5, and 15. Then we note that $2^2 \equiv 1 \pmod{3}$, $2^4 \equiv 1 \pmod{5}$, and $2^4 \equiv 16 \equiv 1 \pmod{15}$. In each case we have shown that $\text{ord}_s 2 \leq 4$. Hence by Theorem 10.4 the maximum period length is 4.

10.1.15. We have $x_0 = 1$ and $x_1 = 24$. Using the definition of the Fibonacci generator, it follows that $x_2 \equiv x_1 + x_0 \equiv 1 + 24 = 25 \pmod{31}$. Hence $x_2 = 25$. Continuing, we find that $x_3 \equiv x_2 + x_1 \equiv 25 + 24 = 49 \equiv 18 \pmod{31}$, so $x_3 = 18$. We compute successive terms in the same manner: $x_4 \equiv x_3 + x_2 \equiv 18 + 25 = 43 \equiv 12 \pmod{31}$, so $x_4 = 12$; $x_5 \equiv x_4 + x_3 \equiv 12 + 18 = 30 \pmod{31}$, so $x_5 = 30$; $x_6 \equiv x_5 + x_4 \equiv 30 + 12 = 42 \equiv 11 \pmod{31}$, so $x_6 = 11$; $x_7 \equiv x_6 + x_5 \equiv 11 + 30 = 41 \equiv 10 \pmod{31}$, so $x_7 = 10$; and $x_8 \equiv x_7 + x_6 \equiv 10 + 11 = 21 \pmod{31}$, so $x_8 = 21$. The terms x_i with $i = 0, 1, 2, \ldots, 8$ are 1, 24, 25, 18, 12, 30, 11, 10, and 21.

10.1.17. Check that 7 has maximal order 1800 modulo $2^{25} - 1$. To make a large enough multiplier, raise 7 to a power relatively prime to $\phi(2^{25} - 1) = 32400000$, for example, to the 11th power.

10.1.19. We must have $313a \equiv 145 \pmod{1000}$. Solving this congruence yields $a = 665$.

10.1.21. a. We compute $x_1 \equiv 2^3 \equiv 8 \pmod{15}$, and $x_2 \equiv 8^3 \equiv 64 \cdot 8 \equiv 4 \cdot 8 \equiv 32 \equiv 2 \pmod{15}$. Since $x_2 = x_0$, the sequence is $8, 2, 8, 2, 8, 2, \ldots$

b. We compute $x_1 \equiv 3^2 \equiv 9 \pmod{23}$, $x_2 \equiv 9^2 \equiv 81 \equiv 12 \pmod{23}$, $x_3 \equiv 12^2 \equiv 6 \pmod{23}$, $x_4 \equiv 6^2 \equiv 13 \pmod{23}$, $x_5 \equiv 13^2 \equiv 8 \pmod{23}$, $x_6 \equiv 8^2 \equiv 18 \pmod{23}$, $x_7 \equiv 18^2 \equiv 2 \pmod{23}$, $x_8 \equiv 2^2 \equiv 4 \pmod{23}$, $x_9 \equiv 4^2 \equiv 16 \pmod{23}$, $x_{10} \equiv 16^2 \equiv 3 \pmod{23}$. Since $x_{10} = x_0$, the sequence is $9, 12, 6, 13, 8, 18, 2, 4, 16, 3, 9, 12, 6, \ldots$

10.2. The ElGamal Cryptosystem

10.2.1. We select $k = 1234$ for our random integer. Converting the plaintext into numerical equivalents results in 0700 1515 2401 0817 1907 0300 2423, where we filled out the last block with an X. Using a calculator or computational software, we find $\gamma \equiv r^k \equiv 6^{1234} \equiv 517 \pmod{2551}$. Then for each block P we compute $\delta \equiv P \cdot b^k \equiv P \cdot 33^{1234} \equiv P \cdot 651 \pmod{2551}$. The resulting blocks are $0700 \cdot 651 \equiv 1622 \pmod{2551}$, $1515 \cdot 651 \equiv 1579 \pmod{2551}$, $2401 \cdot 651 \equiv 1839 \pmod{2551}$, $0817 \cdot 651 \equiv 1259 \pmod{2551}$, $1907 \cdot 651 \equiv 1671 \pmod{2551}$, $0300 \cdot 651 \equiv 1424 \pmod{2551}$ and $2423 \cdot 651 \equiv 855 \pmod{2551}$. Therefore, the ciphertext is $(517, 1622), (517, 1579), (517, 1839), (517, 1259), (517, 1671), (517, 1424), (517, 855)$. To decrypt this ciphertext, we compute $\gamma^{p-1-a} \equiv 517^{2551-1-13} \equiv 517^{2537} \equiv 337 \pmod{2551}$. Then for each block of the cipher text we compute $P \equiv 337 \cdot \delta \pmod{2551}$. For the first block we have $337 \cdot 1622 \equiv 0700 \pmod{2551}$ which was the first block of the plaintext. The other blocks are decrypted the same way.

10.2.3. We start by computing $\overline{\gamma^a} \equiv 2161^{2713-1-17} \equiv 2161^{2695} \equiv 167 \pmod{2713}$. Then multiplying the second number of each block and reducing yields $167 \cdot 660 \equiv 1700 \pmod{2713}$, $167 \cdot 1284 \equiv 0101 \pmod{2713}$, and $167 \cdot 1467 \equiv 0819 \pmod{2713}$. So the plaintext is 170001010819 which is equivalent to RABBIT.

10.2.5. First we compute $\gamma \equiv 3^{101} \equiv 2022 \pmod{2657}$. Using the Euclidean algorithm we can compute $\overline{101} \equiv 973 \pmod{2656}$ then the signature is given by $s \equiv (823 - 211 \cdot 2022)973 \equiv 833 \pmod{2656}$. To verify this signature, we compute $V_1 \equiv 2022^{833} 801^{2022} \equiv 1014 \pmod{2657}$ and $V_2 \equiv 3^{823} \equiv 1014 \pmod{2657}$. Since $V_1 = V_2$, the signature is verified.

10.2.7. Let $\delta_1 = P_1 b^k$ and $\delta_2 = P_2 b^k$ as in the ElGamal cryptosystem. If P_1 is known, it is easy to compute an inverse for P_1 modulo p. Then $b^k \equiv \overline{P_1}\delta_1 \pmod{p}$. Then it is also easy to compute an inverse for b^k $\pmod{p}$. Then $P_2 \equiv \overline{b^k}\delta_2 \pmod{p}$. Hence the plaintext P_2 is recovered.

10.3. An Application to the Splicing of Telephone Cables

10.3.1. a. Since 17 is prime it has a primitive root. Hence the maximal ± 1-exponent of 17 is $\phi(17)/2 = 8$.

b. Since 22 is of the form $2p$ where p is prime it has a primitive root. Hence the maximal ± 1-exponent of 22 is $\phi(22)/2 = 10/2 = 5$.

c. We see that $24 = 2^3 \cdot 3$ does not have a primitive root since it is not a power of a prime nor twice a power of a prime. Hence the maximal ± 1-exponent of 24 is $\lambda(24) = [\lambda(2^3), \phi(3)] = [2, 2] = 2$.

d. We see that $36 = 2^2 \cdot 3^2$ does not have a primitive root since it is not a power of a prime nor twice a power of a prime. Hence the maximal ± 1-exponent of 36 is $\lambda(36) = [\lambda(2^2), \phi(3^2)] = [2, 6] = 6$.

e. We see that $99 = 3^2 \cdot 11$ does not have a primitive root since it is not a power of a prime not twice a power of a prime. Hence the maximal ± 1-exponent of 99 is $\lambda(99) = [\phi(3^2), \phi(11)] = [6, 10] = 30$.

f. We see that $100 = 2^2 5^2$ does not have a primitive root since it is not a power of a prime nor twice a power of a prime. Hence the maximal ± 1-exponent of 100 is $\lambda(100) = [\lambda(2^2), \phi(5^2)] = [2, 20] = 20$.

10.3.3. a. By Theorems 10.6 and 9.23, the maximal ± 1-exponent of 50 is $\lambda_0(50) = \lambda(50) = [\lambda(2), \phi(25)] = 20$. We seek an integer with order 20 $\pmod{50}$ to be the spread. Since 3 is a primitive root for 5, either 3 or 8 is a primitive root for 25. It turns out that $\text{ord}_{25} 3 = 20 = \phi(25)$. So it follows that $\text{ord}_{50} 3 = 20$. So we choose $s = 3$ for our spread.

b. We compute $\lambda_0(76) = [\lambda(4), \phi(19)] = [2, 18] = 18$. Since 2 is a primitive root modulo 19, we consider $s = 19 + 2 = 21$ for our spread. A quick computation shows that $\text{ord}_{76} 21 = 18$.

c. We compute $\lambda_0(125) = \phi(125) = 100$. Since 2 is a primitive root for 5, we start there. Now $2^{50} \equiv -1 \pmod{125}$, so $\text{ord}_{100} 2 = 100$. Thus we use $s = 2$ as our spread.

CHAPTER 11

Quadratic Residues

11.1. Quadratic Residues and Nonresidues

11.1.1. a. We have $1^2 \equiv 2^2 \equiv 1 \pmod{3}$. Hence the quadratic residues of 3 are those integers congruent to 1 modulo 3.

b. We have $1^2 \equiv 4^2 \equiv 1 \pmod 5$ and $2^2 \equiv 3^2 \equiv 4 \pmod 5$. Hence the quadratic residues of 5 are those integers congruent to 1 or 4 modulo 5.

c. We have $1 \equiv 12^2 \equiv 1 \pmod{13}$, $2^2 \equiv 11^2 \equiv 4 \pmod{13}$, $3^2 \equiv 10^2 \equiv 9 \pmod{13}$, $4^2 \equiv 9^2 \equiv 3 \pmod{13}$, $5^2 \equiv 8^2 \equiv 12 \pmod{13}$, and $6^2 \equiv 7^2 \equiv 10 \pmod{13}$. Hence the quadratic residues of 13 are those integers congruent to 1, 3, 4, 9, 10, or 12 modulo 13.

d. We have $1^2 \equiv 18^2 \equiv 1 \pmod{19}$, $2^2 \equiv 17^2 \equiv 4 \pmod{19}$, $3^2 \equiv 16^2 \equiv 9 \pmod{19}$, $4^2 \equiv 15^2 \equiv 16 \pmod{19}$, $5^2 \equiv 14^2 \equiv 6 \pmod{19}$, $6^2 \equiv 13^2 \equiv 17 \pmod{19}$, $7^2 \equiv 12^2 \equiv 11 \pmod{19}$, $8^2 \equiv 11^2 \equiv 7 \pmod{19}$, and $9^2 \equiv 10^2 \equiv 5 \pmod{19}$. Hence the quadratic residues of 19 are those integers congruent to 1, 4, 5, 6, 7, 9, 11, 16, or 17 modulo 19.

11.1.3. From Exercise 1 (b) we have $\left(\frac{1}{5}\right) = \left(\frac{4}{5}\right) = 1$ and $\left(\frac{2}{5}\right) = \left(\frac{3}{5}\right) = -1$.

11.1.5. a. We compute $\left(\frac{7}{11}\right) \equiv 7^{(11-1)/2} \equiv 7^5 \equiv 49^2 \cdot 7 \equiv 5^2 \cdot 7 \equiv 3 \cdot 7 \equiv -1 \pmod{11}$

b. We compute $(7, 14, 21, 28, 35) \equiv (7, 3, 10, 6, 2) \pmod{11}$ and three of these are greater than $11/2$, so $\left(\frac{7}{11}\right) = (-1)^3 = -1$

11.1.7. We know that $\left(\frac{-2}{p}\right) = \left(\frac{-1}{p}\right)\left(\frac{2}{p}\right)$ by Theorem 11.4 Using Theorems 11.5 and 11.6 we have: If $p \equiv 1 \pmod 8$ then $\left(\frac{-2}{p}\right) = (1)(1) = 1$. If $p \equiv 3 \pmod 8$ then $\left(\frac{-2}{p}\right) = (-1)(-1) = 1$. If $p \equiv -1 \pmod 8$ then $\left(\frac{-2}{p}\right) = (-1)(1) = -1$. If $p \equiv -3 \pmod 8$ then $\left(\frac{-2}{p}\right) = (1)(-1) = -1$.

11.1.9. Since $p - 1 \equiv -1, p - 2 \equiv -2, \ldots, (p+1)/2 \equiv (p-1)/2 \pmod p$, we have $((p-1)/2)!^2 \equiv -(p-1)! \equiv 1 \pmod p$ by Wilson's theorem. (since $p \equiv 3 \pmod 4$ the minus signs cancel). By Euler's criterion, $((p-1)/2)!^{(p-1)/2} \equiv \left(\frac{1}{p}\right)\left(\frac{2}{p}\right) \cdots \left(\frac{(p-1)/2}{p}\right) \equiv (-1)^t \pmod p$, by definition of the Legendre symbol. Since $(p-1)/2! \equiv \pm 1 \pmod p$, and $(p-1)/2$ is odd, we have the result.

11.1.11. If $p \equiv 1 \pmod 4$, $\left(\frac{-a}{p}\right) = \left(\frac{-1}{p}\right)\left(\frac{a}{p}\right) = 1 \cdot 1 = 1$. If $p \equiv 3 \pmod 4$, $\left(\frac{-a}{p}\right) = \left(\frac{-1}{p}\right)\left(\frac{a}{p}\right) = (-1) \cdot 1 = -1$.

11.1.13. a. We will use properties of congruence to complete the square. Suppose that $x^2 + x + 1 \equiv 0 \pmod 7$. Adding $-7x + 8$ to both sides give $x^2 - 6x + 9 \equiv -7x + 8 \equiv 1 \pmod 7$. Hence $(x-3)^2 \equiv 1 \pmod 7$. Since the solutions of $y^2 \equiv 1 \pmod 7$ are $y \equiv 1$ and $y \equiv -1 \pmod 7$, this implies that $x - 3 \equiv 1 \pmod 7$ or $x - 3 \equiv -1 \pmod 7$. It follows that the solutions are those x satisfying $x \equiv 4 \pmod 7$ or $x \equiv 2 \pmod 7$.

b. Suppose that $x^2 + 5x + 1 \equiv 0 \pmod 7$. Adding $-7x$ to both sides gives $x^2 - 2x + 1 \equiv 7x \equiv 0 \pmod 7$. Hence $(x-1)^2 \equiv 0 \pmod 7$. It follows that $x - 1 \equiv 0 \pmod 7$, so all solutions are given by $x \equiv 1 \pmod 7$.

c. Suppose that $x^2 + 3x + 1 \equiv 0 \pmod 7$. Then adding $-7x + 3$ to both sides gives $x^2 - 4x + 4 \equiv -7x + 3 \equiv 3 \pmod 7$. Hence $(x - 2)^2 \equiv 3 \pmod 7$. But 3 is a quadratic nonresidue of 7. Hence there are no solutions.

11.1.15. Suppose that p is a prime that is at least 7. At least one of the three incongruent integers 2, 3, and 6 is a quadratic residue of p, because if neither 2 nor 3 is a quadratic residue of p then $2 \cdot 3 = 6$ is a quadratic residue of p. If 2 is a quadratic residue, then 2 and 4 are quadratic residues that differ by 2; if 3 is a quadratic residue, then 1 and 3 are quadratic residues that differ by 2; while if 6 is a quadratic residue then 4 and 6 are quadratic residues that differ by 2.

11.1.17. a. Since $p = 4n + 3$, $2n + 2 = (p+1)/2$. Then $x^2 \equiv (\pm a^{n+1})^2 \equiv a^{2n+2} \equiv a^{(p+1)/2} \equiv a^{(p-1)/2}a \equiv 1 \cdot a \equiv a \pmod p$ using the fact that $a^{(p-1)/2} \equiv 1 \pmod p$ since a is a quadratic residue of p.

   b. From Lemma 11.1, there are exactly two solutions to $y^2 \equiv 1 \pmod p$, namely $y \equiv \pm 1 \pmod p$. Since $p \equiv 5 \pmod 8$, -1 is a quadratic residue of p and 2 is a quadratic nonresidue of p. Since $p = 8n + 5$, we have $4n + 2 = (p-1)/2$ and $2n + 2 = (p+3)/4$. Then $(\pm a^{n+1})^2 \equiv a^{(p+3)/4} \pmod p$ and $(\pm 2^{2n+1}a^{n+1})^2 \equiv 2^{(p-1)/2}a^{(p+3)/4} \equiv -a^{(p+3)/4} \pmod p$ by Euler's criterion. We must show that one of $a^{(p+3)/4}$ or $-a^{(p+3)/4} \equiv a \pmod p$. Now a is a quadratic residue of p, so $a^{(p-1)/2} \equiv 1 \pmod p$ and therefore $a^{(p-1)/4}$ solves $x^2 \equiv 1 \pmod p$. But then $a^{(p-1)/4} \equiv \pm 1 \pmod p$, that is $a^{(p+3)/4} \equiv \pm a \pmod p$ or $\pm a^{(p+3)/4} \equiv a \pmod p$ as desired.

11.1.19. Note $x^2 \equiv 1 \pmod{15}$ if and only if $x^2 \equiv 1 \pmod 3$ and $x^2 \equiv 1 \pmod 5$. The solutions to $x^2 \equiv 1 \pmod 3$ are $x \equiv 1$ and $x \equiv 2 \pmod 3$ and the solutions to $x^2 \equiv 1 \pmod 5$ are $x \equiv 1$ and $x \equiv 4 \pmod 5$. We use the Chinese remainder theorem to solve the four sets of simultaneous congruences: $x \equiv 1 \pmod 3$ and $x \equiv 1 \pmod 5$, $x \equiv 1 \pmod 3$ and $x \equiv 4 \pmod 5$, $x \equiv 2 \pmod 3$ and $x \equiv 1 \pmod 5$, and $x \equiv 2 \pmod 3$ and $x \equiv 4 \pmod 5$. This yields the four incongruence solutions $x \equiv 1, 4, 11,$ and $14 \pmod{15}$.

11.1.21. Note that $1001 = 7 \cdot 11 \cdot 13$, so we solve the congruence modulo each of these primes. First we have $x^2 \equiv 207 \equiv 4 \pmod 7$, so $x \equiv 2$ or $5 \pmod 7$. Next we have $x^2 \equiv 207 \equiv 9 \pmod{11}$, so $x \equiv 3$ or $8 \pmod{11}$. Next we have $x^2 \equiv 207 \equiv -1 \pmod{13}$, so $x \equiv 5$ or $8 \pmod{13}$. There are now 8 systems of three congruences each to solve via the Chinese remainder theorem. The solution of $x \equiv 2 \pmod 7$, $x \equiv 3 \pmod{11}$, $x \equiv 5 \pmod{13}$ is $x \equiv 135 \pmod{1001}$. The solution of $x \equiv 2 \pmod 7$, $x \equiv 3 \pmod{11}$, $x \equiv 8 \pmod{13}$ is $x \equiv 905 \pmod{1001}$. The solution of $x \equiv 2 \pmod 7$, $x \equiv 8 \pmod{11}$, $x \equiv 5 \pmod{13}$ is $x \equiv 954 \pmod{1001}$. The solution of $x \equiv 2 \pmod 7$, $x \equiv 8 \pmod{11}$, $x \equiv 8 \pmod{13}$ is $x \equiv 723 \pmod{1001}$. The solution of $x \equiv 5 \pmod 7$, $x \equiv 3 \pmod{11}$, $x \equiv 5 \pmod{13}$ is $x \equiv 278 \pmod{1001}$. The solution of $x \equiv 5 \pmod 7$, $x \equiv 3 \pmod{11}$, $x \equiv 8 \pmod{13}$ is $x \equiv 47 \pmod{1001}$. The solution of $x \equiv 5 \pmod 7$, $x \equiv 8 \pmod{11}$, $x \equiv 5 \pmod{13}$ is $x \equiv 96 \pmod{1001}$. The solution of $x \equiv 5 \pmod 7$, $x \equiv 8 \pmod{11}$, $x \equiv 8 \pmod{13}$ is $x \equiv 866 \pmod{1001}$. In order, the solutions modulo 1001 are 47, 96, 135, 278, 723, 866, 905, and 954.

11.1.23. If $x_0^2 \equiv a \pmod{p^{e+1}}$ then $x_0^2 \equiv a \pmod{p^e}$. Conversely, if $x_0^2 \equiv a \pmod{p^e}$ then $x_0^2 = a + bp^e$ for some integer b. We can solve the linear congruence $2x_0 y \equiv -b \pmod p$, say $y = y_0$. Let $x_1 = x_0 + y_0 p^e$. Then $x_1^2 \equiv x_0^2 + 2x_0 y_0 p^e = a + p^e(b + 2x_0 y_0) \equiv a \pmod{p^{e+1}}$ since $p \mid 2x_0 y_0 + b$. This is the induction step in showing that $x^2 \equiv a \pmod{p^e}$ has solutions if and only if $\left(\frac{a}{p}\right) = 1$.

11.1.25. a. $75 = 5^2 3$ and $\left(\frac{31}{5}\right) = \left(\frac{1}{5}\right) = 1$ and $\left(\frac{31}{3}\right) = \left(\frac{1}{3}\right) = 1$, so there are $2^2 = 4$ solutions.

   b. $105 = 3 \cdot 5 \cdot 7$ and 16 is a quadratic residue of 3, 5, and 7, so there are $2^3 = 8$ solutions.

   c. $231 = 3 \cdot 7 \cdot 11$ and $\left(\frac{46}{3}\right) = \left(\frac{1}{3}\right) = 1$, $\left(\frac{46}{7}\right) = \left(\frac{4}{7}\right) = 1$, $\left(\frac{46}{11}\right) = \left(\frac{2}{11}\right) = 1$ so there are $2^3 = 8$ solutions.

   d. $\left(\frac{1156}{3}\right) = \left(\frac{1}{3}\right) = 1$, $\left(\frac{1156}{5}\right) = \left(\frac{1}{5}\right) = 1$, $\left(\frac{1156}{7}\right) = \left(\frac{1}{7}\right) = 1$, and $\left(\frac{1156}{11}\right) = \left(\frac{1}{11}\right) = 1$ so there are $2^4 = 16$ solutions.

11.1. QUADRATIC RESIDUES AND NONRESIDUES

11.1.27. Suppose $p_1, p_2, \ldots, p_n$ are the only primes of the form $4k + 1$. Let $N = 4(p_1 p_2 \cdots p_n)^2 + 1$. Let q be an odd prime factor of N. Then $q \neq p_i$, $i = 1, 2, \ldots, n$, but $N \equiv 0 \pmod{q}$, so $4(p_1 p_2 \cdots p_n)^2 \equiv -1 \pmod{q}$ and therefore $\left(\frac{-1}{q}\right) = 1$, so $q \equiv 1 \pmod 4$ by Theorem 11.5.

11.1.29. Let b_1, b_2, b_3, and b_4 be the four modular square roots of a modulo pq. Then each b_i is a solution to exactly one of the four systems of congruences given in the text. For convenience, let the subscripts correspond to the lower case Roman numerals of the systems. Suppose two of the b_i's were quadratic residues modulo pq. Without loss of generality, say $b_1 \equiv y_1^2 \pmod{pq}$ and $b_2 \equiv y_2^2 \pmod{pq}$. Then from systems (i) and (ii), we have that $y_1^2 \equiv b_1 \equiv x_2 \pmod q$ and $y_2^2 \equiv b_2 \equiv -x_2 \pmod q$. Therefore both x_2 and $-x_2$ are quadratic residues modulo q, but this is impossible since $q \equiv 3 \pmod 4$. The other cases are identical.

11.1.31. Let r be a primitive root for p and let $a \equiv r^s \pmod p$ and $b \equiv r^t \pmod p$ with $1 \leq s, t \leq p-1$. If $a \equiv b \pmod p$, then $s = t$ and so s and t have the same parity. By Theorem 11.2, we have part (i). Further, we have $ab \equiv r^{s+t} \pmod p$. Then the right hand side of (ii) is 1 exactly when s and t have the same parity, which is exactly when the left hand side is 1. This proves part (ii). Finally, since $a^2 \equiv r^{2s} \pmod p$ and $2s$ is even, we must have that a^2 is a quadratic residue modulo p, proving part (iii).

11.1.33. If r is a primitive root of q, then the set of all primitive roots is given by $\{r^k : (k, \phi(q)) = (k, 2p) = 1\}$. So the $p - 1$ numbers $\{r^k : k$ is odd and $k \neq p, 1 \leq k < 2p\}$ are all the primitive roots of q. On the other hand, q has $(q - 1)/2 = p$ quadratic residues, which are given by $\{r^2, r^4, \ldots, r^{2p}\}$. This set has no intersection with the first one.

11.1.35. First suppose $p = 2^{2^n} + 1$ is a Fermat prime and let r be a primitive root for p. Then $\phi(p) = 2^{2^n}$. Then an integer a is a nonresidue if and only if $a = r^k$ with k odd. But then $(k, \phi(p)) = 1$, so a is also a primitive root. Conversely, suppose that p is an odd prime and every quadratic nonresidue of p is also a primitive root of p. Let r be a particular primitive root of p. Then, r^k is a quadratic nonresidue and hence a primitive root for p if and only if k is odd. But this implies that every odd number is relatively prime to $\phi(p)$, so $\phi(p)$ must be a power of 2. Thus $p = 2^b + 1$ for some b. If b had a nontrivial odd divisor, then we could factor p as a difference of b powers, contradicting the primality of of p. Therefore b is a power of 2 and so p is a Fermat prime.

11.1.37. a. We have $q = 2p + 1 = 2(4k + 3) + 1 = 8k + 7$, so $\left(\frac{2}{q}\right) = 1$ by Theorem 11.6. Then by Euler's criterion, $2^{(q-1)/2} \equiv 2^p \equiv 1 \pmod q$. Therefore $q \mid 2^p - 1$.

b. $11 = 4(2) + 3$ and $23 = 2(11) + 1$, so $23 \mid 2^{11} - 1 = M_{11}$, by part (a); $23 = 4(5) + 3$ and $47 = 2(23) + 1$, so $47 \mid M_{23}$; $251 = 4(62) + 3$ and $503 = 2(251) + 1$, so $503 \mid M_{251}$.

11.1.39. Let $q = 2k + 1$. Since q does not divide $2^p + 1$, we must have, by Exercise 38, that $k \equiv 0$ or $3 \pmod 4$. That is, $k \equiv 0, 3, 4$ or $7 \pmod 8$. Then $q \equiv 2(0, 3, 4$ or $7) + 1 \equiv \pm 1 \pmod 8$.

11.1.41. Note that $\left(\frac{j(j+1)}{p}\right) = \left(\frac{j \cdot j(1+\bar{j})}{p}\right) = \left(\frac{j^2(1+\bar{j})}{p}\right) = \left(\frac{(1+\bar{j})}{p}\right)$ since j^2 is a perfect square. Then, $\sum_{j=1}^{p-2}\left(\frac{j(j+1)}{p}\right) = \sum_{j=1}^{p-2}\left(\frac{\bar{j}+1}{p}\right) = \sum_{j=2}^{p-1}\left(\frac{j}{p}\right) = \sum_{j=1}^{p-1}\left(\frac{j}{p}\right) - 1 = -1$. Here we have used the method in the solution to Exercise 10 to evaluate the last sum, and the fact that as j runs through the values 1 through $p - 2$, so does $\bar{j}$.

11.1.43. Let r be a primitive root of p. Then $x^2 \equiv a \pmod p$ has a solution if and only if $2\,\mathrm{ind}_r x \equiv \mathrm{ind}_r a \pmod{p-1}$ has a solution in $\mathrm{ind}_r x$. Since $p - 1$ is even, the last congruence is solvable if and only if $\mathrm{ind}_r a$ is even, which happens when $a = r^2, r^4, \ldots, r^{p-1}$, i.e. $(p-1)/2$ times.

11.1.45. $q = 2(4k + 1) + 1 = 8k + 3$, so 2 is a quadratic nonresidue of q. By Exercise 33, 2 is a primitive root.

11.1.47. Check that $q \equiv 3 \pmod{4}$, so -1 is a quadratic nonresidue of q. Since $4 = 2^2$, we have $\left(\frac{-4}{q}\right) = \left(\frac{-1}{q}\right)\left(\frac{2^2}{q}\right) = (-1)(1) = -1$. Therefore -4 is a nonresidue of q. By Exercise 33, -4 is a primitive root.

11.1.49. a. By adding $(\overline{2}b)^2$ to both sides of the congruence $C \equiv P(P+b) \pmod{n}$, we have $C + a \equiv P^2 + Pb + (\overline{2}b)^2 \equiv (P + \overline{2}b)^2 \pmod{n}$.

    b. There are 4 solutions to $x^2 \equiv C + a \pmod{pq}$. From each, subtract $\overline{2}b$, which gives the 4 messages.

    c. First we solve $2x \equiv 1 \pmod{2773}$ to get $\overline{2} = 1387$. Then $\overline{2}b \equiv 1338 \pmod{2773}$, and $(\overline{2}b)^2 \equiv 2082 \pmod{2773}$. For the first block of ciphertext, we have $1819 \equiv P(P+3) \pmod{2773}$. we add 2082 to both sides to get $1128 \equiv (P + 1388)^2 \pmod{2773}$. We solve $x^2 \equiv 1128 \pmod{2773}$ to find the two solutions 1692 and 1081. Subtracting 1388 from both of these and reducing gives us the two possible values for P, 0304 and 2466. Since 66 is not the numerical equivalent of a letter, we know the solution must be 0304 which is equivalent to DE. Similarly we find that 0459 has $P = 0856, 1796, 1914$, or 0974. Of these, only 1914 has letter equivalents, so the second digraph is TO. Finally, we find that 0803 has $P = 2346, 2017, 0424$, or 0753. Two of these have letter equivalents: 0424 is equivalent to EY, which makes the message DETOEY; 2017 is equivalent to UR, which makes the message DETOUR. We guess that the message is DETOUR.

11.1.51. a. By noting this, the second player can tell which cards dealt are quadratic residues, since the ciphertext will also be quadratic residues modulo p.

    b. All ciphers will be quadratic residues modulo p.

11.1.53. The quadratic residues modulo 11 are $1, 3, 4, 5$, and 9. So there are several chains: $1 + 3 = 4, 4 + 5 = 9$, and $9 + 5 = 14 \equiv 3 \pmod{11}$, for example.

11.2. The Law of Quadratic Reciprocity

11.2.1. a. Since $53 \equiv 1 \pmod{4}$ the law of quadratic reciprocity shows that $\left(\frac{3}{53}\right) = \left(\frac{53}{3}\right)$. We have $\left(\frac{53}{3}\right) = \left(\frac{2}{3}\right) = -1$. Hence $\left(\frac{3}{53}\right) = -1$.

    b. Since $79 \equiv 1 \pmod{4}$ the law of quadratic reciprocity shows that $\left(\frac{7}{79}\right) = \left(\frac{79}{7}\right)$. We have $\left(\frac{79}{7}\right) = \left(\frac{2}{7}\right) = 1$ since 2 is a quadratic residue of 7.

    c. We have $\left(\frac{15}{101}\right) = \left(\frac{3}{101}\right)\left(\frac{5}{101}\right)$. Since $101 \equiv 1 \pmod{4}$, the law of quadratic reciprocity shows that $\left(\frac{3}{101}\right) = \left(\frac{101}{3}\right)$ and $\left(\frac{5}{101}\right) = \left(\frac{101}{5}\right)$. We have $\left(\frac{101}{3}\right) = \left(\frac{2}{3}\right) = -1$ and $\left(\frac{101}{5}\right) = \left(\frac{1}{5}\right) = 1$. Hence $\left(\frac{15}{101}\right) = -1 \cdot 1 = 1$.

    d. $\left(\frac{31}{641}\right) = \left(\frac{641}{31}\right) = \left(\frac{21}{31}\right) = \left(\frac{3}{31}\right)\left(\frac{7}{31}\right) = \left(-\left(\frac{31}{3}\right)\right)\left(-\left(\frac{31}{7}\right)\right) = \left(\frac{1}{3}\right)\left(\frac{3}{7}\right) = 1\left(-\left(\frac{7}{3}\right)\right) = -\left(\frac{2}{3}\right) = 1$.

    e. $\left(\frac{111}{991}\right) = \left(\frac{3}{991}\right)\left(\frac{37}{991}\right) = -\left(\frac{991}{3}\right)\left(\frac{991}{37}\right) = -\left(\frac{1}{3}\right)\left(\frac{29}{37}\right) = -\left(\frac{29}{37}\right) = -\left(\frac{37}{29}\right) = -\left(\frac{8}{29}\right) = -\left(\frac{2}{29}\right)\left(\frac{4}{29}\right) = -\left(\frac{2}{29}\right) = -(-1) = 1$.

    f. $\left(\frac{105}{1009}\right) = \left(\frac{3}{1009}\right)\left(\frac{5}{1009}\right)\left(\frac{7}{1009}\right) = \left(\frac{1009}{3}\right)\left(\frac{1009}{5}\right)\left(\frac{1009}{7}\right) = \left(\frac{1}{3}\right)\left(\frac{4}{5}\right)\left(\frac{1}{7}\right) = 1$.

11.2.3. If $p \equiv 1 \pmod{6}$ there are 2 cases: If $p \equiv 1 \pmod{4}$ then $\left(\frac{-1}{p}\right) = 1$ and $\left(\frac{3}{p}\right) = \left(\frac{p}{3}\right) = \left(\frac{1}{3}\right) = 1$. So $\left(\frac{-3}{p}\right) = 1$. If $p \equiv 3 \pmod{4}$ then $\left(\frac{-1}{p}\right) = -1$ and $\left(\frac{3}{p}\right) = -\left(\frac{p}{3}\right)$, so $\left(\frac{-3}{p}\right) = (-1)(-1) = 1$. If $p \equiv -1 \pmod{6}$ and $p \equiv 1 \pmod{4}$, then $\left(\frac{-3}{p}\right) = \left(\frac{-1}{p}\right)\left(\frac{3}{p}\right) = 1 \cdot \left(\frac{p}{3}\right) = \left(\frac{-1}{3}\right) = -1$. If $p \equiv 3 \pmod{4}$, then

$\left(\frac{-3}{p}\right) = \left(\frac{-1}{p}\right)\left(\frac{3}{p}\right) = (-1)\left(-\left(\frac{p}{3}\right)\right) = \left(\frac{p}{3}\right) = \left(\frac{-1}{3}\right) = -1.$

11.2.5. Suppose that p is an odd prime. By the law of quadratic reciprocity it follows that $\left(\frac{7}{p}\right) = \left(\frac{p}{7}\right)$ if $p \equiv 1 \pmod{4}$ and $\left(\frac{7}{p}\right) = -\left(\frac{p}{7}\right)$ if $p \equiv 3 \pmod{4}$. So, 7 is a quadratic residue of a prime p with $p \equiv 1 \pmod{4}$ if $\left(\frac{p}{7}\right) = 1$. This is the case when $p \equiv 1, 2$ or $4 \pmod{7}$. Using the Chinese remainder theorem we see that 7 is a quadratic residue of p when $p \equiv 1, 9$ or $25 \pmod{28}$, and 7 is a quadratic nonresidue of p when $p \equiv 5, 13$ or $17 \pmod{28}$. Also, 7 is a quadratic residue of a prime p with $p \equiv 3 \pmod{4}$ if $\left(\frac{p}{7}\right) = -1$. This is the case when $p \equiv 3, 5$ or $6 \pmod{7}$. Using the Chinese remainder theorem we see that 7 is a quadratic residue of p when $p \equiv 3, 19$ or $27 \pmod{28}$ and 7 is a quadratic nonresidue of p when $p \equiv 11, 15$ or $23 \pmod{28}$. It follows that 7 is a quadratic residue of p if and only if $p \equiv 1, 3, 9, 19, 25$ or $27 \pmod{28}$.

11.2.7. a. We have $F_1 = 2^{2^1} + 1 = 5$. We find that $3^{(F_1-1)/2} = 3^{(5-1)/2} = 3^2 = 9 \equiv -1 \pmod{F_1}$. Hence by Pepin's test we come (to the already obvious) conclusion that $F_1 = 5$ is prime.

b. We have $F_3 = 2^{2^3} + 1 = 257$. We find that $3^{(F_3-1)/2} = 3^{(257-1)/2} = 3^{128} \equiv (3^8)^{16} \equiv 136^{16} \equiv (136^4)^4 \equiv 64^4 \equiv (64^2)^2 \equiv 241^2 \equiv 256 \equiv -1 \pmod{257}$. Hence by Pepin's test we see that $F_3 = 257$ is prime.

c. Using a calculator we find $3^{255} \equiv 94 \pmod{F_4}$. $3^{32768} \equiv 3^{255 \cdot 128} 3^{128} \equiv 94^{128} 3^{128} \equiv -1 \pmod{F_4}$.

11.2.9. a. The lattice points in the rectangle are the points (i, j) where $0 < i < p/2$ and $0 < j < q/2$. There are the lattice points (i, j) with $i = 1, 2, \ldots, (p-1)/2$ and $j = 1, 2, \ldots, (q-1)/2$. Consequently, there are $(p-1)/2 \cdot (q-1)/2$ such lattice points.

b. The points on the diagonal connecting **O** and **C** are the points (x, y) where $y = (q/p)x$. Suppose that x and y are integers with $y = (q/p)x$. Then $py = qx$. Since $(p, q) = 1$ it follows that $p \mid x$ which is impossible if $0 < x < p/2$. Hence there are no lattice points on this diagonal.

c. The number of lattice points in the triangle with vertices **O,A,** and **C** is the number of lattice points (i, j) with $i = 1, 2, \ldots, (p-1)/2$ and $1 \le j \le iq/p$. For a fixed value of i in the indicated range, there are $[iq/p]$ lattice points (i, j) in the triangle. Hence the total number of lattice points in the triangle is $\sum_{i=1}^{(p-1)/2}[iq/p]$.

d. The number of lattice points in the triangle with vertices **O,B,** and **C** is the number of lattice points (i, j) with $j = 1, 2, \ldots, (q-1)/2$ and $1 \le i < jp/q$. For a fixed value of j in the indicated range, there are $[jp/q]$ lattice points (i, j) in the triangle. Hence the total number of lattice points in the triangle is $\sum_{j=1}^{(q-1)/2}[jp/q]$.

e. Since there are $(p-1)/2 \cdot (q-1)/2$ lattice points in the rectangle, and no points on the diagonal **OC**, the sum of the numbers of lattice points in the triangles **OBC** and **OAC** is $(p-1)/2 \cdot (q-1)/2$. By parts (b) and (c) it follows that $\sum_{j=1}^{(p-1)/2}[jq/p] + \sum_{j=1}^{(q-1)/2}[jp/q] = (p-1)/2 \cdot (q-1)/2$. By Lemma 11.3 it follows that $\left(\frac{p}{q}\right) = (-1)^{T(p,q)}$ and $\left(\frac{q}{p}\right) = (-1)^{T(q,p)}$ where $T(p,q) = \sum_{j=1}^{(p-1)/2}[jp/q]$ and $T(q,p) = \sum_{j=1}^{(q-1)/2}[jq/p]$. We conclude that $\left(\frac{p}{q}\right)\left(\frac{q}{p}\right) = (-1)^{(p-1)/2 \cdot (q-1)/2}$. This is the law of quadratic reciprocity.

11.2.11. First suppose $a = 2$. Then we have $p \equiv \pm q \pmod{8}$ and so $\left(\frac{a}{p}\right) = \left(\frac{a}{q}\right)$ by Theorem 11.6 Now suppose a is an odd prime. If $p \equiv q \pmod{4a}$, then $p \equiv q \pmod{a}$ and so $\left(\frac{q}{a}\right) = \left(\frac{p}{a}\right)$. And since $p \equiv q \pmod{4}$, $(p-1)/2 \equiv (q-1)/2 \pmod{2}$. Then by Theorem 11.7, $\left(\frac{a}{p}\right) = \left(\frac{p}{a}\right)(-1)^{(p-1)/2 \cdot (a-1)/2} = \left(\frac{q}{a}\right)(-1)^{(q-1)/2 \cdot (a-1)/2} = \left(\frac{a}{q}\right)$. But if $p \equiv -q \pmod{4a}$, then $p \equiv -q \pmod{a}$ and so $\left(\frac{-q}{a}\right) = \left(\frac{p}{a}\right)$. And since $p \equiv -q \pmod{4}$, $(p-1)/2 \equiv (q-1)/2 + 1 \pmod{2}$. Then by Theorem 11.7, $\left(\frac{a}{p}\right) =$

$\left(\frac{p}{a}\right)(-1)^{(p-1)/2\cdot(a-1)/2} = \left(\frac{-q}{a}\right)(-1)^{((q-1)/2+1)\cdot(a-1)/2} = \left(\frac{-1}{a}\right)(-1)^{(a-1)/2}\left(\frac{a}{q}\right) = \left(\frac{a}{q}\right)$. The general case follows from the multiplicativity of the Legendre symbol.

11.2.13. **a.** Recall that $e^{xi} = 1$ if and only if x is a multiple of 2π. First we compute $(e^{(2\pi i/n)k})^n = e^{(2\pi i/n)nk} = (e^{(2\pi i)})^k = 1^k = 1$, so $e^{(2\pi i/n)k}$ is an nth root of unity. Now if $(k, n) = 1$, then $((2\pi i/n)k)a$ is a multiple of $2\pi i$ if and only if $n | a$. Therefore $a = n$ is the least positive integer for which $(e^{(2\pi i/n)k})^a = 1$. Therefore $e^{(2\pi i/n)k}$ is a primitive nth root of unity. Conversely, suppose $(k, n) = d > 1$. Then $(e^{(2\pi i/n)k})^{(n/d)} = e^{(2\pi i)k/d} = 1$, since k/d is an integer, and so in this case $e^{(2\pi i/n)k}$ is not a primitive nth root of unity.

b. Let $m = l + kn$ where k is an integer. Then $\zeta^m = \zeta^{l+kn} = \zeta^l \zeta^{kn} = \zeta^l$. Now suppose ζ is a primitive nth root of unity and that $\zeta^m = \zeta^l$, and without loss of generality, assume $m \geq l$. From the first part of this exercise, we may take $0 \leq l \leq m < n$. Then $0 = \zeta^m - \zeta^l = \zeta^l(\zeta^{m-l} - 1)$. Hence, $\zeta^{m-l} = 1$. Since n is the least positive integer such that $\zeta^n = 1$, we must have $m - l = 0$.

c. First, $f(z+1) = e^{2\pi i(z+1)} - e^{-2\pi i(z+1)} = e^{2\pi iz}e^{2\pi i} - e^{-2\pi iz}e^{-2\pi i} = e^{2\pi iz}1 - e^{-2\pi iz}1 = f(z)$. Next, $f(-z) = e^{-2\pi iz} - e^{2\pi iz} = -(e^{2\pi iz} - e^{-2\pi iz}) = -f(z)$. Finally, suppose $f(z) = 0$. Then $0 = e^{2\pi iz} - e^{-2\pi iz} = e^{-2\pi iz}(e^{4\pi iz} - 1)$, so $e^{4\pi iz} = 1$. Therefore $4\pi iz = 2\pi in$ for some integer n, and so $z = n/2$.

d. Fix y and consider $g(x) = x^n - y^n$ and $h(x) = (x - y)(\zeta x - \zeta^{-1}y) \cdots (\zeta^{n-1}x - \zeta^{-(n-1)}y)$ as polynomials in x. Both polynomials have degree n. The leading coefficient in $h(x)$ is $\zeta^{1+2+\cdots+n-1} = \zeta^{n(n-1)/2} = (\zeta^n)^{(n-1)/2} = 1$, since $n-1$ is even. So both polynomials are monic. Further, note that $g(\zeta^{-2k}y) = (\zeta^{-2k}y)^n - y^n = y^n - y^n = 0$ for $k = 0, 1, 2, \ldots, n-1$. Also $h(\zeta^{-2k}y)$ has $(\zeta^k \zeta^{-2k}y - \zeta^{-k}y) = (\zeta^{-k}y - \zeta^{-k}y) = 0$ as one of its factors. So g and h are monic polynomials sharing these n distinct zeros (since $-2k$ runs through a complete set of residues modulo n). By the Fundamental Theorem of Algebra, g and h are identical.

e. Let $x = e^{2\pi iz}$ and $y = e^{-2\pi iz}$ in the identity from part (d). Then the right hand side becomes
$$\prod_{k=0}^{n-1} (\zeta^k e^{2\pi iz} - \zeta^{-k}e^{-2\pi iz}) = \prod_{k=0}^{n-1} \left(e^{2\pi i(z+k/n)} - e^{-2\pi i(z+k/n)}\right) = \prod_{k=0}^{n-1} f\left(z + \frac{k}{n}\right) =$$
$$f(z) \prod_{k=1}^{(n-1)/2} f\left(z + \frac{k}{n}\right) \prod_{k=(n+1)/2}^{n-1} f\left(z + \frac{k}{n}\right).$$
From the identities in part (c), this last product becomes
$$\prod_{k=(n+1)/2}^{n-1} f\left(z + \frac{k}{n}\right) = \prod_{k=1}^{(n-1)/2} f\left(z + \frac{n-k}{n}\right) = \prod_{k=1}^{(n-1)/2} f\left(z + 1 - \frac{k}{n}\right) = \prod_{k=1}^{(n-1)/2} f\left(z - \frac{k}{n}\right).$$
So the product above is equal to $f(z) \prod_{k=1}^{(n-1)/2} f\left(z + \frac{k}{n}\right) \prod_{k=1}^{(n-1)/2} f\left(z - \frac{k}{n}\right) =$
$f(z) \prod_{k=1}^{(n-1)/2} f\left(z + \frac{k}{n}\right) f\left(z - \frac{k}{n}\right)$. Then noting that the left side of the identity in part (d) is $(e^{2\pi iz})^n - (e^{-2\pi iz})^n = e^{2\pi inz} - e^{-2\pi inz} = f(nz)$ finishes the proof.

f. For $l = 1, 2, \ldots, (p-1)/2$, let k_l be the least positive residue of la modulo p. Then $\prod_{l=1}^{(p-1)/2} f\left(\frac{la}{p}\right) =$
$\prod_{l=1}^{(p-1)/2} f\left(\frac{k_l}{p}\right)$ by the periodicity of f established in part (c). We break this product into two pieces
$\prod_{k_l < p/2} f\left(\frac{k_l}{p}\right) \prod_{k_l > p/2} f\left(\frac{k_l}{p}\right) = \prod_{k_l < p/2} f\left(\frac{k_l}{p}\right) \prod_{k_l > p/2} -f\left(\frac{-k_l}{p}\right) = \prod_{k_l < p/2} f\left(\frac{k_l}{p}\right) \prod_{k_l > p/2} -f\left(\frac{p-k_l}{p}\right) =$
$\prod_{l=1}^{(p-1)/2} f\left(\frac{l}{p}\right)(-1)^N$, where N is the number of k_l exceeding $p/2$. But by Gauss' lemma, $(-1)^N =$

$\left(\frac{a}{p}\right)$. This establishes the identity.

g. Let $z = l/p$ and $n = q$ in the identities in parts (e) and (f). Then we have $\left(\frac{q}{p}\right) = \prod_{l=1}^{(p-1)/2} f\left(\frac{lq}{p}\right) / f\left(\frac{l}{p}\right) =$
$\prod_{l=1}^{(p-1)/2} \prod_{k=1}^{(q-1)/2} f\left(\frac{l}{p}\right) + \frac{k}{q} f\left(\frac{l}{p}\right) - \frac{k}{q} = \prod_{l=1}^{(p-1)/2} \prod_{k=1}^{(q-1)/2} f\left(\frac{k}{q}\right) + \frac{l}{p} f\left(\frac{k}{q}\right) - \frac{l}{p} (-1)^{(p-1)/2 \cdot (q-1)/2}$, where we
have used the fact that $f(-z) = -f(z)$ and the fact that there are exactly $(p-1)/2 \cdot (q-1)/2$ factors in the double product. But, by symmetry, this is exactly the expression for $\left(\frac{q}{p}\right)(-1)^{(p-1)/2 \cdot (q-1)/2}$ which completes the proof.

11.2.15. Since $p \equiv 1 \pmod{4}$, we have $\left(\frac{q}{p}\right) = \left(\frac{p}{q}\right)$. And since $p \equiv 1 \pmod{q}$ for all primes $q \leq 23$, then $\left(\frac{p}{q}\right) = \left(\frac{1}{q}\right) = 1$. Then if a is an integer with $0 < a < 29$ and prime factorization $a = p_1 p_2 \cdots p_k$, then each $p_i < 29$ and $\left(\frac{a}{p}\right) = \left(\frac{p_1}{p}\right) \cdots \left(\frac{p_k}{p}\right) = 1^k = 1$. So there are no quadratic nonresidues modulo p less than 29. Further, since a quadratic residue must be an even power of any primitive root r, then r^1 can not be less than 29.

11.2.17. a. If $a \in T$ then $a = qk$ for some $k = 1, 2, \ldots (p-1)/2$. So $1 \leq a \leq q(p-1)/2 < (pq-1)/2$. Further, since $k \leq (p-1)/2$, and p is prime, we have $(p, k) = 1$. Since $(q, p) = 1$, then $(a, p) = (qk, p) = 1$, so $a \in S$, and hence $T \subset S$. Now suppose $a \in S - T$. Then $1 \leq a \leq (pq-1)/2$ and $(a, p) = 1$, and since $a \notin T$, then $a \neq qk$ for any k. Thus $(a, q) = 1$, so $(a, pq) = 1$, and so $a \in R$. Thus $S - T \subset R$. Conversely, if $a \in R$, then $1 \leq a \leq (pq-1)/2$ and $(a, pa) = 1$, so certainly $(a, q) = 1$, and so a is not a multiple of q and hence $a \notin T$. Hence $a \in S - T$. Thus $R \subset S - T$. Therefore $R = S - T$.

b. Since by part (a), $R = S - T$ we have $\prod_{a \in S} a = \prod_{a \in R} a \prod_{a \in T} a = A(q \cdot 2q \cdots ((p-1)/2)a) = Aq^{(p-1)/2}((p-1)/2)! \equiv A\left(\frac{q}{p}\right)((p-1)/2)! \pmod{p}$ by Euler's criterion. Note that $(pq-1)/2 = p(q-1)/2 + (p-1)/2$, so that we can evaluate $\prod_{a \in S} a \equiv ((p-1)!)^{(q-1)/2}((p-1)/2)! \equiv (-1)^{(q-1)/2}((p-1)/2)! \pmod{p}$ by Wilson's Theorem. When we set these two expressions congruent to each other modulo p and cancel we get $A \equiv (-1)^{(q-1)/2}\left(\frac{q}{p}\right)$ as desired.

c. Since the roles of p and q are identical in the hypotheses and in parts (a) and (b), the result follows by symmetry.

d. Assume that $(-1)^{(q-1)/2}\left(\frac{q}{p}\right) = (-1)^{(p-1)/2}\left(\frac{p}{q}\right)$. Then $A = \pm 1$ so certainly $A \equiv \pm 1 \pmod{pq}$. Conversely, suppose $A \equiv 1 \pmod{pq}$. Then $A \equiv 1 \pmod{p}$ and $A \equiv 1 \pmod{q}$. Then by parts (b) and (c) we have $(-1)^{(q-1)/2}\left(\frac{q}{p}\right) = A = (-1)^{(p-1)/2}\left(\frac{p}{q}\right)$. The same argument works if $A \equiv -1 \pmod{pq}$.

e. If a is an integer in R, it is in the range $1 \leq a \leq (pq-1)/2$ and therefore its additive inverse modulo pq is in the range $(pq+1)/2 \leq -a \leq pq - 1$ in the set of reduced residue classes. By the Chinese Remainder Theorem, the congruence $a^2 \equiv 1 \pmod{pq}$ has exactly 4 solutions, $1, -1, b,$ and $-b$ $\pmod{pq}$ and the congruence $a^2 \equiv -1 \pmod{pq}$ has solutions if and only $p \equiv q \equiv 1 \pmod{4}$, and in this case it has exactly 4 solutions $i, -i, ib,$ and $-ib \pmod{pq}$. Now for each element $a \in R$, $(a, pq) = 1$, so a has a multiplicative inverse v. By the remark above, exactly one of $v, -v$ is in R. We let U be the set of those elements which are their own inverse or their own negative inverse, that is let $U = \{a \in R | a^2 \equiv \pm 1 \pmod{pq}\}$. Then when we compute A, all other elements will be paired with another element which is either its inverse or the negative of its inverse. Thus we have $A \prod_{a \in R} a \equiv \pm \prod_{a \in U} a \pmod{pq}$. So if $p \equiv q \equiv 1 \pmod{pq}$, then $A \equiv \pm \prod_{a \in U} a \equiv \pm(1 \cdot c \cdot i \cdot ic) \equiv c^2 i^2 \equiv \mp 1 \pmod{pq}$.

100 11. QUADRATIC RESIDUES

Conversely, in the other case, $A \equiv \prod_{a \in U} a \equiv \pm(1 \cdot c) \not\equiv \pm 1 \pmod{pq}$, which completes the proof.

f. By parts (d) and (e) we have that $(-1)^{(q-1)/2} \left(\frac{q}{p}\right) = (-1)^{(p-1)/2} \left(\frac{p}{q}\right)$ if and only if $p \equiv q \equiv 1 \pmod{4}$. So if $p \equiv q \equiv 1 \pmod{4}$ we have $\left(\frac{q}{p}\right) = \left(\frac{p}{q}\right)$. But if $p \equiv 1 \pmod{4}$ while $q \equiv 3 \pmod{4}$ then we must have $-\left(\frac{q}{p}\right) \neq \left(\frac{p}{q}\right)$ which means we must change the sign and have $\left(\frac{q}{p}\right) = \left(\frac{p}{q}\right)$. The case where $p \equiv 3 \pmod{4}$ but $q \equiv 1 \pmod{4}$ is identical. If $p \equiv q \equiv 3 \pmod{4}$, then we must have $-\left(\frac{q}{p}\right) \neq -\left(\frac{p}{q}\right)$ so that we must have $-\left(\frac{q}{p}\right) = \left(\frac{p}{q}\right)$, which concludes the proof.

11.3. The Jacobi Symbol

11.3.1. a. By the reciprocity law for Jacobi symbols, since $5 \equiv 1 \pmod{4}$ we have $\left(\frac{5}{21}\right) = \left(\frac{21}{5}\right) = \left(\frac{1}{5}\right) = 1$.

b. We have $\left(\frac{27}{101}\right) = \left(\frac{3}{101}\right)^3 = \left(\frac{101}{3}\right)^3 = \left(\frac{2}{3}\right)^3 = (-1)^3 = -1$, where we have used the law of quadratic reciprocity to replace $\left(\frac{101}{3}\right)$ by $\left(\frac{3}{101}\right)$ since $101 \equiv 1 \pmod{4}$.

c. Since $1001 \equiv 1 \pmod{4}$, by the reciprocity law of Jacobi symbols we have $\left(\frac{11}{1001}\right) = \left(\frac{1001}{111}\right) = \left(\frac{2}{111}\right) = 1$ since $111 \equiv 7 \pmod{4}$.

d. Since $1009 \equiv 1 \pmod{4}$ by the reciprocity law for Jacobi symbols we have $\left(\frac{1009}{2307}\right) = \left(\frac{2307}{1009}\right) = \left(\frac{289}{1009}\right) = \left(\frac{17}{1009}\right)^2 = 1$.

e. Since $2663 \equiv 3299 \equiv 3 \pmod{4}$ by the reciprocity law for Jacobi symbols we have $\left(\frac{2663}{3299}\right) = \left(\frac{3299}{2663}\right) = \left(\frac{636}{2663}\right) = \left(\frac{4}{2663}\right)\left(\frac{159}{2663}\right) = leg1592663$ since $3229 \equiv 636 \pmod{2663}$. Since $159 \equiv 2663 \equiv 3 \pmod{4}$, by the reciprocity law for Jacobi symbols we have $\left(\frac{159}{2663}\right) = -\left(\frac{2663}{159}\right) = -\left(\frac{85}{159}\right)$ since $2663 \equiv 85 \pmod{159}$. Since $85 \equiv 1 \pmod{4}$, the reciprocity law for Jacobi symbols shows that $\left(\frac{85}{159}\right) = \left(\frac{159}{85}\right) = \left(\frac{-11}{85}\right) = \left(\frac{-1}{85}\right)\left(\frac{11}{85}\right) = \left(\frac{11}{85}\right)$, since $85 \equiv 1 \pmod{4}$. By the reciprocity law for Jacobi symbols we have $\left(\frac{11}{85}\right) = \left(\frac{85}{11}\right) = \left(\frac{8}{11}\right) = \left(\frac{2}{11}\right)^3 = (-1)^3 = -1$ since $11 \equiv 3 \pmod{8}$. It follows that $\left(\frac{2663}{3299}\right) = -1$.

f. Since $10001 \equiv 1 \pmod{4}$ the reciprocity law for Jacobi symbols shows that $\left(\frac{10001}{20003}\right) = \left(\frac{20003}{10001}\right) = \left(\frac{1}{10001}\right) = 1$, where we have used the periodicity of the Jacobi symbol and the congruence $20003 \equiv 1 \pmod{10001}$.

11.3.3. We have $\left(\frac{30}{n}\right) = \left(\frac{2}{n}\right)\left(\frac{15}{n}\right)$. By Theorem 11.10(iv) $\left(\frac{2}{n}\right) = 1$ when $n \equiv \pm 1 \pmod{8}$. From Exercise 2, $\left(\frac{15}{n}\right) = 1$ when $n \equiv 1, 7, 11, 17, 43, 49, 53,$ or $59 \pmod{60}$. By the Chinese remainder theorem, $\left(\frac{2}{n}\right) = \left(\frac{15}{n}\right) = 1$, when $n \equiv 1, 7, 17, 49, 61, 67, 77,$ or $109 \pmod{120}$ and $\left(\frac{2}{n}\right) = \left(\frac{15}{n}\right) = -1$ when $n \equiv 13, 19, 29, 37, 71, 83, 91, 101, 103, 107, 113,$ or $119 \pmod{120}$.

11.3.5. We have $21 = 3 \cdot 7$. The only quadratic nonresidues of 3 is 2. The quadratic nonresidues of 7 are 3, 5, and 6. From Exercise 4, we need to solve each of the systems of congruences $x \equiv a \pmod 3, x \equiv b \pmod 7$ where a is a quadratic nonresidue of 3 and b is a quadratic nonresidue of 7. For $x \equiv 2 \pmod 3, x \equiv 3 \pmod 7$, we have $x = 17$. For $x \equiv 2 \pmod 3, x \equiv 5 \pmod 7$, we have $x = 5$. And for $x \equiv 2 \pmod 3, x \equiv 6 \pmod 7$, we have $x = 20$. So the pseudo-squares modulo 21 are 5, 17 and 20.

11.3.7. We follow the strategy in the solution to Exercise 5. We have $143 = 11 \cdot 13$. The set of quadratic residues modulo 11 is $S = \{2, 6, 7, 8, 10\}$ and the set of quadratic residues modulo 13 is $T = \{2, 5, 6, 7, 8, 11\}$. We form the 30 systems of congruence $x \equiv a \pmod{11}, x \equiv b \pmod{13}$, where $a \in S$ and $b \in T$, and solve them. We find that the pseudo-squares modulo 143 are: 1, 3, 4, 9, 12, 14, 16, 23, 25, 27, 36, 38, 42, 48, 49, 53, 56, 64, 69, 75, 81, 82, 92, 100, 103, 108, 113, 114, 126, and 133.

11.3.9. Since n is odd and square-free, n has prime factorization $n = p_1 p_2 \cdots p_r$. Let b be one of the $(p-1)/2$ quadratic nonresidues of p_1, so that $\left(\frac{b}{p_1}\right) = -1$. By the Chinese Remainder Theorem, let a be a solution to the system of linear congruences:

$$x \equiv b \pmod{p_1}$$
$$x \equiv 1 \pmod{p_2}$$
$$\vdots$$
$$x \equiv 1 \pmod{p_r}$$

Then $\left(\frac{a}{p_1}\right) = \left(\frac{b}{p_1}\right) = -1, \left(\frac{1}{p_2}\right) = 1, \ldots \left(\frac{a}{p_r}\right) = \left(\frac{1}{p_r}\right) = 1$. Therefore $\left(\frac{a}{n}\right) = \left(\frac{a}{p_1}\right)\left(\frac{a}{p_2}\right) \cdots \left(\frac{a}{p_r}\right) = (-1) \cdot 1 \cdots 1 = -1$.

11.3.11. a. Note that $(a, b) = (b, r_1) = (r_1, r_2) = \cdots = (r_{n-1}, r_n) = 1$ and since the q_i are even, the r_i are odd. Since $r_0 = b$ and $a \equiv \epsilon_1 r_1 \pmod{b}$ we have $\left(\frac{a}{b}\right) = \left(\frac{\epsilon_1 r_1}{r_0}\right) = \left(\frac{\epsilon_1}{r_0}\right)\left(\frac{r_1}{r_0}\right) = \left(\frac{\epsilon_1}{r_0}\right)\left(\frac{r_0}{r_1}\right)(-1)^{(r_0-1)/2 \cdot (r_1-1)/2}$ by Theorem 11.11. If $\epsilon_1 = 1$, then $\left(\frac{a}{b}\right) = (-1)^{(r_0-1)/2 \cdot (\epsilon_1 r_1 - 1)/2}\left(\frac{r_0}{r_1}\right)$ If $\epsilon_1 = -1$, then $\left(\frac{\epsilon_1}{r_0}\right) = (-1)^{(r_0-1)/2}$ and we have $\left(\frac{a}{b}\right) = (-1)^{(r_0-1)/2 \cdot (r_1+1)/2}\left(\frac{r_0}{r_1}\right) = -1^{(r_0-1)/2 \cdot (-r_1-1)/2}\left(\frac{r_0}{r_1}\right) = (-1)^{(r_0-1)/2 \cdot (\epsilon_1 r_1 - 1)/2}\left(\frac{r_0}{r_1}\right)$ since $(r_1+1)/2$ and $(-r_1-1)/2$ have the same parity. Similarly, $\left(\frac{r_0}{r_1}\right) = (-1)^{(r_1-1)/2 \cdot (\epsilon_2 r_2 - 1)/2}\left(\frac{r_1}{r_2}\right)$, so $\left(\frac{a}{b}\right) = (-1)^{(r_0-1)/2 \cdot (\epsilon_1 r_1 - 1)/2 + (r_1-1)/2 \cdot (\epsilon_2 r_2 - 1)/2}\left(\frac{r_1}{r_2}\right)$. Proceed inductively until the last step, when $\left(\frac{r_n}{r_{n-1}}\right) = \left(\frac{1}{r_{n-1}}\right) = 1$.

b. If either $r_{i-1} \equiv 1 \pmod{4}$ or $\epsilon_i r_i \equiv 1 \pmod{4}$, then $(r_{i-1} - 1)/2 \cdot (\epsilon_1 r_i - 1)/2$ is even. Otherwise, that is, if $r_{i-1} \equiv \epsilon_i r_i \equiv 3 \pmod{4}$, then $(r_{i-1} - 1)/2 \cdot (\epsilon_i r_i - 1)/2$ is odd. Then $(r_{n-1} - 1)/2 \cdot (\epsilon_n r_n - 1)/2$ the exponent in part (a) is even or odd as T is even or odd.

11.3.13. a. We have $\left(\frac{5}{12}\right) = \left(\frac{5}{2}\right)^2 \left(\frac{5}{3}\right) = (-1)^2 \cdot \left(\frac{2}{3}\right) = 1 \cdot (-1) = -1$.

b. We have $\left(\frac{13}{20}\right) = \left(\frac{13}{2}\right)^2 \left(\frac{13}{5}\right) = (-1)^2 \cdot \left(\frac{3}{5}\right) = 1 \cdot (-1) = -1$.

c. We have $\left(\frac{101}{200}\right) = \left(\frac{101}{2}\right)^3 \left(\frac{101}{5}\right)^2 = (-1)^3 \cdot 1^2 = -1$.

11.3.15. Let $n_1 = p_1^{a_1} p_2^{a_2} \cdots p_r^{a_r}$ and $n_2 = q_1^{b_1} q_2^{b_2} \cdots q_s^{b_s}$ be the prime factorizations of n_1 and n_2. Then by the definition of the Kronecker symbol, we have $\left(\frac{a}{n_1 n_2}\right) = \left(\frac{a}{p_1}\right)^{a_1} \cdots \left(\frac{a}{p_r}\right)^{a_r} \left(\frac{a}{q_1}\right)^{b_1} \cdots \left(\frac{a}{q_s}\right)^{b_s} = \left(\frac{a}{n_1}\right)\left(\frac{a}{n_2}\right)$.

11.3.17. If a is odd, then by Exercise 16, we have $\left(\frac{a}{n_1}\right) = \left(\frac{n_1}{|a|}\right)$. By Theorem 11.10(i), we have $\left(\frac{n_1}{|a|}\right) = \left(\frac{n_2}{|a|}\right) = \left(\frac{a}{n_2}\right)$, using Exercise 16 again. If a is even, say $a = 2^s t$ with t odd, Exercise 16 gives $\left(\frac{a}{n_1}\right) = \left(\frac{2}{n_1}\right)^s (-1)^{(t-1)/2 \cdot (n_1-1)/2}\left(\frac{n_1}{|t|}\right)$ and $\left(\frac{a}{n_2}\right) = \left(\frac{2}{n_2}\right)^s (-1)^{(t-1)/2 \cdot (n_2-1)/2}\left(\frac{n_2}{|t|}\right)$. Since $n_1 \equiv n_2 \pmod{|t|}$, we have $\left(\frac{n_1}{|t|}\right) = \left(\frac{n_2}{|t|}\right)$, and since $4 \mid a$, $m_1 \equiv m_2 \pmod 4$ and so $(-1)^{(t-1)/2 \cdot (n_1-1)/2} = (-1)^{(t-1)/2 \cdot (n_2-1)/2}$. Now $a \equiv 0 \pmod 4$, so $s \geq 2$. If s is 2, then certainly $\left(\frac{2}{n_1}\right)^2 = \left(\frac{2}{n_2}\right)^2$. If $s > 2$, then $8 \mid a$ and $m_1 \equiv m_2 \pmod 8$, so $\left(\frac{2}{n_1}\right) = (-1)^{(n_1^2-1)/8} = (-1)^{(n_2^2-1)/8} = \left(\frac{2}{n_2}\right)$. Therefore $\left(\frac{a}{n_1}\right) = \left(\frac{a}{n_2}\right)$.

11.3.19. If $a \equiv 1 \pmod{4}$, then $|a| \equiv 1 \pmod{4}$ if $a > 0$ and $|a| \equiv -1 \pmod{4}$ if $a < 0$, so by Exercise 16 we have $\left(\frac{a}{|a|-1}\right) = \left(\frac{|a|-1}{|a|}\right) = \left(\frac{-1}{|a|}\right) = (-1)^{(|a|-1)/2} = 1$ if $a > 0$ and $= -1$ if $a < 0$. If $a \equiv 0 \pmod{4}$, $a = 2^s t$ with t odd and $t \geq 2$, then by Exercise 16 $\left(\frac{a}{|a|-1}\right) = \left(\frac{2}{|a|-1}\right)^s (-1)^{(t-1)/2} \left(\frac{|a|-1}{|t|}\right)$. Since $s \geq 2$, check that $\left(\frac{2}{|a|-1}\right)^s = 1$, $(|a|-1 \equiv 7 \pmod 8$ if $s > 2)$. Also $(-1)^{(t-1)/2} \left(\frac{|a|-1}{|t|}\right) = (-1)^{(t-1)/2}\left(\frac{-1}{|t|}\right) = (-1)^{(t-1)/2 + (|t|-1)/2} = 1$ if $t > 0$ and $= -1$ if $t < 0$.

11.4. Euler Pseudoprimes

11.4.1. We find that $2^{(561-1)/2} = 2^{280} = (2^{10})^{28} \equiv (-98)^{28} \equiv (-98^2)^{14} \equiv 67^{14} \equiv (67^2)^7 \equiv 1^7 = 1 \pmod{561}$. Furthermore, we see that $\left(\frac{2}{561}\right) = 1$ since $561 \equiv 1 \pmod 8$.

11.4.3. Suppose that n is an Euler pseudoprime to both the bases a and b. Then $a^{(n-1)/2} \equiv \left(\frac{a}{n}\right)$ and $b^{(n-1)/2} \equiv \left(\frac{b}{n}\right)$. It follows that $(ab)^{(n-1)/2} \equiv \left(\frac{a}{n}\right)\left(\frac{b}{n}\right) = \left(\frac{ab}{n}\right)$. Hence n is an Euler pseudoprime to the base ab.

11.4.5. Suppose that $n \equiv 5 \pmod 8$ and n is an Euler pseudoprime to the base 2. Since $n \equiv 5 \pmod 8$ we have $\left(\frac{2}{n}\right) = -1$. Since n is an Euler pseudoprime to the base 2, we have $2^{(n-1)/2} \equiv \left(\frac{2}{n}\right) = -1 \pmod n$. Write $n - 1 = 2^2 t$ where t is odd. Since $2^{((n-1))/2} \equiv 2^{2t} \equiv -1 \pmod n$, n is a strong pseudoprime to the base 2.

11.4.7. Let n be an Euler pseudoprime to the base 5 such that $n \equiv 5 \pmod{20}$. Then $n = 20k + 5$ and $n - 1 = 20k + 4 = 2^2(5k+1) = 2^2 t$, and $5^{(n-1)/2} \equiv 5^{2(5k+1)} \equiv 5^{2t} \equiv \left(\frac{5}{n}\right) \pmod n$. If $\left(\frac{5}{n}\right) = -1$, then n satisfies Miller's test to the base 5. If $\left(\frac{5}{n}\right) = 1$, then $5^{2t} \equiv 1 \pmod n$ and so $5^t \equiv -1$ and n satisfies Miller's test to the base 5. Therefore n is a strong pseudoprime to the base 5.

11.4.9. Using Exercise 8, we compute $561 = 3 \cdot 11 \cdot 17$. Then $561 = 1 + 2^4 35$, $3 = 1 + 2$, $11 = 1 + 2 \cdot 5$, and $17 = 1 + 2^4$, so $k = 4, k_1 = 1, k_2 = 1$, and $k_3 = 4$. Since $a_1 = 1$ is odd and $k_1 = 1 < k = 4$, we see that $\delta_n = 1/2$. Then the number we seek is $(1/2)((561-1)/2, 3-1)((561-1)/2, 11-1)((561-1)/2, 17-1) = (1/2)(280, 2)(280, 10)(280, 16) = (1/2)2 \cdot 10 \cdot 16 = 80$. So there are 80 different values for b.

11.5. Zero-Knowledge Proofs

11.5.1. We check that both 47 and 67 are congruent to 3 modulo 4. If p is a prime congruent to 3 modulo 4, then $(\pm x^4)^{(p+1)/4} \equiv x^{p+1} \equiv x^2 \pmod p$, by Fermat's little theorem. In this case, we have $x^2 \equiv \pm 2070^{(47+1)/4} \equiv \pm 2^{12} \equiv 7 \pmod{47}$, and $x^2 \equiv \pm 2070^{(67+1)/4} \equiv \pm 7^{17} \equiv \pm 23 \pmod{67}$. Next, since x^2 is a quadratic residue modulo 3149, it must be a quadratic residue modulo each of the factors of 3149. We compute $\left(\frac{7}{47}\right) = 1$, $\left(\frac{-7}{47}\right) = -1$, $\left(\frac{23}{67}\right) = 1$, and $\left(\frac{-23}{67}\right) = -1$. Therefore we solve the system $x^2 \equiv 7 \pmod{47}, x^2 \equiv 23 \pmod{67}$, to find $x^2 = 1229$.

11.5.3. Since $p, q \equiv 3 \pmod 4$, -1 is not a quadratic residue modulo p or q. If the four square roots are found using the method in Example 9.19, then only one of each possibility for choosing $+$ or $-$ can yield a quadratic residue in each congruence, so there is only one system which results in a square.

11.5.5. Paula sends $x = 1226, y = 625$. After receiving a 1, she sends $u\bar{r} = 689$.

11.5.7. The prover sends $x = 1403^2 = 1968409 \equiv 519 \pmod{2491}$. The verifier sends $\{1, 5\}$. The prover sends $y = 1425$. The verifier computes $y^2 z = 1425^2 \cdot 197 \cdot 494 \equiv 519 \equiv x \pmod{2491}$

11.5.9. a. First we find inverses modulo 3953 of the six numbers, getting 3333, 753, 411, 1319, 705, and 1811, respectively. Next we square and reduce these numbers modulo 3953 to get $s_1 = 959, s_2 = 1730, s_3 = $

$2895, s_4 = 441, s_5 = 2900$, and $s_6 = 2684$.

b. Paula sends $y \equiv 403 \cdot 1001 \cdot 21 \cdot 989 \cdot 1039 \equiv 1074 \pmod{3953}$.

c. Vince checks that $1074^2 \cdot 959 \cdot 1730 \cdot 441 \cdot 2684 \equiv 336 \equiv 403^2 \pmod{3953}$.

11.5.11. If Paula sends back a to Vince then $a^2 \equiv w^2 \pmod{n}$, with $a \not\equiv w \pmod{n}$. Then $a^2 - w^2 = (a-w)(a+w) \equiv 0 \pmod{n}$. By computing $(a-w, n)$ and $(a+w, n)$ Vince will likely produce a nontrivial factor of n.

CHAPTER 12

Decimal Fractions and Continued Fractions

12.1. Decimal Fractions

12.1.1. a. Using the recursive formulae from Theorem 12.1. Let $\gamma_0 = 2/5$. Then $c_1 = [10 \cdot (2/5)] = 4$, and $\gamma_1 = 10 \cdot (2/5) - 4 = 0$, so we're done, and the decimal expansion is 0.4.

b. Let $\gamma_0 = 5/12$. Then $c_1 = [10 \cdot (5/12)] = 4$ and $\gamma_1 = 10 \cdot (5/12) - 4 = 1/6$. Then $c_2 = [10 \cdot (1/6)] = 1$, and $\gamma_2 = 10 \cdot (1/6) - 1 = 2/3$. Then $c_3 = [10 \cdot (2/3)] = 6$, and $\gamma_3 = 10 \cdot (2/3) - 6 = 2/3 = \gamma_2$, so the sequence repeats and the decimal expansion is $0.41\overline{6}$.

c. Let $\gamma_0 = 12/13$. Then $c_1 = [10 \cdot (12/13)] = 9$, and $\gamma_1 = 10 \cdot (12/13) - 9 = 3/13$. Then $c_2 = [10 \cdot (3/13)] = 2$, and $\gamma_2 = 10 \cdot (3/13) - 2 = 4/13$. Then $c_3 = [10 \cdot (4/13)] = 3$, and $\gamma_3 = 10 \cdot (4/13) - 3 = 1/13$. Then $c_4 = [10 \cdot (1/13)] = 0$, and $\gamma_4 = 10 \cdot (1/13) - 0 = 10/13$. Then $c_5 = [10 \cdot (10/13)] = 7$, and $\gamma_5 = 10 \cdot (10/13) - 7 = 9/13$. Then $c_6 = [10 \cdot (9/13)] = 6$, and $\gamma_6 = 10 \cdot (9/13) - 6 = 12/13 = \gamma_0$. So the decimal expansion is $.\overline{923076}$.

d. Let $\gamma_0 = 8/15$. Then $c_1 = [10 \cdot (8/15)] = 5$, and $\gamma_1 = 10 \cdot (8/15) - 5 = 5/15$. Then $c_2 = [10 \cdot (5/15)] = 3$, and $\gamma_2 = 10 \cdot (5/15) - 3 = 5/15 = \gamma_1$. So the decimal expansion is $0.5\overline{3}$.

e. Let $\gamma_0 = 1/111$. Then $c_1 = [10 \cdot (1/111)] = 0$, and $\gamma_1 = 10 \cdot (1/111) - 0 = 10/111$. Then $c_2 = [10 \cdot (10/111)] = 0$, and $\gamma_2 = 10 \cdot (10/111) - 0 = 100/111$. Then $c_3 = [10 \cdot (100/111)] = 9$, and $\gamma_3 = 10 \cdot (100/111) - 9 = 1/111 = \gamma_1$. So the decimal expansion is $0.\overline{009}$.

f. Let $\gamma_0 = 1/1001$. Then $c_1 = [10 \cdot (1/1001)] = 0$, and $\gamma_1 = 10 \cdot (1/1001) - 0 = 10/1001$. Then $c_2 = [10 \cdot (10/1001)] = 0$, and $\gamma_2 = 10 \cdot (10/1001) - 0 = 100/1001$. Then $c_3 = [10 \cdot (100/1001)] = 0$, and $\gamma_3 = 10 \cdot (100/1001) - 0 = 1000/1001$. Then $c_4 = [10 \cdot (1000/1001)] = 9$, and $\gamma_4 = 10 \cdot (1000/1001) - 9 = 991/1001$. Then $c_5 = [10 \cdot (991/1001)] = 9$, and $\gamma_5 = 10 \cdot (991/1001) - 9 = 901/1001$. Then $c_6 = [10 \cdot (901/1001)] = 9$, and $\gamma_6 = 10 \cdot (901/1001) - 9 = 1/1001 = \gamma_0$. So the decimal expansion is $0.\overline{000999}$.

12.1.3. a. We reduce $12/100$ to get $3/25$.

b. Note that $.1\overline{2} = .1 + (2/100)\sum_{n=1}^{\infty} 1/10^n = (1/10) + (2/100)(1/(1 - 1/10)) = 11/90$.

c. Let $\alpha = .\overline{12}$. Then $100\alpha = 12.\overline{12}$, so that $99\alpha = 12$. Therefore $\alpha = 12/99 = 4/33$.

12.1.5. All prime divisors of $210 = 2 \cdot 3 \cdot 5 \cdot 7$ must divide b, so $b = 2^r 3^s 5^t 7^u$, with r, s, t, and u nonnegative integers.

12.1.7. a. Since $4 = 2^2$, we have $T = 2^2 | 12^1$, and $U = 1$. So the pre-period length is 1, and $\text{ord}_1 12 = 0$, so the period length is 0.

b. Since $8 = 2^3$, we have $T = 2^3 | 12^2$, and $U = 1$. So the pre-period length is 2, and $\text{ord}_1 12 = 0$, so the period length is 0.

c. Since $10 = 2 \cdot 5$, we have $T = 2 | 12^1$, and $U = 5$. So the pre-period length is 1, and $\text{ord}_5 12 = 4$, so the period length is 4.

105

d. Since $24 = 2^3 3$, we have $T = 2^2 3 | 12^2$, and $U = 1$. So the pre-period length is 2, and $\text{ord}_1 12 = 0$, so the period length is 0.

e. Since $132 = 12 \cdot 11$, we have $T = 12 | 12^1$, and $U = 11$. So the pre-period length is 1, and $\text{ord}_{11} 12 = 1$, so the period length is 1.

f. Since $360 = 2^3 3^2 5$, we have $T = 2^3 3^2 | 12^2$, and $U = 5$. So the pre-period length is 2, and $\text{ord}_5 12 = 4$, so the period length is 4.

12.1.9. If $p = 2$ or 5, the period length is 0. Otherwise, $\text{ord}_p b = n$ is the period length. Now, $\text{ord}_p b = n$ for exactly those primes dividing $10^n - 1$, but not dividing $10^m - 1$ for any $m < n$. Then, (a) $10 - 1 = 3^2, p = 3$ (b) $10^2 - 1 = 3^2 11, p = 11$ (c) $10^3 - 1 = 3 \cdot 11 \cdot 37, p = 37$ (d) $p = 101$ (e) $p = 41$ and 271 (f) $p = 7$ and 13.

12.1.11. Using the construction from Theorem 12.2 and Example 12.1, we use induction to show that $c_k = k - 1$ and $\gamma_k = (kb - k + 1)/(b-1)^2$. The induction step is as follows: $c_{k+1} = [b\gamma_k] = [(kb^2 - bk + b)/(b-1)^2] = [(k(b-1)^2 + b(k+1) - k)/(b-1)^2] = [k + (b(k+1) - k)/(b-1)^2] = k$, and $\gamma_{k+1} = (k+1)b - k$, if $k \neq b - 2$. If $k = b - 2$, we have $c_{b-2} = b$, so we have determined $b - 1$ consecutive digits of the expansion. From the binomial theorem, $(x+1)^a \equiv ax + 1 \pmod{x^2}$, so $\text{ord}_{(b-1)^2} b = b - 1$, which is the period length. Therefore we have determined the entire expansion.

12.1.13. The base b expansion is $(.100100001\ldots)_b$ which is non-repeating and therefore by Theorem 12.4 represents an irrational number.

12.1.15. Let γ be a real number. Set $c_0 = [\gamma]$ and $\gamma_1 = \gamma - c_0$. Then $0 \leq \gamma_1 < 1$ and $\gamma = c_0 + \gamma_1$. From the condition that $c_k < k$ for $k = 1, 2, 3, \ldots$, we must have $c_1 = 0$. Let $c_2 = [2\gamma_1]$ and $\gamma_2 = 2\gamma_1 - c_2$. Then $\gamma_1 = (c_2 + \gamma_2)/2$, so $\gamma = c_0 + c_1/1! + c_2/2! + \gamma_2/2!$ Now let $c_3 = [3\gamma_2]$ and $\gamma_3 = 3\gamma_2 - c_3$. Then $\gamma_2 = (c_3 + \gamma_3)/3$ and so $\gamma = c_0 + c_1/1! + c_2/2! + c_3/3! + \gamma_3/3!$. Continuing in this fashion, for each $k = 2, 3, \ldots$, define $c_k = [k\gamma_{k-1}]$ and $\gamma_k = k\gamma_{k-1} - c_k$. Then $\gamma = c_0 + c_1/1! + c_2/2! + c_3/3! + \cdots + c_k/k! + \gamma_k/k!$. Since each $\gamma_k < 1$, we know that $\lim_{k \to \infty} \gamma_k/k! = 0$, so we conclude that $\gamma = c_0 + c_1/1! + c_2/2! + c_3/3! + \cdots + c_k/k! + \cdots$.

12.1.17. In the proof of Theorem 12.1, the numbers $p\gamma_n$ are the remainders of b^n upon division by p. The process recurs as soon as some γ_i repeats a value. Since $1/p = (.\overline{c_1 c_2 \ldots c_{p-1}})$ has period length $p - 1$, we have by Theorem 12.4 that $\text{ord}_p b = p - 1$, so there is an integer k such that $b^k \equiv m \pmod{p}$. So the remainders of mb^n upon division by p are the same as the remainders of $b^k b^n$ upon division by p. Hence the nth digit of the expansion of m/p is determined by the remainder of b^{k+n} upon division by p. Therefore, it will be the same as the $(k + n)$th digit of $1/p$.

12.1.19. Suppose $n = TU$, with $T = 2^k$ and U odd. Then the period length of the binary expansion of $1/n$ is $\text{ord}_U 2$. If $\text{ord}_U 2 = n - 1$, then $U = n$. So n is prime, and 2 is a primitive root of n.

12.1.21. Suppose $e = h/k$. Then $k!(e - 1 - 1/1! - 1/2! - \cdots - 1/k!)$ is an integer. But this is equal to $k!(1/(k+1)! + 1/(k+2)! + \cdots) = 1/(k+1) + 1/(k+1)(k+2) + \cdots < 1/(k+1) + 1/(k+1)^2 + \cdots = 1/k < 1$. But $k!(e - 1 - 1/1! - 1/2! - \cdots - 1/k!)$ is positive, and therefore cannot be an integer, a contradiction.

12.1.23. Let $\alpha = \sum_{i=1}^{\infty} \frac{(-1)^{a_i}}{10^{i!}}$, and $\frac{p_k}{q_k} = \sum_{i=1}^{k} \frac{(-1)^{a_i}}{10^{i!}}$. Then $\left| \alpha - \frac{p_k}{q_k} \right| = \left| \sum_{i=k+1}^{\infty} \frac{(-1)^{a_i}}{10^{i!}} \right| \leq \sum_{i=k+1}^{\infty} \frac{1}{10^{i!}}$. As in the proof of Corollary 12.5.1, it follows that $\left| \alpha - \frac{p_k}{q_k} \right| < \frac{2}{10^{(k+1)!}}$, which shows that there can be no real number C as in Theorem 12.5. Hence, α must be transcendental.

12.1.25. Suppose $e = h/k$. Then $k!(e - 1 - 1/1! - 1/2! - \cdots 1/k!)$ is an integer. But this is equal to $k!(1/(k+1)! + 1/(k+2)! + \cdots) = 1/(k+1) + 1/(k+1)(k+2) + \cdots < 1/(k+1) + 1/(k+1)^2 + \cdots = 1/k < 1$. But $k!(1/(k+1)! + 1/(k+2)! + \cdots)$ is positive, and therefore can not be an integer, a contradiction.

12.2. Finite Continued Fractions

12.2.1. a. We have $[2;7] = 2 + 1/7 = 15/7$.

b. We have $[1;,2,3] = 1 + \dfrac{1}{2 + 1/3} = 1 + 3/7 = 10/7$.

c. We have $[0'5,6] = \dfrac{1}{5 + (1/6)} = 6/31$.

d. We have $[3;7,15,1] = 3 + \dfrac{1}{7 + \dfrac{1}{15+1}} = 3 + \dfrac{1}{1 + 1/16} = 355/113$. Note that this is a very good approximation for π.

e. We have $[1;1] = 1 + (1/1) = 2$.

f. We have $[1;1,1] = 1 + \dfrac{1}{1+1} = 3/2$.

g. We have $[1;1,1,1] = 1 + \dfrac{1}{1 + \dfrac{1}{1+1}} = 5/3$.

h. We have $[1;1,1,1,1] = 1 + \dfrac{1}{1 + \dfrac{1}{1 + \dfrac{1}{1+1}}} = 8/5$. Note that the numerators and denominators in these last four exercises are Fibonacci numbers.

12.2.3. a. Using the construction in the proof of Theorem 12.8, we let $r_0 = 18$ and $r_1 = 13$. Then $18 = 1 \cdot 13 + 5, 13 = 2 \cdot 5 + 3, 5 = 1 \cdot 3 + 2, 3 = 1 \cdot 2 + 1$, and $2 = 2 \cdot 1$. The sequence of quotient gives us the continued fraction $[1;2,1,1,2]$.

b. We perform the Euclidean algorithm on 32 and 17 to get $32 = 1 \cdot 17 + 15, 17 = 1 \cdot 15 + 2, 15 = 7 \cdot 2 + 1, 2 = 2 \cdot 1$. The sequence of quotients gives us $[1;1,7,2]$.

c. We perform the Euclidean algorithm on 19 and 9 to get $19 = 2 \cdot 9 + 1, 9 = 9 \cdot 1$. The sequence of quotients yields $[2;9]$.

d. We perform the Euclidean algorithm on 310 and 99 to get $310 = 3 \cdot 99 + 13, 99 = 7 \cdot 13 + 8, 13 = 1 \cdot 8 + 5, 8 = 1 \cdot 5 + 3, 5 = 1 \cdot 3 + 2, 3 = 1 \cdot 2 + 1, 2 = 2 \cdot 1$. The sequence of quotients yields $[3;7,1,1,1,1,2]$.

e. We perform the Euclidean algorithm on -931 and 1005 to get $-931 = -1 \cdot 1005 + 74, 1005 = 13 \cdot 74 + 43, 74 = 1 \cdot 43 + 31, 43 = 1 \cdot 31 + 12, 31 = 2 \cdot 12 + 7, 12 = 1 \cdot 7 + 5, 7 = 1 \cdot 5 + 25 = 2 \cdot 2 + 1, 2 = 2 \cdot 1$. The sequence of quotients yields $[-1;13,1,1,2,1,1,2,2]$.

f. We perform the Euclidean algorithm on 831 and 8110 to get $831 = 0 \cdot 8110 + 831, 8110 = 9 \cdot 831 + 631, 831 = 1 \cdot 631 + 200, 631 = 3 \cdot 200 + 31, 200 = 6 \cdot 31 + 14, 31 = 2 \cdot 14 + 3, 14 = 4 \cdot 3 + 2, 3 = 1 \cdot 2 + 1, 2 = 2 \cdot 1$. The sequence of quotients gives us $[0;9,1,3,6,2,4,1,2]$.

12.2.5. a. We compute $p_0 = 1, p_1 = 1 \cdot 2 + 1 = 3, p_2 = 1 \cdot 3 + 1 = 4, p_3 = 1 \cdot 4 + 3 = 7, p_4 = 2 \cdot 7 + 4 = 18$, and $q_0 = 1, q_1 = 2, q_2 = 1 \cdot 2 + 1 = 3, q_3 = 1 \cdot 3 + 2 = 5, q_4 = 2 \cdot 5 + 3 = 13$. Then the convergents are $C_0 = 1/1 = 1, C_1 = 3/2, C_2 = 4/3, C_3 = 7/5, C_4 = 18/13$.

b. . We compute $p_0 = 1, p_1 = 1 \cdot 1 + 1 = 2, p_2 = 7 \cdot 2 + 1 = 15, p_3 = 2 \cdot 15 + 2 = 32$, and $q_0 = 1, q_1 = 1, q_2 = 7 \cdot 1 + 1 = 8, q_3 = 2 \cdot 8 + 1 = 17$. Then the convergents are $C_0 = 1, C_1 = 2, C_2 = 15/8, C_3 = 32/17$.

c. We compute $p_0 = 2, p_1 = 2 \cdot 9 + 1 = 19$, and $q_0 = 1, q_1 = 9$. Then the convergents are $C_0 = 2, C_1 = 19/9$.

d. We compute the sequence of p_i to be 3, 22, 25, 47, 72, 119, 310, and the sequence of q_i to be 1, 7, 8, 15, 23, 38, 99, so the convergents are 3, 22/7, 25/8, 47/15, 72/23, 119/38, 310/99.

e. We compute the sequence of p_i to be $-1, -12, -13, -25, -63, -88, -151, -390, -931$, and the sequence of q_i to be $1, 13, 14, 27, 68, 95, 163, 421, 1005$, so the convergents are $-1, -12/13, -13/14, -25/27, -63/68, -88/95, -151/163, -390/421, -931/1005$.

f. We compute the sequence of p_i to be 0, 1, 1, 4, 25, 54, 241, 295, 831, and the sequence of q_i to be 1, 9, 10, 39, 244, 527, 2352, 2879, 8110, so the convergents are 0, 1/9, 1/10, 4/39, 25/244, 54/527, 241/2352, 295/2879, 831/8110.

12.2.7. For Exercise 5: (a) $3/2 > 7/5$ and $1 < 4/3 < 18/13$ (b) $2 > 32/17$ and $1 < 15/8$ (c) vacuously true (d) $22/7 > 47/15 > 119/38$ and $3 < 25/8 < 72/23, < 310/99$ (e) $-12/13 > -25/27 > -88/95 > -390/421$ and $-1 < -13/14 < -63/68, < -151/163 < -931/1005$ (f) $1/9 > 4/39 > 54/527 > 295/2879$ and $0 < 1/10 < 25/244 < 241/2352 < 831/8110$.

12.2.9. Let $\alpha = r/s$. The Euclidean Algorithm for $1/\alpha = s/r < 1$ gives $s = 0(r) + s; r = a_0(s) + a_1$, and continues just like for r/s.

12.2.11. Proceed by induction. Assume $q_j \geq f_j$ for $j < k$. Then $q_k = a_k q_{k-1} + q_{k-2} \geq a_k f_{k-1} + f_{k-2} \geq f_{k-1} + f_{k-2} = f_k$, as desired.

12.2.13. By Exercise 10, we have $p_n/p_{n-1} = [a_n; a_{n-1}, \ldots, a_0] = [a_0; a_1, \ldots, a_n] = p_n/q_n = r/s$ if the continued fraction is symmetric. Then, $q_n = p_{n-1} = s$ and $p_n = r$, so by Theorem 12.10 we have $p_n q_{n-1} - q_n p_{n-1} = r q_{n-1} - s^2 = (-1)^{n-1}$. Then $r q_{n-1} = s^2 + (-1)^{n-1}$ and so $r | s^2 - (-1)^n$. Conversely, if $r | s^2 + (-1)^{n-1}$, then $(-1)n - 1 = p_n q_{n-1} - q_n p_{n-1} = r q_{n-1} - p_{n-1} s$. So $r | p_{n-1}s + (-1)^{n-1}$ and hence $r | (s^2 + (-1)^{n-1}) - (p_{n-1}s + (-1)^{n-1}) = s(s - p_{n-1})$. Since $s, p_{n-1} < r$ and $(r, s) = 1$, we have $s = p_{n-1}$. Then $[a_n; a_{n-1}, \ldots, a_0] = p_n/p_{n-1} = r/s = [a_0; a_1, \ldots, a_n]$.

12.2.15. Note that the notation $[a_0; a_1, \ldots, a_n]$ makes sense, even if the a_j are not integers. Use induction. Assume the statement is true for k odd and prove it for $k + 2$. Define $a'_k = [a_k; a_{k+1}, a_{k+2}]$ and check that $a'_k < [a_k; a_{k+1}, a_{k+2} + x] = a'_k + x'$. Then $[a_0; a_1, \ldots, a_{k+2}] = [a_0; a_1, \ldots, a'_k] > [a_0; a_1, \ldots, a'_k + x'] = [a_0; a_1, \ldots, a_{k+2} + x]$. Proceed similarly for k even.

12.3. Infinite Continued Fractions

12.3.1. a. We compute $a_0 = [\sqrt{2}] = 1, \alpha_1 = 1/(\sqrt{2} - 1) = \sqrt{2} + 1, a_1 = [\alpha_1] = 2, \alpha_2 = \frac{1}{(\sqrt{2}+1)-2} = \sqrt{2} + 1 = \alpha_1$. Therefore the sequence repeats, and we have $\sqrt{2} = [1; 2, 2, \ldots]$.

b. We compute $a_0 = [\sqrt{3}] = 1, \alpha_1 = 1/(\sqrt{3} - 1) = (\sqrt{3} + 1)/2, a_1 = [\alpha_1] = 1, \alpha_2 = \frac{1}{(\sqrt{3}+1)/2-1} = \sqrt{3} + 1, a_2 = [\alpha_2] = 2, \alpha_3 = \frac{1}{(\sqrt{3}+1)-2} = (\sqrt{3}+1)/2 = \alpha_1$. Therefore the sequence repeats, and we have $\sqrt{3} = [1; 1, 2, 1, 2, \ldots]$.

c. We compute $a_0 = [\sqrt{5}] = 2, \alpha_1 = 1/(\sqrt{5} - 2) = \sqrt{5} + 2, a_1 = [\alpha_1] = 4, \alpha_2 = 1/((\sqrt{5}+2)-4) = \sqrt{5} + 2 = \alpha_1$. Therefore the sequence repeats, and we have $\sqrt{5} = [2; 4, 4, \ldots]$.

d. We compute $a_0 = [(\sqrt{5}+1)/2] = 1, \alpha_1 = \frac{1}{(\sqrt{5}+1)/2-1} = (\sqrt{5}+1)/2 = \alpha_0$. This gives $(\sqrt{5}+1)/2 = [1; 1, 1, \ldots]$.

12.3.3. From Example 12.11, we have $\pi = [3; 7, 15, 1, 292, 1, 1, 1, 2, \ldots]$. We compute the convergents until we have a denominator greater than 100000: 3, 22/7, 333/106, 355/113, 103933/33102, 104348/33215,

208341/66317, 312689/99532, 833719/265381, ... Therefore, the best approximation with denominator less than 100000 is 312689/99532.

12.3.5. If $a_1 > 1$, let $A = [a_2; a_3, \ldots]$. Then $[a_0; a_1, \ldots] + [-a_0 - 1; 1, a_1 - 1, a_2, a_3, \ldots] = a_0 + \dfrac{1}{a_1 + (1/A)} + \left(-a_0 - 1 + \dfrac{1}{1 + \dfrac{1}{a_1 - 1 + (1/A)}}\right) = 0$. Similarly if $a_1 = 1$.

12.3.7. If $\alpha = [a_0; a_1, a_2, \ldots]$, then $1/\alpha = 1/[a_0; a_1, a_2, \ldots] = 0 + \dfrac{1}{a_0 + \dfrac{1}{a_1 + \cdots}} = [0; a_0, a_1, a_2, \ldots]$. Then the kth convergent of $1/\alpha$ is $[0; a_0, a_1, a_2, \ldots, a_{k-1}] = 1/[a_0; a_1, a_2, \ldots, a_{k-1}]$, which is the reciprocal of the $(k-1)$st convergent of α.

12.3.9. By Theorem 12.17, such a p/q is a convergent of α. We have $(\sqrt{5}+1)/2 = [1; 1, 1, \ldots]$, so $q_n = f_n$ (Fibonacci) and $p_n = q_{n+1}$. Then $\lim_{n\to\infty} q_{n-1}/q_n = \lim_{n\to\infty} q_{n-1}/p_{n-1} = 2/(\sqrt{5}+1) = (\sqrt{5}-1)/2$. So $\lim_{n\to\infty} ((\sqrt{5}+1)/2 + (q_{n-1}/q_n)) = (\sqrt{5}+1)/2 + (\sqrt{5}-1)/2 = \sqrt{5}$. So $(\sqrt{5}+1)/2 + (q_{n-1}/q_n) > c$ only finitely often. Whence, $1/((\sqrt{5}+1)/2 + (q_{n-1}/q_n)) q_n^2 < 1/(cq_n^2)$. The following identity finishes the proof. Note that $\alpha_n = \alpha$ for all n. Then $|\alpha - (p_n/q_n)| = |(\alpha_{n+1}p_n + p_{n-1})/(\alpha_{n+1}q_n + q_{n-1}) - (p_n/q_n)| = |(-(p_n q_{n-1} - p_{n-1}q_n))/(q_n(\alpha q_n + q_{n-1}))| = 1/(q_n^2(\alpha + (q_{n-1}/q_n)))$.

12.3.11. If β is equivalent to α, then $\beta = (a\alpha + b)/(c\alpha + d)$. Solving for α gives $\alpha = (-d\beta + b)/(c\beta - a)$, so α is equivalent to β.

12.3.13. If $a \neq 0$, then $r/s = ((rb)a + 0)/((sa)b + 0)$, so r/s and a/b are equivalent. If $a = 0$ then $r/s = (1 \cdot a + r)/(0 \cdot b + s)$.

12.3.15. Note that $p_{k,t}q_{k-1} - q_{k,t}p_{k-1} = t(p_{k-1}q_{k-1} - q_{k-1}p_{k-1}) + (p_{k-2}q_{k-1} - p_{k-1}q_{k-2}) = \pm 1$. Thus $p_{k,t}$ and $q_{k,t}$ are relatively prime.

12.3.17. See, for example, the classic work by O. Perron, *Die Lehre von den Kettenbrüchen*, Leipzig, Teubner (1929).

12.3.19. Using Exercise 17, we test each of the pseudoconvergents in Exercise 18 and find that $|\pi - 179/57| < |\pi - 22/7|$.

12.3.21. (Proof by Rob Johnson.) Note first that if $b < d$, then $|a/b - c/d| < 1/2d^2$ implies that $|ad - bc| < b/2d < 1/2$, but since $b \neq d$, $|ad - bc|$ is a positive integer, and so is greater than $1/2$. Thus $b \geq d$. Now assume that c/d is not a convergent of the continued fraction for a/b. Since the denominators of the convergents increase to b, there must be two successive convergents p_n/q_n and p_{n+1}/q_{n+1} such that $q_n < d < q_{n+1}$. Next, by the triangle inequality we have $1/2d^2 > \left|\dfrac{a}{b} - \dfrac{c}{d}\right| = \left|\dfrac{c}{d} - \dfrac{p_n}{q_n}\right| - \left|\dfrac{a}{b} - \dfrac{p_n}{q_n}\right| \geq \left|\dfrac{c}{d} - \dfrac{p_n}{q_n}\right| - \left|\dfrac{p_{n+1}}{q_{n+1}} - \dfrac{p_n}{q_n}\right|$, since the $n+1$st convergent is on the other side of a/b from the nth convergent. Since the numerator of the first difference is a nonzero integer, and applying Corollary 12.3 to the second difference, we have the last expression greater than or equal to $1/dq_n - 1/q_{n+1}q_n$. If we multiply through by d^2 we get $\dfrac{1}{2} > \dfrac{d}{q_n}\left(1 - \dfrac{d}{q_{n+1}}\right) > 1 - \dfrac{d}{q_{n+1}}$ since $d/q_n > 1$. From which we deduce that $1/2 < d/q_{n+1}$.

Now the convergents p_n/q_n and p_{n+1}/q_{n+1} divide the line into three regions. As c/d could be in any of these, there are three cases. Case 1: If c/d is between the convergents, then $\dfrac{1}{dq_n} \leq \left|\dfrac{c}{d} - \dfrac{p_n}{q_n}\right|$ since the numerator of the fraction is a positive integer and the denominators on both sides of the inequality are the same. This last is less than or equal to $\left|\dfrac{p_{n+1}}{q_{n+1}} - \dfrac{p_n}{q_n}\right| = \dfrac{1}{q_{n+1}q_n}$ since the $n+1$st convergent is farther from the nth convergent than c/d and where we have applied Corollary 12.3. But this implies that $d \geq q_{n+1}$, a contradiction. Case 2. If c/d is closer to p_n/q_n, then again $\dfrac{1}{dq_n} \leq \left|\dfrac{c}{d} - \dfrac{p_n}{q_n}\right| \leq \left|\dfrac{a}{b} - \dfrac{c}{d}\right|$ since a/b is on the other side of the nth convergent from c/d. But this last is less than $1/2d^2$ and if we multiply through by d we have $1/q_n < 1/2d$, which implies that $q_n > d$, a contradiction. Case 3. If c/d is closer to p_{n+1}/q_{n+1}, then with the same reasoning as in Case 2, we have $\dfrac{1}{dq_{n+1}} \leq \left|\dfrac{c}{d} - \dfrac{p_{n+1}}{q_{n+1}}\right| < \left|\dfrac{a}{b} - \dfrac{c}{d}\right| < 1/2d^2$. But this implies that $d/q_{n+1} < 1/2$ contradicting the inequality established above. Having exhausted all the cases, we must conclude that c/d must be a convergent of the continued fraction for a/b.

12.4. Periodic Continued Fractions

12.4.1. a. Using Theorem 12.20, we have $\alpha_0 = \sqrt{7}, a_0 = 2, P_0 = 0, Q_0 = 1, P_1 = 2 \cdot 1 - 0 = 2, Q_1 = \frac{7-2^2}{1} = 3, \alpha_1 = \frac{2+\sqrt{7}}{3}, a_1 = 1, P_2 = 1 \cdot 3 - 2 = 1, Q_2 = \frac{7-1^2}{3} = 2, \alpha_2 = \frac{1+\sqrt{7}}{2}, a_2 = 1, P_3 = 1 \cdot 2 - 1 = 1, Q_3 = \frac{7-1^2}{2} = 3, \alpha_3 = \frac{1+\sqrt{7}}{3}, a_3 = 1, P_4 = 1 \cdot 3 - 1 = 2, Q_4 = \frac{7-2^2}{3} = 1, \alpha_4 = \frac{2+\sqrt{7}}{1}, a_4 = 4, P_5 = 4 \cdot 1 - 2 = 2, Q_5 = \frac{7-2^2}{1} = 3, \alpha_5 = \alpha_1$, so $\sqrt{7} = [2; \overline{1, 1, 1, 4}]$.

b. As in part (a), we find $\sqrt{11} = [3; \overline{3, 6}]$.

c. As in part (a), we find $\sqrt{23} = [4; \overline{1, 3, 1, 8}]$.

d. As in part (a), we find $\sqrt{47} = [6; \overline{1, 5, 1, 12}]$.

e. As in part (a), we find $\sqrt{59} = [7; \overline{1, 2, 7, 2, 1, 14}]$.

f. As in part (a), we find $\sqrt{94} = [9; \overline{1, 2, 3, 1, 1, 5, 1, 8, 1, 5, 1, 1, 3, 2, 1, 18}]$.

12.4.3. a. As in Exercise 1, we find $1 + \sqrt{101} = [2; \overline{2}]$.

b. As in Exercise 1, we find $(2 + \sqrt{5}/3 = [1; \overline{2, 2, 2, 1, 12, 1}]$.

c. As in Exercise 1, we find $(5 - \sqrt{7})/4 = [0; 1, 1, \overline{2, 3, 10, 3}]$.

12.4.5. a. Let $x = [2; 1, \overline{5}]$. Then $x = [2; 1, y]$, where $y = [5; \overline{5}]$. Since $y = [5; y]$, we have $y = 5 + 1/y$, so $y^2 - 5y - 1 = 0$, and since y is positive, $y = (5 + \sqrt{29})/2$. Then $x = 2 + \dfrac{1}{5 + (1/y)} = (3y+2)/(y+1) = (23 + \sqrt{29})/10$.

b. Let $x = [2; \overline{1, 5}]$, then $x = [2; y]$, where $y = [\overline{1; 5}]$. Then $y = [1; 5, y] = 1 + 1/(1 + 1/y)$, so $5y^2 - 5y - 1 = 0$, and y is positive, so $y = (5 + 3\sqrt{5})/10$. Then $x = 2 + 1/y = (-1 + 3\sqrt{5})/2$.

c. Let $x = [\overline{2; 1, 5}]$. Then $x = [2; 1, 5, x] = 2 + \dfrac{1}{1 + 1/(5 + (1/x))} = (17x + 3)/(6x + 1)$, so $6x^2 - 16x - 3 = 0$. Noting that x is positive gives $x = (8 + \sqrt{82})/6$.

12.4.7. a. From Exercise 8, we have $[3; \overline{6}] = \sqrt{3^2 + 1} = \sqrt{10}$.

b. From Exercise 8, we have $[4; \overline{8}] = \sqrt{4^2 + 1} = \sqrt{17}$.

c. From Exercise 8, we have $[5; \overline{10}] = \sqrt{5^2 + 1} = \sqrt{26}$.

d. From Exercise 8, we have $[6; \overline{12}] = \sqrt{6^2 + 1} = \sqrt{37}$.

12.4.9. a. $\alpha_0 = \sqrt{d^2 - 1}, a_0 = d - 1, P_0 = 0, Q_0 = 1, P_1 = (d-1)(1) - 0 = d - 1, Q_1 = ((d^2 - 1) - (d-1)^2)/1 = 2d - 2, \alpha_1 = (d - 1 + \sqrt{d^2 - 1})/(2(d-1)) = 1/2 + 1/2\sqrt{(d+1)/(d-1)}, a_1 = 1, P_2 = 1(2d - 2) - $

$(d-1) = d-1, Q_2 = (d^2 - 1 - (d-1)^2)/(2d-2) = 1, \alpha_2 = (d-1+\sqrt{d^2-1})/1, a_2 = 2d-2, P_3 = 2(d-1)(1)-(d-1) = d-1 = P_1, Q_3 = ((d^2-1)-(d-1)^2)/1 = 2d-2 = Q_1$, so $\alpha = [d-1; \overline{1, 2(d-1)}]$.

b. $\alpha_0 = \sqrt{d^2-d}, a_0 = [\sqrt{d^2-d}] = d-1$, since $(d-1)^2 < d^2-d < d^2$. Then $P_0 = 0, Q_0 = 1, P_1 = d-1, Q_1 = d-1, \alpha_1 = ((d-1)+\sqrt{d^2-d})/(d-1) = 1 + \sqrt{d/(d-1)}, a_1 = 2, P_2 = d-1, Q_2 = 1, \alpha_2 = ((d-1)+\sqrt{d^2-d})/1, a_2 = 2(d-1), P_3 = P_1, Q_3 = Q_1$. Therefore, $\sqrt{d^2-d} = [d-1; \overline{2, 2(d-1)}]$.

c. Applying parts (a) and (b) we compute $\sqrt{99} = \sqrt{10^2 - 1} = [9; \overline{1, 18}]$, $\sqrt{110} = \sqrt{11^2 - 11} = [10; \overline{2, 20}]$, $\sqrt{272} = \sqrt{17^2 - 17} = [16; \overline{2, 32}]$, and $\sqrt{600} = \sqrt{25^2 - 25} = [24; \overline{2, 48}]$.

12.4.11. a. Note that $d < \sqrt{d^2+4} < d+1$. We compute $\alpha_0 = \sqrt{d^2+4}, a_0 = d, P_0 = 0, Q_0 = 1, P_1 = d, Q_1 = 4, \alpha_1 = (d+\sqrt{d^2+4})/4, a_1 = [2d/4] = (d-1)/2$, since d is odd. Then, $P_2 = d-2, Q_2 = d, \alpha_2 = (d-2+\sqrt{d^2+4})/d, ((d-2)+d)/d < \alpha_2 < (d-2+d+1)/d$, so $a_2 = 1, P_3 = 2, Q_3 = d, \alpha_3 = (2+\sqrt{d^2+4})/d, a_3 = 1, P_4 = d-2, Q_4 = 4, \alpha_4 = (d-2+\sqrt{d^2+4})/4, (d-2+d)/4 = (d-1)/2 < \alpha_4 < (d-2+d+1)/4$, so $a_4 = (d-1)/2, P_5 = d, Q_5 = 1, \alpha_5 = (d+\sqrt{d^2+4})/1, a_5 = 2d, P_6 = d = P_1, Q_6 = 4 = Q_1$, so $\alpha = [d; \overline{(d-1)/2, 1, 1, (d-1)/2, 2d}]$.

b. Note that $d-1 < \sqrt{d^2-4} < d$. We compute $\alpha_0 = \sqrt{d^2-4}, a_0 = d-1, P_0 = 0, Q_0 = 1, P_1 = d-1, Q_1 = 2d-5, \alpha_1 = (d-1+\sqrt{d^2-4})/(2d-5), (d-1+d-1)/(2d-5) < \alpha_0 < (d-1+d)/(2d-5)$ and $d > 3$ so $a_1 = 1, P_2 = d-4, Q_2 = 4, \alpha_2 = (d-4+\sqrt{d^2-4})/4, a_2 = (d-3)/2, P_3 = d-2, Q_3 = d-2, \alpha_3 = (d-2+\sqrt{d^2-4})/(d-2), a_3 = 2, P_4 = d-2, Q_4 = 4, \alpha_4 = (d-2+\sqrt{d^2-4})/4, a_4 = (d-3)/2, P_5 = d-4, Q_5 = 2d-5, \alpha_5 = (d-4+\sqrt{d^2-4})/(2d-5), a_5 = 1, P_6 = d-1, Q_6 = 1, \alpha_6 = (d-1+\sqrt{d^2-4})/1, a_6 = 2d-2, P_7 = d-1 = P_1, Q_7 = 2d-5 = Q_1$, so $\alpha = [d-1; \overline{1, (d-3)/2, 2, (d-3)/2, 1, 2d-2}]$.

12.4.13. Suppose $\sqrt{d}$ has period length 2. Then $\sqrt{d} = [a; \overline{c, 2a}]$ from the discussion preceding Example 12.16. Then $\sqrt{d} = [a; y]$ with $y = [\overline{c; 2a}] = [c; 2a, y] = c + 1/(2a + (1/y)) = (2acy + c + y)/(2ay + 1)$. Then $2ay^2 - 2acy - c = 0$, and since y is positive, we have $y = (2ac + \sqrt{(2ac)^2 + 4(2a)c})/(4a) = (ac + \sqrt{(ac)^2 + 2ac})/(2a)$. Then $\sqrt{d} = [a; y] = a + (1/y) = a + 2a/(ac + \sqrt{(ac)^2 + 2ac}) = \sqrt{a^2 + 2a/c}$, so $d = a^2 + 2a/c$, and $b = 2a/c$ is an integral divisor of $2a$. Conversely, let $\alpha = \sqrt{a^2+b}$ and $b|2a$, say $kb = 2a$. Then $a_0 = [\sqrt{a^2+b}] = a$, since $(a^2 < a^2 + b < (a+1)^2)$. Then $P_0 = 0, Q_0 = 1, P_1 = a, Q_1 = b, \alpha_1 = (a+\sqrt{a^2+b})/b, a_1 = 4k, P_2 = a, Q_2 = 1, \alpha_2 = (a+\sqrt{a^2+b})/1, a_2 = 2a, P_3 = a = P_1, Q_3 = b = Q_1$, so $\alpha = [a; \overline{4k, 2a}]$, which has period length 2.

12.4.15. a. We have $1 + \sqrt{5} > 1$, but $(1+\sqrt{5})' = 1 - \sqrt{5} < -1$. Hence, by Theorem 12.21, the continued fraction of $1 + \sqrt{5}$ is not purely periodic.

b. We have $2 + \sqrt{8} > 1$ and $-1 < (2+\sqrt{8})' = 2 - \sqrt{8} < 0$, so by Theorem 12.21 the continued fraction expansion of $2 + \sqrt{8}$ is purely periodic.

c. We have $4 + \sqrt{17} > 1$ and $-1 < (4+\sqrt{17})' = 4 - \sqrt{17} < 0$, so by Theorem 12.21 the continued fraction expansion of $4 + \sqrt{17}$ is purely periodic.

d. We have $(11 - \sqrt{10})/9 < 1$, so by Theorem 12.21, the continued fraction expansion of $(11 - \sqrt{10})/9$ is not purely periodic.

e. We have $(3 + \sqrt{23})/2 > 1$ and $-1 < ((3+\sqrt{23})/2)' = (3 - \sqrt{23})/2 < 0$, so by Theorem 12.21 the continued fraction expansion of $(3 + \sqrt{23})/2$ is purely periodic.

f. We have $(17 + \sqrt{188})/3 > 1$ but $((17+\sqrt{188})/3)' = (17 - \sqrt{188})/3 > 0$, so by Theorem 12.21 the continued fraction expansion of $(17 + \sqrt{188})/3$ is not purely periodic.

12.4.17. Let $\alpha = (a+\sqrt{b})/c$. Then $-1/\alpha' = -(c)/(a-\sqrt{b}) = (ca+\sqrt{bc^2})/(b-a^2) = (A+\sqrt{B})/C$, say. By Exercise 16, $0 < a < \sqrt{b}$ and $\sqrt{b} - a < c < \sqrt{b} + a < 2\sqrt{b}$. Multiplying by c gives $0 < ca < \sqrt{bc^2}$ and $\sqrt{bc^2} - ca <$

$c^2 < \sqrt{bc^2} + ca < 2\sqrt{bc^2}$. That is, $0 < A < \sqrt{B}$ and $\sqrt{B} - A < c^2 < \sqrt{B} + A < 2\sqrt{B}$. Multiply $\sqrt{b} - a < c$ by $\sqrt{b} + a$ to get $C = b - a^2 < \sqrt{bc^2} + ca = A + \sqrt{B}$. Multiply $c < \sqrt{b} + a$ by $\sqrt{b} - a$ to get $\sqrt{B} - A = \sqrt{bc^2} - ac < b - a^2 = C$. So, $-1/\alpha'$ satisfies all the inequalities in Exercise 16, and therefore is reduced.

12.4.19. Start with $\alpha_0 = \sqrt{D_k} + 3^k + 1$ (this will have the same period since it differs from $\sqrt{D_k}$ by an integer) and use induction. Apply the continued fraction algorithm to show $\alpha_{3i} = \sqrt{D_k} + 3^k - 2 \cdot 3^{k-i} + 2/(2 \cdot 3^{k-i})$, $i = 1, 2, \ldots, k$, but $\alpha_{3k+3i} = \sqrt{D_k} + 3^k - 2/(2 \cdot 3^i)$, $i = 1, 2, \ldots, k-1$, and $\alpha_{6k} = \sqrt{D_k} + 3^k + 1 = \alpha_0$. Since $\alpha_i \neq \alpha_0$ for $i < 6k$ we see that the period is $6k$.

12.5. Factoring Using Continued Fractions

12.5.1. We have $19^2 - 2^2 = (19 - 2)(19 + 2) \equiv 0 \pmod{119}$. Then $(19 - 2, 119) = (17, 119) = 17$ and $(19 + 2, 119) = (21, 119) = 7$ are factors of 119.

12.5.3. Using a computer to generate lists $[k, \alpha_k, a_k, P_k, Q_k, \sqrt{Q_k}]$, we have $[1, \sqrt{13290059}, 3645, 0, 1, 1]$, $[2, (3645 + \sqrt{13290059})/4034, 1, 3645, 4034, \sqrt{4034}]$, $[3, (389 + \sqrt{13290059})/3257, 1, 389, 3257, \sqrt{3257}]$, $[4, (2868 + \sqrt{13290059})/1555, 4, 2868, 1555, \sqrt{1555}]$, $[5, (3352 + \sqrt{13290059})/1321, 5, 3352, 1321, \sqrt{1321}]$, $[6, (3253 + \sqrt{13290059})/2050, 3, 3253, 2050, 5\sqrt{82}]$, $[7, (2897 + \sqrt{13290059})/2389, 2, 2897, 2389, \sqrt{2389}]$, $[8, (1881 + \sqrt{13290059})/4082, 1, 1881, 4082, \sqrt{4082}]$, $[9, (2201 + \sqrt{13290059})/2069, 2, 2201, 2069, \sqrt{2069}]$, $[10, (1937 + \sqrt{13290059})/4610, 1, 1937, 4610, \sqrt{4610}]$, $[11, (2673 + \sqrt{13290059})/1333, 4, 2673, 1333, \sqrt{1333}]$, $[12, (2659 + \sqrt{13290059})/4666, 1, 2659, 4666, \sqrt{4666}]$, $[13, (2007 + \sqrt{13290059})/1985, 2, 2007, 1985, \sqrt{1985}]$, $[14, (1963 + \sqrt{13290059})/4754, 1, 1963, 4754, \sqrt{4754}]$, $[15, (2791 + \sqrt{13290059})/1157, 5, 2791, 1157, \sqrt{1157}]$, $[16, (2994 + \sqrt{13290059})/3739, 1, 2994, 3739, \sqrt{3739}]$, $[17, (745 + \sqrt{13290059})/3406, 1, 745, 3406, \sqrt{3406}]$, $[18, (2661 + \sqrt{13290059})/1823, 3, 2661, 1823, \sqrt{1823}]$, $[19, (2808 + \sqrt{13290059})/2965, 2, 2808, 2965, \sqrt{2965}]$, $[20, (3122 + \sqrt{13290059})/1195, 5, 3122, 1195, \sqrt{1195}]$, $[21, (2853 + \sqrt{13290059})/4310, 1, 2853, 4310, \sqrt{4310}]$, $[22, (1457 + \sqrt{13290059})/2591, 1, 1457, 2591, \sqrt{2591}]$, $[23, (1134 + \sqrt{13290059})/4633, 1, 1134, 4633, \sqrt{4633}]$, $[24, (3499 + \sqrt{13290059})/226, 31, 3499, 226, \sqrt{226}]$, $[25, (3507 + \sqrt{13290059})/4385, 1, 3507, 4385, \sqrt{4385}]$, $[26, (878 + \sqrt{13290059})/2855, 1, 878, 2855, \sqrt{2855}]$, $[27, (1977 + \sqrt{13290059})/3286, 1, 1977, 3286, \sqrt{3286}]$, $[28, (1309 + \sqrt{13290059})/3523, 1, 1309, 3523, \sqrt{3523}]$, $[29, (2214 + \sqrt{13290059})/2381, 2, 2214, 2381, \sqrt{2381}]$, $[30, (2548 + \sqrt{13290059})/2855, 2, 2548, 2855, \sqrt{2855}]$, $[31, (3162 + \sqrt{13290059})/1153, 5, 3162, 1153, \sqrt{1153}]$, $[32, (2603 + \sqrt{13290059})/5650, 1, 2603, 5650, 5\sqrt{226}]$, $[33, (3047 + \sqrt{13290059})/709, 9, 3047, 709, \sqrt{709}]$, $[34, (3334 + \sqrt{13290059})/3067, 2, 3334, 3067, \sqrt{3067}]$, $[35, (2800 + \sqrt{13290059})/1777, 3, 2800, 1777, \sqrt{1777}]$, $[36, (2531 + \sqrt{13290059})/3874, 1, 2531, 3874, \sqrt{3874}]$, $[37, (1343 + \sqrt{13290059})/2965, 1, 1343, 2965, \sqrt{2965}]$, $[38, (1622 + \sqrt{13290059})/3595, 1, 1622, 3595, \sqrt{3595}]$, $[39, (1973 + \sqrt{13290059})/2614, 2, 1973, 2614, \sqrt{2614}]$, $[40, (3255 + \sqrt{13290059})/1031, 6, 3255, 1031, \sqrt{1031}]$, $[41, (2931 + \sqrt{13290059})/4558, 1, 2931, 4558, \sqrt{4558}]$, $[42, (1627 + \sqrt{13290059})/2335, 2, 1627, 2335, \sqrt{2335}]$, $[43, (3043 + \sqrt{13290059})/1726, 3, 3043, 1726, \sqrt{1726}]$, $[44, (2135 + \sqrt{13290059})/5059, 1, 2135, 5059, \sqrt{5059}]$, $[45, (2924 + \sqrt{13290059})/937, 7, 2924, 937, \sqrt{937}]$, $[46, (3635 + \sqrt{13290059})/82, 88, 3635, 82, \sqrt{82}]$, $[47, (3581 + \sqrt{13290059})/5689, 1, 3581, 5689, \sqrt{5689}]$, $[48, (2108 + \sqrt{13290059})/1555, 3, 2108, 1555, \sqrt{1555}]$, $[49, (2557 + \sqrt{13290059})/4342, 1, 2557, 4342, \sqrt{4342}]$, $[50, (1785 + \sqrt{13290059})/2327, 2, 1785, 2327, \sqrt{2327}]$, $[51, (2869 + \sqrt{13290059})/2174, 2, 2869, 2174, \sqrt{2174}]$, $[52, (1479 + \sqrt{13290059})/5107, 1, 1479, 5107, \sqrt{5107}]$, $[53, (3628 + \sqrt{13290059})/25, 290, 3628, 25, 5]$. So we have $Q_{53} = 5^2$. Using a computer again, we find that $p_{52} = 35270108682248129250002106 \equiv 2467124 \pmod{13290059}$. Then $(13290059, 2467124 - 5) = 4261$ and $(13290059, 2467124 + 5) = 3119$, so we have $13290059 = 3119 \cdot 4261$.

12.5.5. We have $17^2 = 289 \equiv 3 \pmod{143}$ and $19^2 = 361 \equiv 3 \cdot 5^2 \pmod{143}$. Combining these, we have $(17 \cdot 19)^2 \equiv 3^2 5^2 \pmod{143}$. Hence, $323^2 \equiv 15^2 \pmod{143}$. It follows that $323^2 - 15^2 = (323 - 15)(323 + 15) \equiv 0 \pmod{143}$. This produces the two factors $(323 - 15, 143) = (308, 143) = 11$ and $(323 + 15, 143) = (338, 143) = 13$ of 143.

12.5.7. We use a computer to find $p_0 \equiv 3465, p_{11} \equiv 1211442, p_{27} \equiv 6764708, p_{33} \equiv 6363593$, and $p_{40} \equiv 8464787 \pmod{12007001}$. The product of these reduces to $P = 9815310 \pmod{12007001}$. Then $Q = \sqrt{Q_1 Q_{12} Q_{28} Q_{34} Q_{40}} = 1247455$. Then the factors of 12007001 are $(12007001, P - Q) = 3001$ and

$(12007001, P + Q) = 4001.$

CHAPTER 13

Some Nonlinear Diophantine Equations

13.1. Pythagorean Triples

13.1.1. a. Since $z = m^2 + n^2 \leq 40$, we have $m \leq 6$. The triples we seek are those in Table 13.1 with $z \leq 40$: (3,4,5), (5,12,13), (15,8,17), (7,24,25), (21,20,29), and (35,12,37).

b. These would be triples which are multiples of the primitive triples. In addition to those in part (a), we have (6,8,10), (9,12,15), (12,16,20), (15,20,25), (18,24,30), (21,28,35), (24,32,40), (10,24,26), (15,36,39), and (30,16,34).

13.1.3. By Lemma 13.1, 5 divides at most one of x, y, and z. If $5 \nmid x$ or y, then $x^2 \equiv \pm 1 \pmod 5$ and $y^2 \equiv \pm 1 \pmod 5$. Then, $z^2 \equiv 0, 2$, or $-2 \pmod 5$. But ± 2 is not a quadratic residue modulo 5, so $z^2 \equiv 0 \pmod 5$, whence $5 \mid z$.

13.1.5. Let k be an integer ≥ 3. If $k = 2n+1$, let $m = n+1$. Then m and n have opposite parity, $m > n$ and $m^2 - n^2 = 2n+1 = k$, so m and n define the desired triple. If k has an odd divisor $d > 1$, then use the construction above for d and multiply the result by k/d. If k has no odd divisors, then $k = 2^j$ for some integer $j > 1$. Let $m = 2^{j-1}$ and $n = 1$. Then $k = 2mn$, $m > n$, and m and n have opposite parity, so m and n define the desired triple.

13.1.7. Substituting $y = x+1$ into the Pythagorean equation gives us $2x^2 + 2x + 1 = z^2$, which is equivalent to $m^2 - 2z^2 = -1$ where $m = 2x+1$. Dividing by z^2 yields $m^2/z^2 - 2 = -1/z^2$. Note that $m/z \geq 1$, $1/z^2 = 2 - m^2/z^2 = (\sqrt{2} + m/z)(\sqrt{2} - m/z) < 2(\sqrt{2} - m/z)$. So by Theorem 12.18, m/z must be a convergent of the continued fraction expansion of $\sqrt{2}$. Further, by the proof of Theorem 12.13, it must be one of the even-subscripted convergents. Therefore each solution is given by the recurrence $m_{n+1} = 3m_n + 2z_n$, $z_{n+1} = 2m_n + 3m_n$. (See, e.g., Theorem 13.11.) Substituting x back in yields the recurrences of Exercise 6.

13.1.9. See Exercise 15 with $p = 3$.

13.1.11. We must find all primitive triples containing a divisor of 12: 2, 3, 4, 6, or 12. Such a triple must have $x = m^2 - n^2, y = 2mn, z = m^2 + n^2$, and $(m,n) = 1$. So only y is even. If $y = 2mn = 2$, then $m = n = 1$, and $x = 0$, which is not allowed. If $y = 2mn = 4$, then $m = 2$, and $n = 1$, so $x = 3$ and $z = 5$. If $y = 6 = 2mn$ then $m = 3, n = 1$, which are not of opposite parity. If $y = 12 = 2mn$, then either $m = 6, n = 1, x = 35$, and $z = 37$; or $m = 3, n = 2, x = 5$, and $z = 13$. Now $z \neq 3$ since 9 is not the sum of two squares. If $x = 3 = m^2 - n^2 = (m+n)(m-n)$, then $m = 2, n = 1, y = 4$, and $z = 5$. Multiples of these triples containing 12 are (9,12,15), (35,12,37), (5,12,13), and (12,16,20).

13.1.13. If m is positive, then all solutions are given by $x = 2m, y = m^2 - 1, z = m^2 + 1$.

13.1.15. Check that if $m > \sqrt{p}n$ then $x = (m^2 - pn^2)/2, y = mn, z = (m^2 + pn^2)/2$ is a solution. Conversely, if x, y, z is a primitive solution, then $y^2 = (z^2 - x^2)/p$, so $p \mid (z \pm x)$. Take $m^2 = z \mp x$ and $n^2 = (z \pm x)/p$.

13.1.17. Substituting $f_n = f_{n+2} - f_{n+1}$ and $f_{n+3} = f_{n+2} + f_{n+1}$ into $(f_n f_{n+3})^2 + (2f_{n+1}f_{n+2})^2$ yields $(f_{n+2} - f_{n+1})^2(f_{n+2}+f_{n+1})^2 + 4f_{n+1}^2 f_{n+2}^2 = (f_{n+2}^2 - f_{n+1}^2)^2 + 4f_{n+1}^2 f_{n+2}^2 = f_{n+2}^4 - 2f_{n+1}^2 f_{n+2}^2 + f_{n+1}^4 + 4f_{n+1}^2 f_{n+2}^2 = f_{n+2}^4 + 2f_{n+1}^2 f_{n+2}^2 + f_{n+1}^4 = (f_{n+2}^2 + f_{n+1}^2)^2$, which proves the result.

13.2. Fermat's Last Theorem

13.2.1. Assume without loss of generality that $x < y$. Then $x^n + y^n = x^2 x^{n-2} + y^2 y^{n-2} < (x^2 + y^2) y^{n-2} = z^2 y^{n-2} < z^2 z^{n-2} = z^n$.

13.2.3. a. If $p \mid x, y,$ or z, then certainly $p \mid xyz$. If not, then by Fermat's Little Theorem, $x^{p-1} \equiv y^{p-1} \equiv z^{p-1} \equiv 1 \pmod{p}$. Hence, $1 + 1 \equiv 1 \pmod{p}$, which is impossible.

b. We know $a^p \equiv a \pmod{p}$ for every integer a. Then $x^p + y^p \equiv z^p \pmod{p}$ implies $x + y \equiv z \pmod{p}$, so $p \mid x + y - z$.

13.2.5. Let x and y be the lengths of the legs and z be the hypotenuse. Then $x^2 + y^2 = z^2$. If the area is a perfect square, we have $A = \frac{1}{2}xy = r^2$. Then, if $x = m^2 - n^2$, and $y = 2mn$, we have $r^2 = mn(m^2 - n^2)$. All of these factors are relatively prime, so $m = a^2$, $n = b^2$, and $m^2 - n^2 = c^2$, say. Then, $a^4 - b^4 = c^2$, which contradicts Exercise 4.

13.2.7. We use the method of infinite descent. Assume there is a nonzero solution with where $|x|$ is minimal. Then $(x, y) = 1$. Also x and z cannot both be even, because then y would be odd and then $z^2 \equiv 8 \pmod{18}$, but 8 is not a quadratic residue modulo 16. Therefore x and z are both odd, since $8y^4$ is even. From here it is easy to check that $(x, z) = 1$. We may also assume (by negating if necessary) that $x \equiv 1 \pmod{4}$ and $z \equiv 3 \pmod{4}$. Clearly $x^2 > |z|$. We have $8y^4 = x^4 - z^2 = (x^2 - z)(x^2 + z)$. Since $z \equiv 3 \pmod{4}$, we have $x^2 - z \equiv 2 \pmod{4}$, so $m = (x^2 - z)/2$ is odd, and $n = (x^2 + z)/4$ is an integer. Since no odd prime can divide both m and n, we have $(m, n) = 1$, $m, n > 0$ and $mn = y^4$, whence $m = r^4$ and $n = s^4$, with $(r, s) = 1$. So now $r^4 + 2s^4 = m + 2n = x^2$. This implies $(x, r) = 1$, since no odd prime divides r and x but not s, and r and x are both odd. Also, $|x| > r^2 > 0$. Now consider $2s^4 = (x^2 - r^4) = (x - r^2)(x + r^2)$. Then, s must be even since a difference of squares is not congruent to 2 (mod 4), so $s = 2t$ and $32t^4 = (x - r^2)(x + r^2)$. Recalling $x \equiv 1 \pmod{4}$ and r is odd, we have $U = (x + r^2)/2$ is odd and $V = (x - r^2)/16$ is an integer. Again $(U, V) = 1$ and $UV = t^4$, but we don't know the sign of x. So $U = \pm u^4$ and $V = \pm v^4$, depending on the sign of x. Now $r^2 = \pm(u^4 - 8v^4)$. But since u is odd, the sign can't be $-$ (or else $r^2 \equiv 7 \pmod 8$.) So the sign is $+$ (hence x is positive), and we have $u^4 - 8v^4 = r^2$. Finally, $|v| > 0$ because $|x + r^2| > 0$. So we haven't reduced to a trivial case. Then, $u^4 = U < |x + r^2|/2 < x$, so $|u| < x$, and so $|x|$ was not minimal. This contradiction shows that there are no nontrivial solutions.

13.2.9. (Solution by John R. Ramsden.) First we mimic the construction of the solutions to the Pythagorean equation to solve the 2nd-order diophantine equation $x^2 + 3y^2 = z^2$ and find that all solutions are given by $\pm 2^e x = m^2 - 3n^2$, $2^e y = 2mn$, $2^e = m^2 + 3n^2$ for relatively prime integers m and n, where (i) x and y are odd and z is even if and only if $e = 1$ and m and n are both odd, and (ii) x and z are odd and y is even if and only if $e = 0$ and m and n have opposite parity.

Now consider the diophantine equation $x^4 + 3y^4 = z^4$ modulo 8. If x and y are odd and z is even, we get a contradiction, so case (i) above doesn't happen for the 4th-order specialization. Assume we have a nontrivial solution with x, y and z pairwise relatively prime and z as small as possible. Then as above, since we must be in case (ii), $x^2 = \pm(m^2 - 3n^2)$, $y^2 = 2mn$ and $z^2 = m^2 + 3n^2$. If m were even and n odd, then $z^2 \equiv 4Z + 3 \pmod 4$ a contradiction, therefore m is odd and n is even. Then if $x^2 = -(m^2 - 3n^2) \equiv -1 \pmod 4$, also a contradiction, so we have $x^2 = m^2 - 3n^2$.

Now, since $(m, n) = 1$ and n is even and $y^2 = 2mn$, we have $m = U^2$, $n = 2V^2$ and so $y = 2UV$, for some integers U and V. Then $x^2 = U^4 - 12V^4$ and $z^2 = U^4 + 12V^4$, both of which are diophantine equations which can be considered as 2nd-order of the type solved above. If we do, we find integers $(p, q) = (r, s) = 1$, with p and q of opposite parity and r and s of opposite parity such that $U^2 = p^2 - 3q^2 = r^2 + 3s^2$ and $V^2 = pq = rs$. This last equation shows that $p = P^2$, $q = Q^2$, $r = R^2$, and $s = S^2$ for some integers P, Q, R and S. Since $(p, q) = (r, s) = 1$ and $pq = rs$, there must be integers $(a, d) = (b, c) = 1$ such that $p = ab$, $q = cd$, $r = ac$, $s = bd$ (this is the so-called "Lucas Lemma".) Then since $p = P^2 = ab$, we let $a = fA^2$, $b = fB^2$ where f is squarefree. Then since $r = R^2 = ac = fA^2c$, we see that $c = fC^2$ for some C. But since $(b, c) = 1$, we have that $f = 1$, so $V^2 = pq = rs$ and $(A, D) = (B, C) = 1$. So we have $p = A^2 B^2$, $q = C^2 D^2$, $r = A^2 C^2$, and $s = B^2 D^2$. When we substitute these in for U^2 in the two solutions above we get $(A^4 - 3D^4)B^4 = (A^4 + 3D^4)C^4$. Since $(B, C) = 1$ we must have, for some integer E, $A^4 - 3D^4 = EC^4$ and $A^4 + 3D^4 = EB^4$. Now, since $U^2 = r^2 + 3s^2$, it's easy to check that $(r, 6) = 1$ and hence $(A, 6) = 1$. If we add and subtract the last two equations we get $2A^4 = E(B^4 + C^4)$ and

$6D^4(B^4 - C^4)$. Since $(A, 6) = 1$, we see that E divides 2 and 6 and since it's positive, we conclude $E = 1$ or 2. If $E = 1$, then the second equation $A^4 + 3D^4 = EB^4$ is an equation of the same type but with smaller positive value for z, completing the descent argument. If $E = 2$, then we add the equations and divide by 2 to get $A^4 = B^4 + C^4$ and note that Fermat showed this equations has no nontrivial solutions.

13.2.11. If x were even, the $y^2 = x^3 + 23 \equiv 3 \pmod{4}$, which is impossible, so x must be odd, making y even, say $y = 2v$. If $x \equiv 3 \pmod{4}$, then $y^2 \equiv 3^3 + 23 \equiv 2 \pmod{4}$ which is also impossible, so $x \equiv 1 \pmod{4}$. Add 4 to both sides of the equation to get $y^2 + 4 = 4v^2 + 4 = x^3 + 27 = (x + 3)(x^2 - 3x + 9)$. Then $z = x^2 - 3x + 9 \equiv 1 - 3 + 9 \equiv 3 \pmod{4}$, so a prime $p \equiv 3 \pmod{4}$ must divide z. Then $4v^2 + 4 \equiv 0 \pmod{p}$ or $v^2 \equiv -1 \pmod{p}$. But this shows that a prime congruent to 3 modulo 4 has -1 as a quadratic residue, which contradicts Theorem 11.5. Therefore, the equation has no solutions.

13.2.13. If there were two perfect squares in a Pythagorean triple, then we would have a solution of either the equation in Theorem 13.3 or the equation in Exercise 4, both of which have no nontrivial solutions.

13.2.15. Assume $n \nmid xyz$, and $(x, y, z) = 1$. Now $(-x)^n = y^n + z^n = (y + z)(y^{n-1} - y^{n-2}z + \cdots + z^{n-1})$, and these factors are relatively prime, so they are nth powers, say $y + z = a^n$, and $y^{n-1} - y^{n-2}z + \cdots + z^{n-1} = \alpha^n$, whence $x = a\alpha$. Similarly, $z + x = b^n$, and $(z^{n-1} - z^{n-2}x + \cdots + x^{n-1}) = \beta^n$, $-y = b\beta$, $x + y = c^n$, and $(x^{n-1} - x^{n-2}y + \cdots + y^{n-1}) = \gamma^n$, and $-z = c\gamma$. Since $x^n + y^n + z^n \equiv 0 \pmod{p}$, we have $p \mid xyz$, say $p \mid x$. Then $\gamma^n = (x^{n-1} - x^{n-2}y + \cdots + y^{n-1}) \equiv y^{n-1} \pmod{p}$. Also $2x \equiv b^n + c^n + (-a)^n \equiv 0 \pmod{p}$, so by the condition on p, we have $p \mid abc$. If $p \mid b$ then $y = -b\beta \equiv 0 \pmod{p}$, but then $p \mid x$ and y, a contradiction. Similarly, p cannot divide c. Therefore, $p \mid a$, so $y \equiv -z \pmod{p}$, and so $\alpha^n \equiv (y^{n-1} - y^{n-2}z + \cdots + z^{n-1}) \equiv ny^{n-1} \equiv n\gamma^n \pmod{p}$. Let g be the inverse of $\gamma \pmod{p}$, then $(ag)^n \equiv n \pmod{p}$, which contradicts the condition that there is no solution to $w^n \equiv n \pmod{p}$.

13.2.17. Note that $3^3 + 4^3 + 5^3 = 27 + 64 + 125 = 216 = 6^3$.

13.2.19. If $m \geq 3$ then modulo 8 we have $3^n \equiv -1 \pmod{8}$ which is impossible, so $m = 1$ or 2. If $m = 1$, then $3^n = 2 - 1 = 1$ which implies that $n = 0$ which is not a positive integer, so we have no solutions in this case. If $m = 2$, then $3^n = 2^2 - 1 = 3$, which implies that $n = 1$, and this is the only solution.

13.2.21. a. Substituting the expressions into the left-hand side of the equation yields $a^2 + b^2 + (3ab - c)^2 = a^2 + b^2 + 9a^2b^2 - 6abc + c^2 = (a^2 + b^2 + c^2) + 9a^2 + b^2 - 6abc$. Since (a, b, c) is a solution to Markoff's equation, we substitute $a^2 + b^2 + c^2 = 3abc$ to get the last expression equal to $3abc + 9a^2 + b^2 - 6abc = 9a^2 + b^2 - 3abc = 3ab(3ab - c)$, which is the right-hand side of Markoff's equation evaluated at these expressions.

b. Case 1: If $x = y = z$, then Markoff's equations becomes $3x^2 = 3xyz$ so that $1 = yz$. Then $y = z = 1$ and then $x = 1$ so the only solution in this case is $(1, 1, 1)$.

Case 2: If $x = y \neq z$, then $2x^2 + z^2 = 3x^2z$ which implies that $x^2 \mid z^2$ or $x \mid z$, say $dx = z$. Then $2x^2 + d^2x^2 = 3dx^3$ or $2 + d^2 = 3dx$ or $2 = d(3d - x)$. So $d \mid 2$, but since $x \neq z$, we must have $d = 2$. Then $3d - x = 1$ so that $x = 1 = y$ and $z = 2$, so the only solution in this case is $(1, 1, 2)$.

Case 3: Assume $x < y < z$. From $z^2 - 3xyz + x^2 + y^2 + z^2$ we apply the quadratic formula to get $2z = 3xy \pm \sqrt{9x^2y^2 - 4(x^2 + y^2)}$. Note that $8x^2y^2 - 4x^2 - 4y^2 = 4x^2(y^2 - 1) + 4y^2(x^2 - 1) > 0$ so in the "minus" case of the quadratic formula, we have $2z < 3xy - \sqrt{9x^2y^2 - 8x^2y^2} = 3xy - xy = 2xy$, or $z < xy$. But $3xyz = x^2 + y^2 + z^2 < 3z^2$ so that $xy < z$, a contradiction, therefore we must have the "plus" case in the quadratic formula and $2z = 3xy + \sqrt{9x^2y^2 - 4(x^2 + y^2)} > 3xy$, so that $z > 3xy - z$. This last expression is the formula for the generation of z in part (a). Therefore, by successive use of the formula in part (a), we will reduce the value of $x + y + z$ until it is one of the solutions in Case 1 or Case 2.

13.2.23. Let $\epsilon > 0$ be given then the abc Conjecture gives us $\max(|a|, |b|, |c|) \leq K(\epsilon)\text{rad}(abc)^{1+\epsilon}$ for integers $(a, b) = 1$ and $a + b = c$. Set $M = \log K / \log 2 + (3 + 3\epsilon)$. Suppose x, y, x, a, b, c are positive integers with $(x, y) = 1$ and $x^a + y^b = c^z$, so that we have a solution to Beal's equation. Assume $\min(a, b, c) > M$.

From the abc Conjecture, and since $\text{rad}(x^a y^b y^c) = \text{rad}(xyz)$, we have $\max(x^a, y^b, y^c) \leq K\text{rad}(xyz)^{1+\epsilon} \leq (xyz)^{1+\epsilon}$. If $\max(x,y,z) = x$, then we would have $x^a \leq Kx^{3(1+\epsilon)}$. Taking log's of both sides yields $a \leq \log K/\log x + (3+3\epsilon) < \log K/\log 2 + (3+3\epsilon) = M$, a contradiction. Similarly if the maximum is y or z. Therefore, if the abc Conjecture is true, there are no solutions to the Beal conjecture for sufficiently large exponents.

13.2.25. a. If 1 is a congruent number, then there exist rational numbers r, s and t such that $r^2 + s^2 = t^2$ and $rs/2 = 1$. Let $r = a/d$, $s = b/d$ and $t = c/d$, where a, b, c, and d are integers and d is the least common denominator of the rational numbers r, s and t. Then $a^2 + b^2 = (rd)^2 + (sd)^2 = d^2 t^2 = c^2$, so (a,b,c) is a Pythagorean triple and represents a right triangle whose area is $ab/2 = (rd)(sd)/2 = (d^2)(rs/2) = d^2$, a perfect square. Conversely, if there is a right triangle whose area is a perfect square, d^2, then it is represented by a Pythagorean triple (a,b,c), and $a^2 + b^2 = c^2$. We can divide through by d^2 to get $(a/d)^2 + (b/d)^2 = (c/d)^2$ and so this represents a right triangle with sides $(a/d, b/d, c/d)$ and area $1/2(a/d)(b/d) = (ab/2)(1/d^2) = d^2/d^2 = 1$.

b. Suppose 1 is a congruent number. Then by part (a), there exist integers a, b, c and d, such that $a^2 + b^2 = c^2$ and $ab/2 = d^2$. If we add and subtract 4 times the second equation from the first we get $a^2 + 2ab + b^2 = (a+b)^2 = c^2 + (2d)^2$ and $a^2 - 2ab + b^2 = (a-b)^2 = c^2 - (2d)^2$. Since the right hand sides of both equations are squares, then so is their product, and we have $(c^2 + (2d)^2)(c^2 - (2d)^2) = c^4 - (2d)^4 = (a+b)^2(a-b)^2$, but this is a solution to $x^4 - y^4 = z^2$, which contradicts Exercise 4. Therefore 1 is not a congruent number.

13.3. Sums of Squares

13.3.1. a. We compute $377 = 13 \cdot 29 = (3^2 + 2^2)(5^2 + 2^2) = (3 \cdot 5 + 2 \cdot 2)^2 + (3 \cdot 2 - 2 \cdot 5)^2 = 19^2 + 4^2$.

b. We compute $650 = 13 \cdot 50 = (3^2 + 2^2)(7^2 + 1^2) = (3 \cdot 7 + 2 \cdot 1)^2 + (3 \cdot 1 - 2 \cdot 7)^2 = 23^2 + 11^2$.

c. We compute $1450 = 29 \cdot 50 = (5^2 + 2^2)(7^2 + 1^2) = (5 \cdot 7 + 2 \cdot 1)^2 + (5 \cdot 1 - 2 \cdot 7)^2 = 37^2 + 9^2$.

d. We compute $18850 = 377 \cdot 50 = (19^2 + 4^2)(7^2 + 1^2) = (19 \cdot 7 + 4 \cdot 1)^2 + (19 \cdot 1 - 4 \cdot 7)^2 = 137^2 + 9^2$.

13.3.3. a. We compute $34 = 5^2 + 3^2$.

b. . We compute $90 = 3^2 10 = 3^2(3^2 + 1) = 9^2 + 3^2$.

c. We compute $100 = 10^2 + 0^2$.

d. We compute $490 = 7^2 10 = 7^2(3^2 + 1) = 21^2 + 7^2$.

e. We compute $21658 = 7^2 \cdot 2 \cdot 13 \cdot 17 = 7^2(1^2 + 1^2)(3^2 + 2^2)(4^2 + 1^2) = 7^2(1^2 + 1^2)((3 \cdot 4 + 2 \cdot 1)^2 + (3 \cdot 1 - 2 \cdot 4)^2) = 7^2(1^2 + 1^2)(14^2 + 5^2) = 7^2((1 \cdot 14 + 1 \cdot 5)^2 + (1 \cdot 5 - 1 \cdot 14)^2) = 7^2(19^2 + 9^2) = 133^2 + 63^2$.

f. We compute $324608 = 2^{10} \cdot 317 = 32^2(14^2 + 11^2) = 448^2 + 352^2$.

13.3.5. a. We have $3 = 1^2 + 1^2 + 1^2$.

b. We have $90 = 8^2 + 5^2 + 1^2$.

c. We have $11 = 3^2 + 1^2 + 1^2$.

d. We have $18 = 3^2 + 3^2 + 0^2$.

e. There are no solutions since $23 \equiv 7 \pmod{8}$. See Exercise 6.

13.3. SUMS OF SQUARES

f. There are no solutions since $28 = 4 \cdot 7$. See Exercise 7.

13.3.7. Let $n = x^2 + y^2 + z^2 = 4^m(8k+7)$. If $m = 0$, see Exercise 6. If $m \geq 1$, then n is even, so 0 or 2 of x, y, z are odd. If 2 are odd, $x^2 + y^2 + z^2 \equiv 2$ or 6 (mod 8), but then $4 \nmid n$, a contradiction, so all of x, y, z are even. Then $4^{m-1}(8k+7) = (\frac{x}{2})^2 + (\frac{y}{2})^2 + (\frac{z}{2})^2$ is the sum of 3 squares. Repeat until $m = 0$ and use Exercise 6 to get a contradiction.

13.3.9. **a.** We compute $105 = 7 \cdot 15 = (2^2 + 1^2 + 1^2 + 1^2)(3^2 + 2^2 + 1^2 + 1^2) = (2 \cdot 3 + 1 \cdot 2 + 1 \cdot 1 + 1 \cdot 1)^2 + (2 \cdot 2 - 1 \cdot 3 + 1 \cdot 1 - 1 \cdot 1)^2 + (2 \cdot 1 - 1 \cdot 1 - 1 \cdot 3 + 1 \cdot 2)^2 + (2 \cdot 1 + 1 \cdot 1 - 1 \cdot 2 - 1 \cdot 3)^2 = 10^2 + 1^2 + 0^2 + 2^2$.

b. We compute $510 = 15 \cdot 34 = (3^2 + 2^2 + 1^2 + 1^2)(4^2 + 4^2 + 1^2 + 1^2) = (3 \cdot 4 + 2 \cdot 4 + 1 \cdot 1 + 1 \cdot 1)^2 + (3 \cdot 4 - 2 \cdot 4 + 1 \cdot 1 - 1 \cdot 1)^2 + (3 \cdot 1 - 2 \cdot 1 - 1 \cdot 4 + 1 \cdot 4)^2 + (3 \cdot 1 + 1 \cdot 2 - 1 \cdot 4 - 1 \cdot 4)^2 = 22^2 + 4^2 + 1^2 + 3^2$.

c. We compute $238 = 7 \cdot 34 = (2^2 + 1^2 + 1^2 + 1^2)(4^2 + 4^2 + 1^2 + 1^2) = (2 \cdot 4 + 1 \cdot 4 + 1 \cdot 1 + 1 \cdot 1)^2 + (2 \cdot 4 - 1 \cdot 4 + 1 \cdot 1 - 1 \cdot 1)^2 + (2 \cdot 1 - 1 \cdot 1 - 1 \cdot 4 + 1 \cdot 4)^2 + (2 \cdot 1 + 1 \cdot 1 - 1 \cdot 4 - 1 \cdot 4)^2 = 14^2 + 4^2 + 1^2 + 5^2$.

d. We compute $3570 = 15 \cdot 238 = (3^2 + 2^2 + 1^2 + 1^2)(14^2 + 4^2 + 1^2 + 5^2) = (3 \cdot 14 + 2 \cdot 4 + 1 \cdot 1 + 1 \cdot 5)^2 + (3 \cdot 4 - 2 \cdot 14 + 1 \cdot 5 - 1 \cdot 1)^2 + (3 \cdot 1 - 5 \cdot 2 - 1 \cdot 14 + 1 \cdot 4)^2 + (3 \cdot 5 + 2 \cdot 1 - 1 \cdot 4 - 1 \cdot 14)^2 = 56^2 + 12^2 + 17^2 + 1^2$.

13.3.11. Let $m = n - 169$. Then m is the sum of four squares: $m = x^2 + y^2 + z^2 + w^2$. If, say, x, y, z are 0, then $n = w^2 + 169 = w^2 + 10^2 + 8^2 + 2^2 + 1^2$. If, say, x, y are 0, then $n = z^2 + w^2 + 169 = z^2 + w^2 + 12^2 + 4^2 + 3^2$. If, say, x is 0, then $n = y^2 + z^2 + w^2 + 169 = y^2 + z^2 + w^2 + 12^2 + 5^2$. If none are 0, then $n = x^2 + y^2 + z^2 + w^2 + 13^2$.

13.3.13. If k is odd, then 2^k is not the sum of four positive squares. Suppose $k \geq 3$, and $2^k = x^2 + y^2 + z^2 + w^2$. Then either 0, 2 or 4 of the squares are odd. Modulo 8, we have $0 \equiv x^2 + y^2 + z^2 + w^2$, and since an odd square is congruent to 1 modulo 8, the only possibility is to have x, y, z, w all even. But then we can divide by 4 to get $2^{k-2} = (\frac{x}{2})^2 + (\frac{y}{2})^2 + (\frac{z}{2})^2 + (\frac{w}{2})^2$. Either $k - 2 \geq 3$ and we can repeat the argument, or $k - 2 = 1$, in which case we have 2 equal to the sum of four positive squares, a contradiction.

13.3.15. If $p = 2$ the theorem is obvious. Else, $p = 4k + 1$, whence -1 is a quadratic residue modulo p, say $a^2 \equiv -1 \pmod{p}$. Let x and y be as in Thue's Lemma. Then $x^2 < p$ and $y^2 < p$ and $-x^2 \equiv (ax)^2 \equiv y^2 \pmod{p}$. Thus $p \mid x^2 + y^2 < 2p$; therefore $p = x^2 + y^2$ as desired.

13.3.17. The left sum runs over every pair of integers $i < j$, for $1 \leq i < j \leq 4$, so there are six terms. Each integer subscript 1, 2, 3, and 4 appears in exactly three pairs, so

$$\sum_{1 \leq i < j \leq 4} [(x_i + x_j)^4 + (x_i - x_j)^4] = \sum_{1 \leq i < j \leq 4} (2x_i^4 + 12x_i^2 x_j^2 + 2x_j^4)$$

$$= \sum_{k=1}^{4} 6x_k^4 + \sum_{1 \leq i < j \leq 4} 12x_i^2 x_j^2 = 6\left(\sum_{k=1}^{4} x_k^2\right)^2.$$

13.3.19. If m is positive, then $m = \sum_{k=1}^{4} x_k^2$, for some x_k's. Then $6m = 6\sum_{k=1}^{4} x_k^2 = \sum_{k=1}^{4} 6x_k^2$. Each term of the last sum is the sum of twelve fourth powers by Exercise 18. Therefore $6m$ is the sum of forty-eight fourth powers.

13.3.21. For $n = 1, 2, \ldots, 50$, $n = \sum_{1}^{n} 1^4$. For $n = 51, 52, \ldots, 81$, $n - 48 = n - 3(2^4) = \sum_{1}^{n-48} 1^4$, so $n = 2^4 + 2^4 + 2^4 + \sum_{1}^{n-48} 1^4$ is the sum of $(n - 45)$ 4th powers, and $n - 45 \leq 36 \leq 50$. This result, coupled with the result from Exercise 20, shows that all positive integers can be written as the sum of 50 or fewer 4th powers. That is, $g(4) \leq 50$.

13.3.23. The only quartic residues modulo 16 are 0 and 1. Therefore, the sum of fewer than 15 fourth powers must have a least nonnegative residue between 0 and 14 (mod 16), which excludes any integer congruent to 15 (mod 16).

13.4. Pell's Equation

13.4.1. a. Clearly $|x| \leq 2$. Checking all possibilities gives $(\pm 2, 0)$ and $(\pm 1, \pm 1)$ for solutions.

b. Clearly $|x| < 3$. Checking all possibilities gives no solution.

c. Clearly $|x| < 4, |y| \leq 2$. Checking all possibilities gives the solutions $(\pm 1, \pm 2)$.

13.4.3. We have $\sqrt{31} = [5; \overline{1, 1, 3, 5, 3, 1, 1, 10}]$, which has period 8. The first few convergents are 5/1, 6/1, 11/2, 39/7, ... For part (a), there are solutions by Theorem 13.11. For part (b), there are no solutions by Theorem 13.11. Trying the convergents p/q in the equation with $x = p, y = q$ gives us the values $-6, 5, -3, 2, \ldots$, so we have solutions for parts (c), (d), and (e). Then for part (f), reduce modulo 4 to get $x^2 + y^2 \equiv 3 \pmod 4$ which has no solution.

13.4.5. We have $\sqrt{37} = [6; \overline{12}]$ of period 1. Theorem 13.11 gives the first 3 solutions as $x = 73, y = 12$; $x = 10657, y = 1752$; $x = 1555849, y = 255780$.

13.4.7. We have $x_1 = 1766319049, y_1 = 226153980$. We apply Theorem 13.12 to get $x_2 + y_2\sqrt{61} = (x_1 + y_1\sqrt{61})^2$, which gives $x_2 = 6239765965720528801, y_2 = 798920165762330040$. We used *MAPLE* to do these calculations.

13.4.9. Reduce modulo p to get $x^2 \equiv -1 \pmod p$. Since -1 is a quadratic nonresidue modulo p if $p = 4k + 3$, there is no solution.

13.4.11. Following the hint, we solve $a^2 - 2b^2 = \pm 1$. By Theorem 13.10, we find that every convergent p_k/q_k of $\sqrt{2}$ is a solution. Note that $p_1 = 0, p_1 = 3, p_k = 2p_{k-1} + 2_{k-2}, q_0 = 1, q_1 = 1$, and $q_k = 2q_{k-1} + q_{k-2}$. Then solving $s - t = a, t = b$ yields $s = a + b = p_k + q_k$ and $t = q_k$, whence $x = p_k^2 + 2p_kq + k$ and $y = 2p_kq_k + 2q_k^2$. The first few solutions are $p_0 = 1, q_0 = 1$ corresponding to $x = 1^2 + 2 \cdot 1 \cdot 1 = 3$ and $y = 2 \cdot 1 \cdot 1 + 2 \cdot 1^2 = 4$; $p_1 = 3, q_1 = 2$ corresponding to $x = 3^2 + 2 \cdot 3 \cdot 2 = 21$ and $y = 2 \cdot 3 \cdot 2 + 2 \cdot 2^2 = 20$; $p_2 = 7, q_2 = 5$ corresponding to $x = 7^2 + 2 \cdot 7 \cdot 5 = 119$ and $y = 2 \cdot 7 \cdot 5 + 2 \cdot 5^2 = 120$; $p_3 = 17, q_3 = 12$ corresponding to $x = 17^2 + 2 \cdot 17 \cdot 12 = 697$ and $y = 2 \cdot 17 \cdot 12 + 2 \cdot 12^2 = 696$.

13.4.13. Suppose there is a solution (x, y). Then x must be odd. Note that $(x^2 + 1)^2 = x^4 + 2x^2 + 1 = 2y^2 + 2x^2$ and $(x^2 - 1)^2 = x^4 - 2x^2 + 1 = 2y^2 - 2x^2$. Multiplying these two equations together yields $(x^4 - 1)^2 = 4(y^4 - x^4)$, or since $x^4 \equiv 1 \pmod 4$, $((x^4 - 1)/2)^2 = y^4 - x^4$. But this is a violation of Exercise 4 in Section 13.2.

CHAPTER 14

The Gaussian Integers

14.1. Gaussian Integers and Gaussian Primes

14.1.1. a. First, $(2+i)(2+i) = 4 + 2i + 2i + i^2 = 4 + 4i - 1 = 3 + 4i$. Then we have $(2+i)^2(3+i) = (3+4i)(3+i) = 9 + 12i + 3i + 4i^2 = 9 + 15i - 4 = 5 + 15i$.

b. First, $(2-3i)(2-3i) = 4 - 6i - 6i + 9i^2 = 4 - 12i - 9 = -5 - 12i$. Then we have $(2-3i)^3 = (-5-12i)(2-3i) = -10 - 24i + 15i + 36i^2 = -10 - 9i - 36 = -46 - 9i$.

c. First, $(-i+3)(-i+3) = i^2 - 3i - 3i + 9 = -1 - 6i + 9 = 8 - 6i$. Next, $-i(-i+3) = i^2 - 3i = -1 - 3i$. Finally, we have $-i(-i+3)^3 = (8-6i)(-1-3i) = -8 + 6i - 24i + 18i^2 = -8 - 18i - 18 = -26 - 18i$.

14.1.3. a. We evaluate the fraction $\dfrac{\beta}{\alpha} = \dfrac{5+5i}{2-i} = \dfrac{(5+5i)(2+i)}{(2-i)(2+i)} = \dfrac{5+15i}{5} = 1 + 3i$, which is a Gaussian integer. Therefore, α divides β, since $\alpha(1+3i) = \beta$.

b. We evaluate the fraction $\dfrac{8}{1-i} = \dfrac{8(1+i)}{(1-i)(1+i)} = \dfrac{8+i}{2} = 4 + 4i$, which is a Gaussian integer. Therefore $8 = (1-i)(4+4i)$ and so α divides β.

c. Since $N(\alpha) = N(5) = 25$ and $N(\beta) = N(2+3i) = 4 + 9 = 13$, we observe that $25 \nmid 13$. Therefore, α can not divide β.

d. We evaluate the fraction $\dfrac{26}{3+2i} = \dfrac{26(3-2i)}{(3+2i)(3-2i)} = \dfrac{78-52i}{13} = 6 - 4i$, which is a Gaussian integer. Therefore, α, divides β.

14.1.5. Since a Gaussian integer must be of the form $a + bi$ with a and b rational integers, then for a Gaussian integer α to be divisible by $4 + 3i$, we must have $\alpha = (4+3i)(a+bi) = (4a - 3b) + (4b + 3a)i$ and this gives us a formula for all Gaussian integers divisible by $4 + 3i$ in terms of rational integers a and b. To the right is a display of the pattern of this set in the plane.

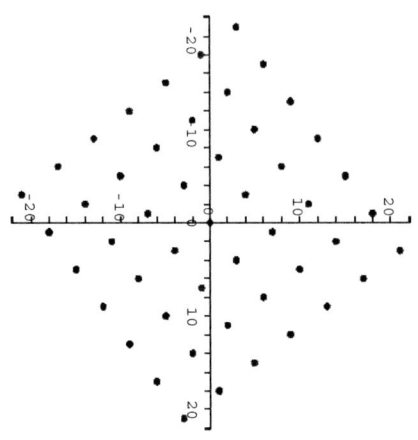

14.1.7. Since $\alpha | \beta$ and $\beta | \gamma$, there exist Gaussian integers μ and ν such that $\mu\alpha = \beta$ and $\nu\beta = \gamma$. Since the product of Gaussian integers is again a Gaussian integer, we have that $\nu\mu$ is also a Gaussian integer. Then $\gamma = \nu\beta = \nu\mu\alpha$ and so $\alpha | \gamma$.

121

14.1.9. Consider the equation $x^5 = x$ or $x^5 - x = 0$. The left side factors over the Gaussian integers as $x(x-1)(x+1)(x-i)(x+i) = 0$, so the solutions of the equation are $0, 1, -1, i,$ and $-i$. Since this includes all of the units for the Gaussian integers, this proves the result.

14.1.11. Since $\alpha|\beta$ and $\beta|\alpha$, there exist Gaussian integers μ and ν such that $\alpha\mu = \beta$ and $\beta\nu = \alpha$. Then $\alpha = \alpha\mu\nu$. Taking norms of both sides yields $N(\alpha) = N(\alpha\mu\nu) = N(\alpha)N(\mu\nu)$ by Theorem 14.1. So that $N(\mu)N(\nu) = 1$. Since μ and ν are Gaussian integers their norms must be nonnegative rational integers. Therefore $N(\mu) = N(\nu) = 1$, and so μ and ν are units, and hence, α and β are associates.

14.1.13. Note that $N(1+2i) = N(2+i) = 5$, so the condition on norms holds, but $(1+2i)/(2+i) = 4/5 + 3/5i$, so this is a counterexample.

14.1.15. First we show existence. If $a > 0$ and $b \geq 0$ we're done. If $a \leq 0$ and $b > 0$ then we multiply by $-i$ to get $-i\alpha = b - ai = c + di$ which has $c > 0$ and $d \geq 0$. If $a < 0$ and $b \leq 0$ then we multiply by -1 to get $-\alpha = -a - bi = c + di$ which has $c > 0$ and $d \geq 0$. If $a \geq 0$ and $b < 0$ then we multiply by i to get $i\alpha = -b + ai = c + di$ which has $c > 0$ and $d \geq 0$. (We have covered the quadrants in the plane in counter-clockwise order.) Having found the associate $c + di$ in the first quadrant, we observe that it is unique, since if we multiply by any unit other than one we get, respectively $-c - di$, which has $-c < 0$, $-d + ci$, which has $-d \leq 0$, or $d - ci$, which has $-c < 0$.

14.1.17. a. We have $\alpha/\beta = (24 - 9i)/(3 + 3i) = 5/2 - 11/2i$. Rounding to the nearest integer in each part, and going up in each case, since we have half integers, yields $\gamma = 3 - 5i$. Then $\rho = \alpha - \beta\gamma = -3i$. Then $N(\rho) = 3^2 + 0^2 = 9 < N(\beta) = 3^2 + 3^2 = 18$.

b. We have $\alpha/\beta = (18 + 15i)/(3 + 4i) = 144/25 - 27/25i$. Rounding to the nearest integer in each part yields $\gamma = 5 - i$. Then we compute $\rho = \alpha - \beta\gamma = -1 - 2i$, so that $N(\rho) = (-1)^2 + (-2)^2 = 5 < N(\beta) = 25$.

c. We have $\alpha/\beta = 87i/(11 - 2i) = -174/125 + 957/125i$. Rounding to the nearest integer in each part yields $\gamma = -1 + 8i$. Then we compute $\rho = \alpha - \beta\gamma = -5 - 3i$, so that $N(\rho) = 5^2 + 3^2 = 34 < N(\beta) = 11^2 + 2^2 = 125$.

14.1.19. a. We have $\alpha/\beta = (24 - 9i)/(3 + 3i) = 5/2 - 11/2i$. Instead of rounding up in each part, we round $5/2$ down to 2, which yields $\gamma = 2 - 5i$. Then $\rho = \alpha - \beta\gamma = 3$. Then $N(\rho) = 3^2 + 0^2 = 9 < N(\beta) = 3^2 + 3^2 = 18$.

b. We have $\alpha/\beta = (18 + 15i)/(3 + 4i) = 144/25 - 27/25i$. Instead of rounding $144/25$ to the nearest integer, we round it down to 4, which yields $\gamma = 4 - i$. Then we compute $\rho = \alpha - \beta\gamma = 2 + 2i$, so that $N(\rho) = 2^2 + 2^2 = 8 < N(\beta) = 25$.

c. We have $\alpha/\beta = 87i/(11 - 2i) = -174/125 + 957/125i$. Instead of rounding $957/125$ to the nearest integer, we round it down to 7, which yields $\gamma = -1 + 7i$. Then we compute $\rho = \alpha - \beta\gamma = -3 + 8i$, so that $N(\rho) = (-3)^2 + 8^2 = 73 < N(\beta) = 11^2 + 2^2 = 125$.

14.1.21. If $\beta|\alpha$ then there is only one pair $\gamma = \alpha/\beta$ and $\rho = 0$. If not, then the complex number α/β can be plotted in the complex plane and lies in a unit square whose vertices are lattice points. If $\alpha = \beta\gamma + \rho$, then $\alpha/\beta - \gamma = \rho/\beta$. Then taking absolute values, we see that $|\alpha/\beta - \gamma| = |\rho/\beta| < 1$. We conclude that the possible values for γ are those Gaussian integers inside a unit circle centered at α/β, each of which generates a unique ρ.

14.1.23. If a and b are both even then the Gaussian integer is divisible by 2. Since $(1 + i)(1 - i) = 2$, then $1 + i$ is a divisor of 2 which is in turn a divisor of $a + bi$. If a and b are both odd we may write $a + bi = (1 + i) + (a - 1) + (b - 1)i$, and $a - 1$ and $b - 1$ are both even. Since both of theses Gaussian integers are multiples of $1 + i$, so is there sum. If a is odd and b is even, then $a - 1 + bi$ is a multiple of $1 + i$ and hence $(a + bi) - (a - 1 + bi) = 1$ is a multiple of $1 + i$ if $a + bi$ is, a contradiction. A similar argument shows

14.1. GAUSSIAN INTEGERS AND GAUSSIAN PRIMES

that if a is even and b is odd then $1+i$ does not divide $a+bi$.

14.1.25. Let $\alpha = a+bi$, and suppose $\alpha^2 + 1$ is a Gaussian prime. Since we can factor $\alpha^2 + 1 = (\alpha+i)(\alpha-i) = (a + (b+1)i)(a+(b-1)i)$, we must have one of these factors a unit. One way is for $b = 0$, so that $a = 0$ and then the first factor is i. But then the second factor is $-i$ and $\alpha^2 + 1 = 0$ which is not prime. The only other way is for $b = \pm 1$, which forces $a = \pm 1$ and leads to $\alpha^2 + 1 = \pm 1 \pm 2i$. Since $N(\pm 1 \pm 2i) = 5$ is prime, we know that $\pm 1 \pm 2i$ are Gaussian primes, and these four are the only ones of this form.

14.1.27. Suppose $7 = (a+bi)(c+di)$ where $a+bi$ and $c+di$ are nonunit Gaussian integers. Taking norms of both sides yields $49 = (a^2+b^2)(c^2+d^2)$. Since $a+bi$ and $c+di$ are not units, we have that the factors on the right are not equal to 1, so we must have $a^2 + b^2 = 7$, a contradiction, since 7 is not the sum of 2 squares.

14.1.29. Since α in not a unit or a prime, it has a nontrivial factors $\alpha = \beta\gamma$ with β and γ nonunits, so that $1 < N(\beta)$ and $1 < N(\gamma)$. Then $N(\alpha) = N(\beta)N(\gamma)$. If $N(\beta) > \sqrt{N(\alpha)}$ then $N(\gamma) = N(\alpha)/N(\beta) < N(\alpha)/\sqrt{N(\alpha)} = \sqrt{N(\alpha)}$. So if β doesn't satisfy the conditions, then γ does.

14.1.31. Following the procedure in Exercise 30, we note that $1+i$ is a Gaussian prime. Its multiples in the 1st Quadrant on or below the line $y = x$ are those Gaussian integers $a+bi$ where a and b are both even or both odd, so we cross these out. The closest integer to the origin not crossed out is $2+i$, so we circle it and cross out its multiples. The new numbers crossed out with norm less than 10 are $5, 6+3i, 9+2i$, and $7+6i$. We also cross out multiples of its conjugate $2-i$, which eliminates $4+3i, 8+i$ and $7+4i$. The next closest integer not crossed out is 3 and the only multiple not crossed out is 9, which we cross out. The next closest number to the origin which is not crossed out is $3+2i$, but its norm is 13, which is greater than $\sqrt{100}$ so we are done with the sieving process. This leaves the following numbers as Gaussian primes with norm less than 100: $3, 7, 1+i, 2+i, 4+i, 6+i, 3+2i, 5+2i, 7+2i, 8+3i, 5+4i, 9+4i, 6+5i$, and $8+5i$, plus their conjugates and associates.

14.1.33. a. Note that $\alpha - \alpha = 0 = 0 \cdot \mu$, so $\mu | \alpha - \alpha$. Thus, $\alpha \equiv \alpha \pmod{\mu}$.

b. Since $\alpha \equiv \beta \pmod{\mu}$, we have $\mu | \alpha - \beta$, so there exists a Gaussian integer γ such that $\mu\gamma = \alpha - \beta$. But then $\mu(-\gamma) = \beta - \alpha$, so $\mu | \beta - \alpha$. Therefore $\beta \equiv \alpha \pmod{\mu}$.

c. Since $\alpha \equiv \beta \pmod{\mu}$ and $\beta \equiv \gamma \pmod{\mu}$, there exist Gaussian integers δ and ϵ such that $\mu\delta = \alpha - \beta$ and $\mu\epsilon = \beta - \gamma$. Then $\alpha - \gamma = \alpha - \beta + \beta - \gamma = \mu\delta + \mu\epsilon = \mu(\delta + \epsilon)$. Therefore $\alpha \equiv \gamma \pmod{\mu}$.

14.1.35. If $\alpha = a_1 + ib_1$ and $\beta = a_2 + ib_2$ let $p = (a_1+b_1)(a_2+b_2)$. Then the real part of $\alpha\beta$ is given by the two multiplications $R = a_1a_2 - b_1b_2$ and the imaginary part is given by $p - R$ which requires only one more multiplication. The second way in the hint goes as follows. Let $m_1 = b_2(a_1+b_1)$, $m_2 = a_2(a_1-b_1)$, and $m_3 = b_1(a_2-b_2)$. These are the three multiplications. Then the real part of $\alpha\beta$ is given by $m_2 + m_3$ and the imaginary part by $m_1 + m_3$.

14.1.37. a. We have $G_0 = 0+i$, $G_1 = 1+i$, $G_2 = 1+2i$, $G_3 = 2+3i$, $G_4 = 3+5i$, $G_5 = 5+8i$.
b. Using the definition of G_k and the properties of the Fibonacci sequence we have $G_k = f_k + if_{k+1} = (f_{k-1}+f_{k-2}) + (f_k+f_{k-1}i) = (f_{k-1}+f_ki) + (f_{k-2}+f_{k-1}i) = G_{k-1} + G_{k-2}$.

14.1.39. We proceed by induction. For the basis step note that $G_2G_1 - G_3G_0 = (1+2i)(1+i) - (2+3i)(i) = 2+i$, so the basis step holds. Now assume the identity holds for values less than n. We compute, using the identity in Exercise 37, $G_{n+2}G_{n+1} - G_{n+3}G_n = (G_{n+1}+G_n)G_{n+1} - (G_{n+2}+G_{n+1})G_n = G_{n+1}^2 - G_{n+2}G_n = G_{n+1}^2 - (G_{n+1}+G_n)G_n = G_{n+1}^2 - G_n^2 - G_{n+1}G_n = (G_{n+1}+G_n)(G_{n+1}-G_n) - G_{n+1}G_n = G_{n+2}G_{n-1} - G_{n+1}G_n = -(-1)^{n-1}(2+i) = (-1)^n(2+i)$, which completes the induction step.

14.1.41. Since the coefficients of the polynomial are real, the other root is $r-si$, and over the complex numbers the polynomial must factor as $(z-(r+si))(z-(r-si)) = z^2 - 2rz + r^2 + s^2$. The z-coefficients, $a = 2r$ and $b = r^2 + s^2$ are integers. Then $r = a/2$ and $s^2 = (4b-r^2)/4$, which shows that $s = c/2$ for some integer c. Multiplying by 4 we have $a^2 + c^2 \equiv 0 \pmod 4$ which can be true only if both a and c are even, hence r and s are integers and $r + si$ is a Gaussian integer.

14.1.43. Let $\beta = 1 + 2i$ so that $N(\beta) = 5$. From the proof of the Division algorithm, we have for a Gaussian integer α, that there exist Gaussian integers γ and ρ such that $\alpha = \gamma\beta + \rho$ with $N(\rho) \leq N(\beta)/2 = 5/2$. Therefore the only possible remainders upon division by $1 + 2i$ are $0, 1, i, 1 + i$ and their associates. Further, if $\alpha = \beta\gamma + (1 + i) = \beta(\gamma + 1) + (1 + i) - (1 + 2i) = \beta(\gamma + 1) - i$. So we may take the entire set of remainders to be $0, 1, -1, i$ and $-i$. Consider dividing each of the Gaussian primes $\pi_1, \ldots, \pi_4$, by β. If any two left the same remainder ρ, then β divides the difference between the two primes. But all these differences are either 2 or $1 \pm i$, which are not divisible by β. Further, since these are all prime, none of the remainders are 0. Therefore, the remainders are exactly the set $1, -1, i$ and $-i$. Now divide $a + bi$ by β and let the remainder be ρ. If ρ is not zero, then it is one of $1, -1, i$ or $-i$. But then one of $\pi_1, \ldots, \pi_4$ leaves the same remainder upon division by β, say π_k. Then β divides $\pi_k - (a + bi)$ which is a unit, a contradiction. Therefore $\rho = 0$. Therefore $1 + 2i$ divides $a + bi$. A similar argument shows that $1 - 2i$ also divides $a + bi$. Therefore the product of these primes $(1 - 2i)(1 + 2i) = 5$ also divides $a + bi$, and hence each of the components.

14.1.45. Taking norms of the equation $\alpha\beta\gamma = 1$ shows that all three numbers must be units in the Gaussian integers, which restricts our choices to $1, -1, i$ and $-i$. Choosing three of these in the equation $\alpha + \beta + \gamma = 1$, we have the possible solutions, up to permutation, $(1, 1, -1)$, $(1, i, -i)$, but only the second solution works in the first equation, leaving $(1, i, -i)$ as the only solution.

14.2. Greatest Common Divisors and Unique Factorization

14.2.1. Certainly $1 | \pi_1$ and $1 | \pi_2$. Suppose $\delta | \pi_1$ and $\delta | \pi_2$. Since π_1 and π_2 are Gaussian primes, δ must be either a unit or an associate of the primes. But since π_1 and π_2 are not associates, then they can not have an associate in common, so δ is a unit and so $\delta | 1$. Therefore 1 satisfies the definition of a greatest common divisor for π_1 and π_2.

14.2.3. Since γ is a greatest common divisor of α and β, we have $\gamma | \alpha$ and $\gamma | \beta$, so there exist Gaussian integers μ and ν such that $\mu\gamma = \alpha$ and $\nu\gamma = \beta$. So that $\overline{\mu\gamma} = \overline{\mu} \cdot \overline{\gamma} = \overline{\alpha}$ and $\overline{\nu\gamma} = \overline{\nu} \cdot \overline{\gamma} = \overline{\beta}$ so that $\overline{\gamma}$ is a common divisor of $\overline{\alpha}$ and $\overline{\beta}$. Further if $\overline{\delta} | \overline{\alpha}$ and $\overline{\delta} | \overline{\beta}$ then $\overline{\overline{\delta}} | \alpha$ and $\overline{\overline{\delta}} | \beta$ and so $\overline{\overline{\delta}} | \gamma$ by the definition of greatest common divisor. But then $\overline{\overline{\overline{\delta}}} = \delta | \overline{\gamma}$, which shows that $\overline{\gamma}$ is a greatest common divisor for $\overline{\alpha}$ and $\overline{\beta}$.

14.2.5. Let $\epsilon\gamma$, where ϵ is a unit, be an associate of γ. Since $\gamma | \alpha$ there is a Gaussian integer μ such that $\mu\gamma = \alpha$. Since ϵ is a unit, $1/\epsilon$ is also a Gaussian integer. Then $(1/\epsilon)\mu(\epsilon\gamma) = \alpha$, so $\epsilon\gamma | \alpha$. Similarly, $\epsilon\gamma | \beta$. If $\delta | \alpha$ and $\delta | \beta$ then $\delta | \gamma$ by definition of greatest common divisor, so there exists a Gaussian integer ν such that $\nu\delta = \gamma$. Then $\epsilon\nu\delta = \epsilon\gamma$, and since $\epsilon\nu$ is a Gaussian integer, we have $\delta | \epsilon\gamma$, so $\epsilon\gamma$ satisfies the definition of a greatest common divisor.

14.2.7. Good examples are the factors of rational primes which factor in the Gaussian integers, such as $13 = (3 - 2i)(3 + 2i)$. Then $\gcd(3 + 2i, 3 - 2i) = 1$, but $N(3 + 2i) = N(3 - 2i) = 13$.

14.2.9. Since a and b are relatively prime rational integers, there exist rational integers m and n such that $am + bn = 1$. Let δ be a greatest common divisor of the Gaussian integers a and b. Then δ divides $am + bn = 1$. Therefore δ is a unit in the Gaussian integers and hence a and b are relatively prime Gaussian integers.

14.2.11. a. We have $44 + 18i = (12 - 16i)(1 + 2i) + 10i$. Then $12 - 16i = (10i)(-2 - i) + (2 + 4i)$. Then $10i = (2 + 4i)(2 + i) + 0$. Since the last nonzero remainder is $2 + 4i$, this is a greatest common divisor.

b. From the equations in part (a) we have $2 + 4i = (12 - 16i) - (10i)(-2 - i) = (12 - 16i) - ((44 + 18i) - (12 - 16i)(1 + 2i))(-2 - i) = (2 + i)(44 + 18i) + (1 + (1 + 2i)(-2 - i))(12 - 16i) = (2 + i)(44 + 18i) + (1 - 5i)(12 - 16i)$. So we take $\mu = 2 + i$ and $\nu = 1 - 5i$.

14.2.13. We proceed by induction. We have $G_0 = i$ and $G_1 = 1 + i$. Since G_0 is a unit, these are relatively prime and this completes the basis step. Assume we know that G_k and G_{k-1} are relatively prime. Suppose $\delta | G_k$ and $\delta | G_{k+1}$. Then $\delta | (G_{k+1} - G_k) = (G_k + G_{k-1} - G_k) = G_{k-1}$, so δ is a common divisor of G_k and

G_{k-1} which are relatively prime. Hence $\delta|1$ and so 1 is a greatest common divisor of G_{k+1} and G_k.

14.2.15. Let k be the smallest rational integer such that $N(\alpha) < 2^k$. Dividing $\beta = \rho_0$ by $\alpha = \rho_1$ in the first step of the Euclidean Algorithm gives us $\beta = \gamma_2 \alpha + \rho_2$ with $N(\rho_2) < N(\alpha) < 2^{k-1}$. The next step of the Euclidean Algorithm, gives us $\alpha = \gamma_3 \rho_2 + \rho_3$ with $N(\rho_3) < N(\rho_2) < 2^{k-2}$. Continuing with the algorithm shows us that $N(\rho_k) < 2^{k-(k-1)} = 2$, so that the Euclidean Algorithm must terminate in no more than $k = [\log_2 N(\alpha)] + 1$ steps. And thus we have $k = O(\log_2(N(\alpha)))$.

14.2.17. a. We compute $N(7 + 6i) = 85 = 5 \cdot 17$. Since $1 \pm 2i$ and their associates have norm 5 and $1 \pm 4i$ and their associates have norm 17, we try these and discover that $7 + 6i = (-1)(1 - 2i)(1 - 4i)$.

b. We compute $N(3 - 13i) = 178 = 2 \cdot 89$. Only $1 + i$ has norm 2 and it divides $3 - 13i$ only once, leaving $-5 - 8i$ which has norm 89, which is a rational prime. Therefore $5 + 8i$ is a Gaussian prime and we have $3 - 13i = (-1)(1 + i)(5 + 8i)$.

c. By Exercise 7 in Section 1, we know 7 is a Gaussian prime and since $4 = 2^2 = (i(1+i)^2)^2 = -(1+i)^4$, we have $28 = (-1)(1 + i)^4 7$.

d. We have $400i = 16 \cdot 25i = (i(1+i)^2)^4(5^2)i = (1+i)^8((1+2i)(1-2i))^2 i = i(1+i)^8(1+2i)^2(1-2i)^2$.

14.2.19. a. We find that $10 = -i(1+i)^2(1+2i)(1-2i)$, so a divisor of 10 must have one of the three Gaussian primes to a power less than or equal to the power to which it appears in this factorization. So the possible number of factors, ignoring associates is $(2+1)(1+1)(1+1) = 12$. Since there are 4 units, when we count associates, there are a total of $4 \cdot 12 = 48$ divisors of 10.

b. We have $128 + 256I = i(1 + i)^{14}(1 + 2i)$, so the number of divisors is $4(14 + 1)(1 + 1) = 120$.

c. We have $27000 = i(1+i)^6(1+2i)^3(1-2i)^3 3^3$, so the number of divisors is $4(6+1)(3+1)(3+1)(3+1) = 1792$.

d. We have $5040 + 40320i = (1+i)^8(1+2i)(1-2i)^2 3^2 7(-3+2i)$, so the number of divisors is $4(8+1)(1+1)(2+1)(2+1)(1+1)(1+1) = 2592$.

14.2.21. Assume n and $a + bi$ are relatively prime. Then there exist Gaussian integers μ and ν such that $\mu n + \nu(a + bi) = 1$. If we take conjugates of both sides and recall that the conjugate of a rational integer is itself, we have $\overline{\mu} n + \overline{\nu}(a - bi) = 1$, so n is also relatively prime to $a - bi$. Since $a - bi$ is an associate of $b + ai$ (multiply by i), we have the result. The converse is true by symmetry.

14.2.23. Suppose that $\pi_1, \pi_2, \ldots, \pi_k$ are all of the Gaussian primes and form the Gaussian integer $Q = \pi_1 \pi_2 \cdots \pi_k + 1$. From Theorem 14.10, we know that Q has a unique factorization into Gaussian primes, and hence is divisible by some Gaussian prime ρ. Then $\rho|Q$ and $\rho|\pi_1 \pi_2 \cdots \pi_k$, so ρ divides their difference, which is 1, a contradiction, unless ρ is a prime different from $\pi_1, \pi_2, \ldots, \pi_k$, proving that we did not have all the Gaussian primes.

14.2.25. Since $2 + 3i$ and $1 + 2i$ are necessarily relatively prime, we perform the Euclidean algorithm to express 1 as a linear combination of the two numbers to get $1 = i(2 + 3i) - 2i(1 + 2i)$. Then we have that $-2i$ is an inverse for $1 + 2i \pmod{2 + 3i}$.

14.2.27. Since α and μ are relatively prime, there exist Gaussian integers σ and τ such that $\sigma \alpha + \tau \mu = 1$. If we multiply through by β we get $\beta \sigma \alpha + \beta \tau \mu = \beta$, so that we know $\alpha(\beta \sigma) \equiv \beta \pmod{\mu}$ and thus $x \equiv \beta \sigma \pmod{\mu}$ is the solution.

14.2.29. a. From the Euclidean algorithm we get $1 = (-4)3 + (1)13$. We multiply by $(2 + i)$ to get $2 + i = (-4)(2+i)3 + 12$, so that we see $x \equiv -8 - 4i \equiv 5 - 4i \pmod{13}$ is the solution.

b. From the Euclidean algorithm we get $1 = (-1 - 2i)(5) + (2 + 2i)(4 + i)$. Then we must have $x \equiv (-1 - 2i)(3 - 2i) \equiv -7 - 4i \equiv 1 + i \pmod{4 + i}$.

c. From the Euclidean algorithm we get $1 = (1-i)(3+i) + i(2+3i)$. Then we must have $x \equiv 4(1-i) \equiv 3i \pmod{2+3i}$.

14.2.31. Statement: Let $\mu_1, \mu_2, \ldots, \mu_r$ be pairwise relatively prime Gaussian integers and let $\alpha_1, \alpha_2, \ldots, \alpha_r$ be Gaussian integers. Then the system of congruences $x \equiv \alpha_i \pmod{\mu_i}, i = 1, \ldots, r$ has a unique solution modulo $M = \mu_1 \mu_2 \cdots \mu_r$.

Proof: To construct a solution, for each $k = 1, \ldots, r$, let $M_k = M/\mu_k$. Then M_k and μ_k are relatively prime, since μ_k is relatively prime to all of the factors of M_k. Then from Exercise 24, we know M_k has an inverse λ_k modulo μ_k, so that $M_k \lambda_k \equiv 1 \pmod{\mu_k}$. Now let $x = \alpha_1 M_1 \lambda_1 + \cdots + \alpha_r M_r \lambda_r$. We will show x is the solution to the system.

Since $\mu_k | M_j$ whenever $j \neq k$, we have $\alpha_j M_j \lambda_k \equiv 0 \pmod{\mu_k}$ whenever $j \neq k$. Therefore $x \equiv \alpha_k M_k \lambda_k \pmod{\mu_k}$ Also, since λ_k is an inverse for M_k modulo μ_k, we have $x \equiv \alpha_k \pmod{\mu_k}$ for every k, as desired.

Now suppose there is another solution y to the system. Then $x \equiv \alpha_k \equiv y \pmod{\mu_k}$ and so $\mu_k | (x-y)$ for every k. Since the μ_k are pairwise relatively prime, no Gaussian prime appears in more than one of their prime factorizations. Therefore, if a Gaussian prime power $\pi^e | (x-y)$ then it divides exactly one of the μ_k's. Therefore, the product M of the μ_k's also divides $x - y$ and so $x \equiv y \pmod{M}$ showing that x is unique modulo M.

14.2.33. Using Exercise 31, we let $M = (2+5i)(3-4i) = 26 + 7i$, so that $M_1 = 3 - 4i$ and $M_2 = 2 + 5i$. An inverse for M_1 modulo $2 + 5i$ is $\lambda_1 = -1 + 2i$. An inverse for M_2 modulo $3 - 4i$ is $\lambda_2 = -2$. Then the solution is $x = (1+3i)(3-4i)(-1+2i) + (2-i)(2+5i)(-2) = -43 + 9i \equiv 9 + 23i \pmod{26 + 7i}$.

14.2.35. a. Using the construction in the solution to Exercise 37, we note that $N(1-i) = 2$ and $(1,1) = 1 = d$, so that $S = \{0, 1\}$ which is a complete residue system.

b. Using the construction in Exercise 37, we note that $N(2) = 4$ and $(2, 0) = 2 = d$, so that $S = \{0, 1, i, 1+i\}$, which is a complete residue system.

c. Using the construction in Exercise 37, we note that $N(2+3i) = 13$ and $(2,3) = 1 = d$, so that $S = \{0, 1, 2, 3, 4, 5, 6, 7, 8, 9, 10, 11, 12\}$. Reducing each of these modulo $2 + 3i$ gives us $\{0, 1, 2, 2i, -1-i, -i, 1-i, -1+i, i, 1+i, -2i, -2, -1\}$ for a complete residue system.

14.2.37. Let $\alpha = a + bi$ and $d = \gcd(a, b)$. We assert that the set $S = \{p + qi | 0 \leq p < N(\alpha)/d, 0 \leq q < d\}$ is a complete residue system. Note that this represents a rectangle of lattice points in the plane. We create two multiples of α. First, $N(\alpha)/d = \alpha(\overline{\alpha}/d)$ is a real number and a multiple of α. Second, there exist rational integers r and s such that $ra + sb = d$. So we have the multiple of α given by $v = (s + ir)\alpha = (s+ir)(a+bi) = (as - br) + di$. Now it is clear that any Gaussian integer is congruent modulo α to an integer in the rectangle S, since first we can add or subtract multiples of v until the imaginary part is between 0 and $d - 1$ and then add and subtract multiples of $N(\alpha)/d$ until the real part is between 0 and $N(\alpha)/d - 1$. It remains to show the elements of S are incongruent to each other modulo α. Suppose β and γ are in S and congruent to each other modulo α. Then the imaginary part of $\beta - \gamma$ must be divisible by d, but since these must lie in the interval from 0 to $d-1$, they must be equal. Therefore the difference between β and γ is real and divisibly by α, hence by $\overline{\alpha}$ and hence by $\alpha \overline{\alpha}/d = N(\alpha)/d$, which proves they are equal. Since S has $N(\alpha)$ elements, we are done.

14.2.39. a. From Exercise 37, we find a complete residue system modulo $2 + 2i$ to be $S = \{0, i, 2i, 3i, 1, 1+i, 1+2i, 1+3i\}$. Also, we have $2 + 2i = -i(1+i)^3$, so every element in S with the same parity in real and imaginary parts is not relatively prime to $2 + 2i$. Deleting these gives us $\{i, 3i, 1, 1+2i\}$. Reducing modulo $2 + 2i$ gives us $\{i, -i, 1, -1\}$ for a reduced residue system.

b. From Exercise 37, we find a complete residue system modulo 4 to be $S = \{0, i, 2i, 3i, 1, 1+i, 1+2i, 1+3i, 2, 2+i, 2+2i, 2+3i, 3, 3+i, 3+2i, 3+3i\}$. Also, we have $4 = -(1+i)^4$, so every element in S with the same parity in real and imaginary parts is not relatively prime to 4. Deleting these gives us $\{i, 3i, 1, 1+2i, 2+i, 2+3i, 3, 3+2i\}$. Reducing modulo 4 gives us $\{i, -i, 1, 1+2i, 2+i, 2-

$I, -1, -1 + 2i\}$ for a reduced residue system.

c. From Exercise 37, we find a complete residue system modulo $4+2i$ to be $S = \{0, i, 2i, 3i, 4i, 5i, 6i, 7i, 8i, 9i, 1, 1 + i, 1 + 2i, 1 + 3i, 1 + 4i, 1 + 5i, 1 + 6i, 1 + 7i, 1 + 8i, 1 + 9i\}$. Also, we have $4 + 2i = (1 + i)^2(1 - 2i)$, so every element in S with the same parity in real and imaginary parts is not relatively prime to $4 + 2i$. Deleting these gives us $\{i, 3i, 5i, 7i, 9i, 1, 1 + 2i, 1 + 4i, 1 + 6i, 1 + 8i\}$. Reducing modulo $4 + 2i$ gives us $\{i, 2 - i, 2 + i, -2 + i, -i, 1, 1 + 2i, -1 - 2i, -1, -1 + 2i\}$. Note that $2 + i$ and $-1 + 2i$ are associates of $1 - 2i$ which is a prime divisor of $4 + 2i$, so we delete them, leaving $\{i, 2 - i, -2 + i, -i, 1, 1 + 2i, -1 - 2i, -1\}$ for a reduced residue system.

14.2.41. From the properties of the norm function and Exercise 37, we know that there are $N(\pi^e) = N(\pi)^e$ residue classes modulo π^e. Let $\pi = r + si$, and $d = \gcd(r, s)$. Also, by Exercise 37, a complete residue system modulo π^e is given by the rectangle $S = \{p + qi | 0 \leq p < N(\pi^e)/d, 0 \leq q < d\}$, while a complete residue system modulo π is given by the rectangle $T = \{p + qi | 0 \leq p < N(\pi)/d, 0 \leq q < d\}$. Note that in T there is exactly one element not relatively prime to π, and that there are $N(\pi)^{e-1}$ copies of T, congruent modulo π, inside of S. Therefore, there are exactly $N(\pi)^{e-1}$ elements in S not relatively prime to π. Thus there are $N(\pi)^e - N(\pi)^{e-1}$ elements in a reduced residue system modulo π^e.

14.2.43. a. A polynomial is called *primitive* if the greatest common divisor of its coefficients is 1. We require a result from algebra called Gauss' Lemma, which states that the product of primitive polynomials is primitive. To prove this, suppose $f(x) = a_0 + a_1 x + \cdots + a_n x^n$ and $g(x) = b_0 + b_1 x + \cdots + b_m x^m$ are primitive integer polynomials. Let p be any prime. Let a_j be the first coefficient of $f(x)$ which p doesn't divide. Likewise, let b_k be the first coefficient of $g(x)$ which p doesn't divide. Then $f(x)g(x) = c_0 + c_1 x + \cdots + c_{j+k} + \cdots + c_{n+m} x^{n+m}$, where $c_{j+k} = b_0 a_{j+k} + b_1 a_{j+k-1} + \cdots + b_k a_j + \cdots + b_{j+k} a_0$. Since every term is divisible by p except $b_k a_j$, we see that c_{j+k} is not divisible by p. We conclude that no prime can divide all the coefficients of $f(x)g(x)$ and so it is primitive.

Now suppose $\alpha = r + s\sqrt{-5}$ is an algebraic integer. Then it is a root of a monic polynomial $f(x)$ with integer coefficients. We may assume $f(x)$ has smallest positive degree of all such polynomials. If $f(x) = x + b$, then $f(\alpha) = r + s\sqrt{-5} + b$ so that $s = 0$ and $r = b$, which are both integers. So assume that $\deg(f) \geq 2$. Note that $f(x)$ is necessarily irreducible over the integers, since if $f(x) = g(x)h(x)$ is a nontrivial factorization of f, then $g(\alpha)h(\alpha) = 0$ and so α satisfies one of g or h which contradicts the minimality of f.

Note that α is a root of $g(x) = (x - \alpha)(x - \overline{\alpha}) = (x^2 - 2rx + r^2 + 5s^2)$. If we divide $f(x)$ by $g(x)$ we get $f(x) = q(x)g(x) + r(x)$, with $\deg(r) < \deg(g) = 2$ or $r(x) = 0$. Then we have $f(\alpha) = q(\alpha)g(\alpha) + r(\alpha)$, so that $r(\alpha) = 0$. But α can not be the root of a polynomial of degree 1 or 0, so $r(x) = 0$ and we have $f(x) = q(x)g(x)$, where $q(x)$ and $g(x)$ have rational coefficients. We can factor out any common factors of the coefficients of q and g and write $f(x) = (a/b)q_1(x)g_1(x)$, where q_1 and g_1 are primitive integer polynomials and $(a, b) = 1$. But by Gauss' Lemma, $q_1 g_1$ is primitive, so no prime factor of b can divide all of the coefficients. Therefore $b = 1$, and since $f(x)$ is monic, we have $a = 1$. Further, since f is irreducible, we must have $q_1 = 1$ and so $f(x) = g(x) = x^2 - 2rx + r^2 + 5s^2$ and we know that $2r$ and $r^2 + 5s^2$ are integers. Then $r = b/2$ and $5s^2 = (4c - b^2)/4$ for some integers b and c. So $s = e/2$ for some integer e. (5 can not appear in the denominator of s, else when we square it, the single factor of 5 in the expression leaves a remaining factor in the denominator, which does not appear on the right side of the equation.) Substituting these expressions in for r and s, we have $(b/2)^2 + 5(e/2)^2 = c$, or, upon multiplication by 4, $b^2 + 5e^2 = 4c \equiv 0 \pmod{4}$ which has solutions only when b and e are even. Therefore r and s are rational integers.

b. First we seek rational integers a and b such that $(2 + 3\sqrt{-5})(a + b\sqrt{-5}) = -9 + 11\sqrt{-5}$. Multiplying out the left side yields $(2a - 15b) + (3a + 2b)\sqrt{-5} = -9 + 11\sqrt{-5}$. So we must have $2a - 15b = -9$ and $3a + 2b = 11$. Solving this system of equations gives us $a = 3$ and $b = 1$. Since these are rational integers, we have $(2 + 3\sqrt{-5})(3 + 1\sqrt{-5}) = -9 + 11\sqrt{-5}$.

Next, we seek rational integers a and b such that $(1 + 4\sqrt{-5})(a + b\sqrt{-5}) = (a - 20b) + (4a + b)\sqrt{-5} = 8 + 13\sqrt{-5}$. We must have $a - 20b = 8$ and $4a + b = 13$, but this system leads to $b =$

$-19/81$, which is not an integer, so we conclude that $1 + 4\sqrt{-5}$ does not divide $8 + 13\sqrt{-5}$.

c. Let $\alpha = a + b\sqrt{-5}$ and $\beta = c + d\sqrt{-5}$. Then $\alpha + \beta = (a+c) + (b+d)\sqrt{-5}$ and $\alpha - \beta = (a-c) + (b-d)\sqrt{-5}$, and $\alpha\beta = (ac - 5bd) + (ad + bc)\sqrt{-5}$. Since the rational integers are closed under addition, subtraction and multiplication, all of the results are again of the form $p + q\sqrt{-5}$ with p and q rational integers.

d. Let $\alpha = a + b\sqrt{-5}$ and $\beta = c + d\sqrt{-5}$. Then $N(\alpha)N(\beta) = (a^2 + 5b^2)(c^2 + 5d^2) = a^2c^2 + 5a^2d^2 + 5b^2c^2 + 25b^2d^2$. On the other hand, $\alpha\beta = (ac - 5bd) + (ad + bc)\sqrt{-5}$ and $N(ac - 5bd) + (ad + bc)\sqrt{-5}) = (ac - 5bd)^2 + 5(ad + bc)^2 = a^2c^2 - 10acbd + 25b^2d^2 + 5(a^2d^2 + 2adbc + b^2c^2) = a^2c^2 + 5a^2d^2 + 5b^2c^2 + 25b^2d^2$, which is equal to the expression above, proving the assertion.

e. If ϵ is a unit in $\mathbb{Z}[\sqrt{-5}]$, then there exists an η such that $\epsilon\eta = 1$. From part (d) we have $N(\epsilon\eta) = N(\epsilon)N(\eta) = N(1) = 1$, so $N(\epsilon) = 1$. Suppose $\epsilon = a + b\sqrt{-5}$, then $N(\epsilon) = a^2 + 5b^2 = 1$, which shows that $b = 0$, and hence $a^2 = 1$, so that we know $a = \pm 1$. Therefore the only units are 1 and -1.

f. If an integer α in $\mathbb{Z}[\sqrt{-5}]$ is not a unit and not prime, then it must have two non unit divisors β and γ such that $N(\beta)N(\gamma) = N(\alpha)$. To see that 2 is prime, note that a divisor $\beta = a + b\sqrt{-5}$ has norm $a^2 + 5b^2$, while $N(2) = 4$, which forces $b = 0$. If β is not a unit, then $a = \pm 2$, but then this forces γ to be a unit, hence 2 is prime. To see that 3 is prime, we seek divisors of $N(3) = 9$ among $a^2 + 5b^2$. We see that b can be only 0 or ± 1 or else the norm is too large. And if $b = \pm 1$, then the only possible divisor is 9 itself, forcing the other divisor to be a unit. If $b = 0$ then $a = \pm 3$, and hence 3 is prime. To see that $1 \pm \sqrt{-5}$ is prime, note that its norm is 6. A divisor $a + bi$ can have b take on the values 0 and ± 1 else the norm is too large. If $b = 0$, then $a^2 | 6$ a contradiction, so $b = \pm 1$. But then $a^2 + 5 | 6$ forcing $a = \pm 1$. But $N(\pm 1 \pm \sqrt{-5}) = 6$ so the other divisor is a unit, and so $1 \pm \sqrt{5}$ is also prime. Note then that $2 \cdot 3 = 6$ and $(1 - \sqrt{-5})(1 + \sqrt{-5}) = 6$, so that we do not have unique factorization into primes in $\mathbb{Z}[\sqrt{-5}]$.

g. Suppose γ and ρ exist. Note first that $(7 - 2\sqrt{-5})/(1 + \sqrt{-5}) = -1/2 - 3/2\sqrt{-5}$, so $\rho \neq 0$. Let $\gamma = a + b\sqrt{-5}$ and $\rho = c + d\sqrt{-5}$. Then from $7 - 2\sqrt{-5} = (1 + \sqrt{-5})(a + b\sqrt{-5}) + (c + d\sqrt{-5}) = (a - 5b + c) + (a + b + d)\sqrt{-5}$ we get $7 = a - 5b + c$ and $-2 = a + b + d$. If we subtract the second equation from the first we have $9 = -6b + c - d$ or $c - d = 6b + 9$. Therefore, $3 | c - d$, and since $\rho \neq 0$, $c - d \neq 0$, so $|c - d| \geq 3$. We consider $N(\rho) = c^2 + 5d^2$. If $d = 0$, then $N(\rho) \geq c^2 \geq 3^2 > 6$. If $d = \pm 1$, then $|c| \geq 2$ and $N(\rho) = c^2 + 5d^2 \geq 4 + 5 > 6$. If $|d| \geq 2$, then $N(\rho) \geq 5d^2 \geq 5 \cdot 2^2 = 20 > 6$, so in every case the norm of ρ is greater than 6. So no such γ and ρ exist, and there is no analog for the division algorithm in $\mathbb{Z}[\sqrt{-5}]$.

h. Suppose $\mu = a + b\sqrt{-5}$ and $\nu = c + d\sqrt{-5}$ is a solution to the equation. Then $3(a + b\sqrt{-5}) + (1 + \sqrt{-5})(c + d\sqrt{-5}) = (3a + c - 5d) + (3b + c + d)\sqrt{-5} = 1$. So we must have $3a + c - 5d = 1$ and $3b + c + d = 0$. If we subtract the second equation from the first, we get $3a - 3b - 6d = 1$ which implies that $3 | 1$, an absurdity. Therefore no such solution exists.

14.3. Gaussian Integers and Sums of Squares

14.3.1. a. Since the prime factorization for 5 is 5^1 and $5 \equiv 1 \pmod{4}$, we have, by Theorem 14.13, that the number of ways to write 5 as the sum of two squares is $4(1 + 1) = 8$.

b. The prime factorization of 20 is $2^2 5$, and $5 \equiv 1 \pmod{4}$. So by Theorem 14.13, the number of ways to write 20 as the sum of two squares is $4(1 + 1) = 8$.

c. We have $120 = 2^3 5 \cdot 3$, where $5 \equiv 1 \pmod 4$ but $3 \equiv 3 \pmod 4$. So by Theorem 14.3, there is no way to write 120 as the sum of two squares.

d. We have $1000 = 2^3 5^3$, so the number of ways to write 1000 as the sum of two squares is $4(3+1) = 16$.

14.3. GAUSSIAN INTEGERS AND SUMS OF SQUARES

14.3.3. We first check that a greatest common divisor δ of α and β divides γ, otherwise no solution exists. If a solution exists, we use the Euclidean algorithm and back substitution to express δ as a linear combination of α and β: $\alpha\mu + \beta\nu = \delta$. Since δ divides γ there is a Gaussian integer η such that $\delta\eta = \gamma$. If we multiply the last equation by η we have $\alpha\mu\eta + \beta\nu\eta = \delta\eta = \gamma$, so we may take $x_0 = \mu\eta$ and $y_0 = \nu\eta$ as a solution. The set of all solutions is given by $x = x_0 + \beta\tau/\delta$, $y = y_0 - \alpha\tau/\delta$, where τ ranges over the Gaussian integers.

14.3.5. a. We find that a greatest common divisor of $3 + 4i$ and $3 - i$ is $2 + i$. Then we compute $7i/(2+i) = 7/5 + 14/5i$, which is not a Gaussian integer. Therefore there are no solutions to the diophantine equation.

b. We find that a greatest common divisor of $7 + i$ and $7 - i$ is $1 + i$ which does not divide 1. Therefore the diophantine equation has no solutions.

14.3.7. Suppose x, y, z is a primitive Pythagorean triple with y even, so that x and z are necessarily odd. Then $z^2 = x^2 + y^2 = (x+iy)(x-iy)$ in the Gaussian integers. If a rational prime p divides $x + iy$, then it must divide both x and y, which contradicts the fact that the triple is primitive. Therefore, the only Gaussian primes which divide $x + iy$ are of the form $m + in$ with $n \neq 0$. Also, if $1 + i | x + iy$, then we have the conjugate relationship $1 - i | x - iy$, which implies that $2 = (1-i)(1+i)$ divides z^2, which is odd, a contradiction. Therefore we conclude that $1 + i$ does not divide $x + iy$, and hence neither does 2. Suppose δ is a common divisor of $x + iy$ and $x - iy$. Then δ divides the sum $2x$ and the difference $2iy$. Since we know that 2 is not a common factor, δ must divide both x and y, which we know are relatively prime. Hence δ is a unit and $x + iy$ and $x - iy$ are also relatively prime. Then we know that every prime which divides $x + iy$ is of the form $\pi = u + iv$ and so $\overline{\pi} = u - iv$ divides $x - iy$. Since their product equals a square, each factor is a square. Thus $x + iy = (m + in)^2$ and $x - iy = (m - in)^2$ for some Gaussian integer $m + in$ and its conjugate. But then $x + iy = m^2 - n^2 + 2mni$ so $x = m^2 - n^2$ and $y = 2mn$. And $z^2 = (m+ni)^2(m-ni)^2 = (m^2+n^2)^2$, so $z = m^2 + n^2$. Further, if m and n were both odd or both even, we would have z even, a contradiction, so we may conclude that m and n have opposite parity. Finally, having found m and n which work, if $m < n$ we can multiply by i and reverse their roles to get $m > n$. The converse is exactly as in Section 13.1.

14.3.9. By Lemma 14.3, there is a unique rational prime p such that $\pi | p$. Let $\alpha = a + bi$ and consider 3 cases.
Case 1: If $p = 2$, then π is an associate of $1 + i$ and $N(\pi) - 1 = 1$. Since there are only two congruence classes modulo $1 + i$ and since α and $1 + i$ are relatively prime, we have $\alpha^{N(\pi)-1} = \alpha \equiv 1 \pmod{1+i}$.
Case 2: If $p \equiv 3 \pmod 4$, then $\pi = p$ and $N(\pi) - 1 = p^2 - 1$. Also $(-i)^p = -i$. By the Binomial theorem, we have $\alpha^p = (a + bi)^p \equiv a^p + (bi)^p \equiv -ib^p \equiv a - bi \equiv \overline{\alpha} \pmod p$, using Fermat's little theorem. Similarly $\overline{\alpha}^p \equiv \alpha \pmod p$, so that $\alpha^{p^2} \equiv \overline{\alpha}^p \equiv \alpha \pmod p$ and since $p = \pi$ and α and π are relatively prime, we have $\alpha^{N(\pi)-1} \equiv 1 \pmod p$.
Case 3: If $p \equiv 1 \pmod 4$, then $\pi\overline{\pi} = p$, $i^p = i$, and $N(\pi) - 1 = p - 1$. By the Binomial theorem, we have $\alpha^p = (a + bi)^p \equiv a^p + (bi)^p \equiv a + bi \equiv \alpha \pmod p$, using Fermat's little theorem. Cancelling an α gives us $\alpha^{p-1} \equiv 1 \pmod p$, and since $\pi | p$ we have $\alpha^{N(\pi)-1} \equiv 1 \pmod \pi$, which concludes the proof.

14.3.11. Let π be a Gaussian prime. If $\alpha^2 \equiv 1 \pmod \pi$, then $\pi | \alpha^2 - 1 = (\alpha - 1)(\alpha + 1)$, so that either $\alpha \equiv 1$ or $\alpha \equiv -1 \pmod \pi$. Therefore only 1 and -1 can be their own inverses modulo π. Now let $\alpha_1 = 1, \alpha_2, \ldots, \alpha_{r-1}, \alpha_r = -1$ be a reduced residue system modulo π. For each α_k, $k = 2, 3, \ldots, r - 1$, there is a multiplicative inverse modulo π α'_k such that $\alpha_k \alpha'_k \equiv 1 \pmod \pi$. If we group all such pairs in the reduced residue system together, then the product is easy to evaluate: $\alpha_1 \alpha_2 \cdots \alpha_r = 1(\alpha_2\alpha'_2)(\alpha_3\alpha'_3)\cdots(\alpha_{r-1})(\alpha'_{r-1})(-1) \equiv -1 \pmod \pi$, which proves the theorem.

APPENDIX A

Axioms for the Set of Integers

A.0.1. a. By the commutative law, $a(b+c) = (b+c)a$. Now, using the distributive law, $a(b+c) = (b+c)a = ba + ca = ab + ac$.

    b. By the distributive law, $(a+b)^2 = (a+b)(a+b) = a(a+b) + b(a+b) = a^2 + ab + ba + b^2$. By the law of commutativity, this is equal to $a^2 + 2ab + b^2$.

    c. From the commutative law of addition, $a + (b+c) = a + (c+b)$. This is equal to $(a+c) + b$ by associativity. With a final application of commutativity, we see that $a + (b+c) = (c+a) + b$.

    d. Using the definition of subtraction and additive commutativity, $(b-a) + (c-b) + (a-c) = (-a+b) + (-b+c) + (-c+a)$. By associativity, this is equal to $-a + (b-b) + (c-c) + a$. Using the definition of an additive inverse, this is 0.

A.0.3. By the definition of the inverse of an element, $0 + (-0) = 0$. But since 0 is an identity element, we have $0 + (-0) = -0$. It follows that $-0 = 0$.

A.0.5. Let x be a positive integer. Since $x = x - 0$ is positive, $x > 0$. Now let $x > 0$. Then $x - 0 = x$ is positive.

A.0.7. We have $a - c = a + (-b+b) - c = (a-b) + (b-c)$, which is positive from our hypothesis and the closure of the positive integers.

APPENDIX B

Binomial Coefficients

B.0.1. a. We have $\binom{100}{0} = 100!/(0!100!) = 1$.

  b. We have $\binom{50}{1} = 50!/(1!49!) = 50$.

  c. We have $\binom{20}{3} = 20!/(3!17!) = 1140$.

  d. We have $\binom{11}{5} = 11!/(5!6!) = 462$.

  e. We have $\binom{10}{7} = 10!/(7!3!) = 120$.

  f. We have $\binom{70}{70} = 70!/(70!0!) = 1$.

B.0.3. a. We compute $(a+b)^5 = a^5 + 5a^4b + 10a^3b^2 + 10a^2b^3 + 5ab^4 + b^5$.

  b. We compute $(x+y)^{10} = x^{10} + 10x^9y + 45x^8y^2 + 120x^7y^3 + 210x^6y^4 + 252x^5y^5 + 210x^4y^6 + 120x^3y^7 + 45x^2y^8 + 10xy^9 + y^{10}$.

  c. We compute $(m-n)^7 = m^7 - 7m^6n + 21m^5n^2 - 35m^4n^3 + 35m^3n^4 - 21m^2n^5 + 7mn^6 - n^7$.

  d. We compute $(2a+3b)^4 = 16a^4 + 96a^3b + 216a^2b^2 + 216ab^3 + 81b^4$.

  e. We compute $(3x-4y)^5 = 243x^5 - 1620x^4y + 4320x^3y^2 - 5760x^2y^3 + 3840xy^4 - 1024y^5$.

  f. We compute $(5x+7)^8 = 390625x^8 + 4375000x^7 + 21437500x^6 + 60025000x^5 + 105043750x^4 + 117649000x^3 + 82354300x^2 + 32941720x + 5764801$.

B.0.5. On the one hand, $(1+(-1))^n = 0^n = 0$. On the other hand, by the binomial theorem, $\sum_{k=0}^{n}(-1)^k\binom{n}{k} = (1+(-1))^n$.

B.0.7. We have $\binom{n}{r}\binom{r}{k} = n!/(r!(n-r)!) \cdot r!/(k!(r-k)!) = n!(n-k)!/(k!(n-k)!(n-r)!(n-k-n+r)!) = \binom{n}{k}\binom{n-k}{n-r}$.

B.0.9. We fix r and proceed by induction on n. It is easy to check the cases when $n = r$ and $n = r+1$. Suppose the identity holds for all values from r to $n-1$. Then consider the sum $\binom{r}{r} + \binom{r+1}{r} + \cdots + \binom{n}{r} = \binom{r-1}{r-1} + \left(\binom{r}{r} + \binom{r}{r-1}\right) + \left(\binom{r+1}{r} + \binom{r+1}{r-1}\right) + \cdots + \left(\binom{n-1}{r} + \binom{n-1}{r-1}\right)$, where we have used $\binom{r}{r} = \binom{r-1}{r-1}$ and Pascal's identity. Regrouping this sum gives us $\left(\binom{r-1}{r-1} + \binom{r}{r-1} + \cdots + \binom{n-1}{r-1}\right) + \left(\binom{r}{r} + \binom{r+1}{r} + \cdots + \binom{n-1}{r}\right)$. By our induction hypothesis, these two sums are equal to $\binom{n}{r+1} + \binom{n+1}{r+1} = \binom{n+1}{r+1}$ which concludes the induction step.

B.0.11. Using Exercise 10, $\binom{x}{n} + \binom{x}{n+1} = x!/(n!(x-n)!) + x!/((n+1)!(x-n-1)!) = (x!(n+1))/((n+1)!(x-n)!) + (x!(x-n))/((n+1)!(x-n)!) = (x!(x-n+n+1))/((n+1)!(x-n)!) = (x+1)!/((n+1)!(x-n)!) = \binom{x+1}{n+1}$.

B.0.13. Let S be a set of n copies of $x+y$. Consider the coefficient of $x^k y^{n-k}$ in the expansion of $(x+y)^n$. Choosing the x from each element of a $k-$element subset of S, we notice that the coefficient of $x^k y^{n-k}$ is the number of $k-$element subsets of S, $\binom{n}{k}$.

B.0.15. By counting elements with exactly 0, 1, 2, and 3 properties, we see that only elements with 0 properties are counted in $n - [n(P_1) + n(P_2) + n(P_3)] + [n(P_1, P_2) + n(P_1, P_3) + n(P_2, P_3)] - [n(P_1, P_2, P_3)]$, and those only once.

B.0.17. A term of the sum is of the form $ax_1^{k_1} x_2^{k_2} \cdots x_m^{k_m}$ where $k_1 + k_2 + \cdots + k_m = n$ and $a = \frac{n!}{k_1! k_2! \cdots k_m!}$.

B.0.19. From Exercise 17 it follows that the coefficient is $\frac{12!}{3!4!5!} 2^3 (-3^4) 5^5 = 27720 \cdot 8 \cdot 81 \cdot 3125 = 56,133,000,000$.